Punches, Dies and Tools
FOR
Manufacturing in Presses

A Cyclopaedia of Die-Making, Punch-Making, Die-Sinking, Sheet Metal Working, and Making of Special Tools, Devices and Mechanical Combinations for Piercing, Punching, Cutting, Bending, Forming, Drawing, Compressing, Embossing, Forging and Assembling Metal Parts, and also Articles of Other Materials in Machine Tools, including Special Sections Illustrating and Explaining the Making of Cartridge Shells, Wire and Bar Steel Drawing Dies, Press Tools for Making Hydraulic Leather Packing, the Making of Paint and Chemical Tablets, Manufacture of Pens, Pins and Needles, Jewelry and Eye-Glass Die-Making, Spoon and Fork-Making Dies, Sub-Press Die-Making for Watch and Clock Work, Drop Dies and Drop Forging.

By

JOSEPH V. WOODWORTH, M. E.
Author of "American Tool Making," etc.

A COMPANION AND REFERENCE VOLUME TO THE AUTHOR'S ELEMENTARY WORK ENTITLED "DIES, THEIR CONSTRUCTION AND USE FOR THE MODERN WORKING OF SHEET METALS"

Copyright © 2017 Read Books Ltd.
This book is copyright and may not be reproduced or copied in any way without the express permission of the publisher in writing

British Library Cataloguing-in-Publication Data
A catalogue record for this book is available from the British Library

Metal Work

Metalworking is the process of working with metals to create individual parts, assemblies, or large-scale structures. The term covers a wide range of work from large ships and bridges to precise engine parts and delicate jewellery. It therefore includes a correspondingly wide range of skills, processes, and tools. The oldest archaeological evidence of copper mining and working was the discovery of a copper pendant in northern Iraq from 8,700 BC, and the oldest gold artefacts in the world come from the Bulgarian Varna Necropolis and date from 4450BC. As time progressed, metal objects became more common, and ever more complex. The need to further acquire and work metals grew in importance. Fates and economies of entire civilizations were greatly affected by the availability of metals and metalsmiths. The metalworker depends on the extraction of precious metals to make jewellery, buildings, electronics and industrial applications, such as shipping containers, rail, and air transport. Without metals, goods and services would cease to move around the globe with the speed and scale we know today.

One of the more common types of metal worker, is an iron worker – who erect (or even dismantle) the structural steel framework of pre-engineered metal buildings. This can even stretch to gigantic stadiums and arenas, hospitals, towers, wind turbines and bridges. Historically ironworkers mainly worked with wrought iron, but today they utilize many different materials including ferrous and non-ferrous metals, plastics, glass, concrete and composites. Ironworkers also unload, place and tie reinforcing steel bars (rebar) as well as install post-tensioning systems, both of which give strength to the concrete used in piers, footings, slabs, buildings and bridges. Such labourers are also likely to finish buildings by erecting curtain wall and window wall systems, pre-cast concrete and stone, stairs and handrails, metal doors, sheeting and elevator fronts – performing any maintenance necessary.

During the early twentieth century, steel buildings really gained in popularity. Their use became more widespread during the Second World War and significantly expanded after the war when steel became more available. This construction method has been widely accepted, in part due to cost efficiency, yet also because of the vast range of application – expanded with improved materials and computer-aided design. The main advantages of steel over wood, are that steel is a 'green' product, structurally sound and manufactured to strict specifications and tolerances, and 100% recyclable. Steel also does not warp, buckle, twist or bend, and is therefore easy to modify and maintain, as well as offering design flexibility. Whilst these advantages are substantial, from aesthetic as well as financial points of view, there are some down-sides to steel construction. It conducts heat 310 times more efficiently than wood, and faulty aspects of the design process can lead to the corrosion of the iron and steel components – a costly problem.

Sheet metal, often used to cover buildings in such processes, is metal formed by an industrial process into thin, flat pieces. It is one of the fundamental forms used in metalworking and it can be cut and bent into a variety of shapes. Countless everyday objects are constructed with sheet metal, including bikes, lampshades, kitchen utensils, car and aeroplane bodies and all manner of industrial / architectural items. The thickness of sheet metal is commonly specified by a traditional, non-linear measure known as its gauge; the larger the gauge number, the thinner the metal. Commonly used steel sheet metal ranges from 30 gauge to about 8 gauge. There are many different metals that can be made into sheet metal, such as aluminium, brass, copper, steel, tin, nickel and titanium, with silver, gold and platinum retaining their importance for decorative uses. Historically, an important use of sheet metal was in plate armour worn by cavalry, and sheet metal continues to have many ornamental uses, including in horse tack. Sheet metal workers are also known as 'tin bashers' (or 'tin knockers'), a name derived from the hammering of panel seams when installing tin roofs.

There are many different forming processes for this type of metal, including 'bending' (a manufacturing process that produces a V-shape, U-shape, or channel shape along a straight axis in ductile materials), 'decambering' (a process of removing camber, or horizontal bend, from strip shaped materials), 'spinning' (where a disc or tube of metal is rotated at high speed and formed into an axially symmetric part) and 'hydroforming.' This latter technique is one of the most commonly used industrial methods; a cost-effective method of shaping metals into lightweight, structurally stiff and strong pieces. One of the largest applications of hydroforming is in the automotive industry, which makes use of the complex shapes possible, to produce stronger, lighter, and more rigid body-work, especially with regards to the high-end sports car industry.

One of the most important, and widely incorporating roles in metalwork, comes with the welding of all this steel, iron and sheet metal together. 'Welders' have a range of options to accomplish such welds, including forge welding (where the metals are heated to an intense yellow or white colour) or more modern methods such as arc welding (which uses a welding power supply to create an electric arc between an electrode and the base material to melt the metals at the welding point). Any foreign material in the weld, such as the oxides or 'scale' that typically form in the fire, can weaken it and potentially cause it to fail. Thus the mating surfaces to be joined must be kept clean. To this end a welder will make sure the fire is a reducing fire: a fire where at the heart there is a great deal of heat and very little oxygen. The expert will also carefully shape the mating faces so that as they are brought together foreign material is squeezed out as the metal is joined. Without the proper precautions, welding and metalwork more generally can be a dangerous and unhealthy practice, and therefore only the most skilled practitioners are usually employed.

As is evident from this incredibly brief introduction, metalwork, and metalworkers more broadly, have been, and still are – integral to society as we know it. Most of our modern buildings are constructed using metal. The boats, aeroplanes, ships, trains and bikes that we travel on are constructed via metalwork, and mining, metal forming and welding have provided jobs for thousands of workers. It is a tough, often dangerous, but incredibly important field. We hope the reader enjoys this book.

PREFACE

THIS book has been written and compiled by a practical man for the use of all practical men who are interested in the working of sheet metals, the designing and constructing of punches and dies, and the manufacturing of repetition parts and articles in presses. The designer, the machinist, the toolmaker, the die-maker, and the manufacturer of sheet-metal goods—or the producer of any other articles that may be manufactured advantageously by means of dies in presses—will find in this work a book of reference on the art of punch- and die-making, tool designing, and sheet-metal working, that will fill a place that has heretofore been vacant in the field of literature devoted to practical mechanics.

The favorable reception accorded my first volume on this subject—"Dies, Their Construction and Use, for the Modern Working of Sheet Metals,"—has been sufficient encouragement to induce me to offer the public a new work—or a second volume on this interesting mechanical art.

The present volume is broader and more comprehensive than the first. It is a work which, instead of dealing with and treating of the fundamental principles of construction, and exhaustively detailing the numerous methods of procedure for constructing punches and dies in machine tools, deals with many more branches of the art and treats in a less detailed manner of the methods of construction. It is a cyclopedia and work of reference for the mechanic, the designer, and the manufacturer who has mastered the requirements for successful tool-making as set down in the first volume. It is also well to state that it is my further purpose eventually to give this branch of practical mechanics a complete and exhaustive literature distinctly its own, covering in succeeding volumes other special branches of die-making and die-sinking, together with the working of heavy sheet metals and other materials in presses.

Throughout this entire volume it has been my endeavor to eliminate all obsolete processes, designs, and methods; to confine myself exclusively to the design, use, adaptation, and operation of the numerous sets of tools illustrated, and to make as brief as possible the descriptions relating to the most modern and approved methods for constructing such tools. The fundamentals have been fully and comprehensively covered in the first volume.

When it is considered that all other branches of practical mechanics enjoy a wide and extensive literature, it is indeed strange that until lately this branch possessed comparatively none. For many years all mechanics interested in the use and operation of dies, presses, and other sheet-metal working tools in modern use, were without any work that treated of these subjects in a thoroughly practical manner. The wide circulation of my first volume demonstrated the widespread interest in the subjects and the need for other works that would cover the art carefully, authoritatively, and entirely. Therefore, I beg to state that in this work I have striven to deal with the subjects which come within the scope of the book in a thoroughly practical manner, to the exclusion of all that is technical or that might appear in any way ambiguous to the practical man. A large number of valuable and interesting processes, rules, formulas, and designs have been embodied in the work, and it is believed will be found of inestimable value in connection with the construction, use, and adaptation of dies and press tools which form the subject matter of the book.

A large number of the tools illustrated and described herein were designed by myself; others have been selected from published articles written by myself for the technical press. For many other descriptions and illustrations I am indebted to the kind courtesy of the Editors of *American Machinist* and *Machinery*, respectively; who readily signified their permission to the use of extracts and engravings taken from innumerable articles on die-making and sheet-metal working contributed to the columns of those publications. It is indeed a pleasure to state here that it is largely to the enthusiastic manner in which those two journals have, editorially, constantly advocated the discussion of these subjects, and the great amount of space they have weekly and monthly given over to illustrating and describing processes and methods of die-making, that we have any literature to-day at all on this great branch of mechanical art. I therefore also beg to extend to the writers of those articles my thanks and sense of deep obligation, and trust that they will also take pleasure in learning of the assembling together in permanent and readily accessible form of the "*meat*" of their many valuable communications.

Trusting that the reception to be accorded this book will be such as to cause me to persevere to complete this series of works, I hopefully surrender it for the perusal of my fellow-workers.

JOSEPH V. WOODWORTH.

JANUARY, 1907.

CONTENTS.

SECTION I.

SIMPLE BENDING AND FORMING DIES; THEIR CONSTRUCTION, USE, AND OPERATION.

PAGES

Construction of Bending Dies—Copper Clip Bending Dies—Making a Set of Hinge Dies—Die for Punching, Forming, and Cutting Off Pipe Straps—Dies for Making a Spring Latch—Dies for Making Joint Ferrules—Piercing, Shearing, and Bending Die—Special Bending Die and its Work, ... 13–32

SECTION II.

INTRICATE COMBINATION BENDING AND FORMING DIES, FOR ACCURATE AND RAPID PRODUCTION.

Novel and Ingenious Cutting and Bending Dies—Dies for Making Shuttle Carriers for Sewing Machines—Dies for Crimped Box-Corner Fasteners—Blanking, Drawing, and Bending in One Die—Die for Finishing Buttons with Celluloid Tops—Stove-Rim Curving Dies—Press Tools for Forming Steel Range Bases—Die for Forming Corners for Stoves and Ranges—Dies for Cutting-off, Bending, Forming, and Driving Double-Pointed Tacks—Bending an Odd-Shaped Piece—Press Tools for Forming a Brass Stud—A Cutting, Bending, and Forming Die, . . 33–66

SECTION III.

AUTOMATIC FORMING, BENDING, AND TWISTING DIES AND PUNCHES, FOR DIFFICULT AND NOVEL SHAPING.

An Automatic Bending Die—Dies for Bending Eyes of Various Shapes—An Ingenious Bending Arrangement—Accurate Automatic Bending and Shaping Die—Punch and Die for Forming Hinge Springs for Novelty Boxes—Five Operations of Bending and Forming—Hot Forming and Bending Odd-Shaped Steel

CONTENTS.

Springs—Die with Automatic Feed for Punching, Shearing, and Drawing—Combination Shearing, Piercing, Bending, and Forming Die with Automatic Feed Attachment, ... 67–93

SECTION IV.

CUT, CARRY, AND FOLLOW DIES, TOGETHER WITH TOOL COMBINATIONS FOR PROGRESSIVE SHEET-METAL WORKING.

Automatic Slide Die for Piercing, Blanking, and Drawing in One Operation—Cut and Carry Dies for Bicycle-Rim Washers—A Follow Die for Five Operations—Dies for Chain Purse Bodies—Blanking and Forming a Sheet-Metal Roller—A Three-Operation Follow Die for Thin Stock—Gang or Multiple Blanking Tools—Blanking Die with Guide Pins for the Punch—Piercing, Blanking, and Bending Dies—Combination Cutting-off, Bending, and Forming Die—Forming and Embossing Die for Spring Clip—Compound Punch and Die for Leather Washers—Combination Die for Punching, Piercing, and Splitting Labels, ... 94–130

SECTION V.

NOTCHING, PERFORATING, AND PIERCING PUNCHES, DIES, AND TOOLS.

Punching Small Holes in Tool-Steel Blanks—Special Attachment for Notching Armature Disks—Armature Blanking and Piercing Punch and Die—Slot Index Die for Armature Punchings—Construction of Multiple Punch for Thin Metal—A Compound Piercing Die—Piercing, Blanking, and Bending Dies—Punching Four Holes at Right Angles—Sheet-Metal Perforating, . . 131–154

SECTION VI.

COMPOSITE, SECTIONAL, COMPOUND, AND ARMATURE DISK AND SEGMENT PUNCHES AND DIES.

An Accurate Sectional Blanking Punch and Die—A Double Sectional Blanking Die and Punch—A Sectional Trimming Die for Tool-Steel Parts—Compound Armature Segment Die—Double Sectional Blanking Die for Tool-Steel Blanks—Making a Com-

CONTENTS.

pound Armature-Disk Die for Small Disks—Another Compound Die for Small Armature Disks—Compound Die for Accurate Piercing and Blanking—Compound Armature-Disk Die with Special Stripper, ... 155–181

SECTION VII.

PROCESSES AND TOOLS FOR MAKING RIFLE CARTRIDGES, CARTRIDGE CASES OF QUICK-FIRING GUNS, AND NICKEL BULLET JACKETS.

Progress in Development of Drawing Processes—Set of Punches and Dies for Drawing Brass Cartridge Shells—Making Cupro-Nickel Bullet Jackets for 30-Calibre Cartridges—Assembling Cartridges in Clips—The Drawing of Cartridge Cases for Quick-Firing Guns—Cupping the Shells—First Drawing—Second Drawing—Third Drawing—Fourth Drawing—Fifth Drawing—Indenting for Primer—Sixth Drawing—Seventh Drawing—Eighth Drawing—Ninth Drawing—Tenth Drawing—Heading—Other Mechanical Operations, ... 182–210

SECTION VIII.

THE MANUFACTURE AND USE OF DIES FOR DRAWING WIRE AND BAR STEEL.

Making Dies for Drawing Wire—Diamond, Sapphire, and Agate Dies—Chilled-Iron Dies—The Making of Draw-Plates by Casting Iron Steel with Diamonds—The Drawing of Round and Rectangular Bar Steel—Pickling the Stock—Making Drawing Dies of Car Wheels—Chilled-Iron Dies vs. Tool-Steel Dies, . . . 211–220

SECTION IX.

PENS, PINS AND NEEDLES; THEIR EVOLUTION AND MANUFACTURE.

Processes for Gold-Pen Making—Manufacture of Steel Pens—Preliminary Processes—Forming and Shaping the Pen—Hardening the Pens—Tempering the Pens—Splitting the Pens—Processes of Pin Manufacture—The Making of Needles—Making Steel Spring and Latch Needles—Manufacture of Sewing-Machine Needles—Methods of Production—Cold-Swaging Process of Pointing Needles, ... 221–236

CONTENTS.

SECTION X.

PUNCHES, DIES, AND PROCESSES FOR MAKING HYDRAULIC PACKING LEATHERS; TOGETHER WITH TOOLS FOR PAINT AND CHEMICAL TABLETS.

PAGES

Efficiency of Leather Packing—Three Kinds of Hydraulic Leathers—The Selection and Preparation of Leather for Working Up—Making of Cup Leathers—Table of Blanks for Cup Leathers—Table of Blanks for Hat Leathers—Making Hat Leathers—Making of U Leathers—Table of Blanks for U Leathers—Practice to Adopt for Working Hydraulic Leathers—Re-Leathering a Machine—Making Punches and Dies for Chemical Tablets—The Making of Paint Tablets on Punch Presses, . . . 237–252

SECTION XI.

DRAWING, RE-DRAWING, REDUCING, FLANGING, FORMING, REVERSING, AND CUPPING PROCESSES, PUNCHES AND DIES FOR CIRCULAR AND RECTANGULAR SHEET-METAL ARTICLES.

A Quintuple Combination Die for Producing Five Drawn and Embossed Tin Shells at One Operation—Set of Press Tools for Drawing an Odd-Shaped Cup—Drawing and Forming Tools for Double-Acting Press—Combination Blanking and Forming Die—Die for Redrawing Large Shells—Drawing Die with Blank-Holder Attachment—Combined Cutting, Drawing, and Knurling Die—The Blanking and Drawing of Rectangular Shells—Punch and Die for Drawing a Tin Ferrule—Drawing Flanged Cups—Blanking, Drawing, and Hole-Cutting Die with Positive Knock-out—Die for Rapidly Blanking and Drawing Small Shells—Set of Drawing and Forming Dies for Making Tin Nozzles—Cupping Tools for a Double-Acting Cam Press—Economic Construction of Redrawing Dies—Punch and Die for Reducing Brass Tubing—Inside-Out Shell Drawing—Reversing Formed Sheet-Metal Shells—Drawing a Central Hub in Heavy Sheet-Metal Blank, ... 253–301

SECTION XII.

BEADING, WIRING, CURLING, AND SEAMING PUNCHES AND DIES FOR CLOSING AND SEAMING METAL PARTS.

An Expansion Punch for Beading a Shell—A Simple Wiring Die and Its Work—Expanding a Double Bead in a Brass Cup—A Punching and Half-Wiring Operation—Details of Horning and Seaming Operation—Tapering Tools for Hollow Screw Tops, . 302–312

CONTENTS.

SECTION XIII.

JEWELRY DIE-MAKING, EYEGLASS LENS AND MEDAL DIES; AND CONSTRUCTION OF SPOON AND FORK MAKING TOOLS.

PAGES

The Working a Gold-Filled Material—Jewelry Cutters and Cutter Plates—A Bit of Jewelry Die Work—Methods for Jewelry Die-Making—Cutting, Bending, and Piercing Gold Stock—The Making of Dies for Eyeglass Lens Trial Rims—Another Set of Punches and Dies Used in Making Rims for Lenses—Eyeglass Strap Punches and Dies and Their Making—The Stamping of Small Medallions—Manufacturing Methods of Spoon Making—The Making of German-Silver Forks—The Making of Souvenir Spoons, ... 313–360

SECTION XIV.

DESIGN, CONSTRUCTION, AND USE OF SUB-PRESS DIES FOR WATCH AND CLOCK WORK, AND ACCURATELY PERFORATED BLANKS OF IRREGULAR SHAPE.

The Making of a Sub-Die and Punch—Making a Compound Sub-Die for Punching an Irregular Piece—Making the Upper Die—Making the Small Punches—Making the Lower Punch—Making the Lower Shedder—Grinding the Upper Die—Locating and Fitting the Parts—Aligning the Punches and Dies—Making a Set of Sub-Dies for Clock-Wheels—Making the Chuck for the Five Segmental Punches—Finishing the Segments—Making the Filling-In Pieces—Making and Finishing the Bottom Punch—Steel for Sub-Press Plungers—A Departure from Established Sub-Press Design—Proportions of Sub-Presses, etc.—Table of Dimensions of Sub-Presses—Indexing Sub-Presses—Fine Power-Feed Punching Without Sub-Press, . . . 361–388

SECTION XV.

DROP FORGING AND DIE SINKING, TOGETHER WITH MAKING OF DROP DIES, STEAM-HAMMER DIES, NUMBER-PLATE DIES, AND DIES FOR BOLT MACHINE.

Drop Forgings—Characteristic of Drop Forgings—Practical Applications of Art of Drop Forging—Dies for Making Drop Forgings—Methods of Drop Forging—The Making of Sets of Drop Forging Dies—The Hardening of Drop-Forge Dies—Method of Reproducing Drop Dies—Set of Forging Dies for Pneumatic and Steam Hammer Work—Striking of Forming Dies for Glove Fasteners, etc.—Set of Dies for Striking up Number Plates—Special Dies for Bolt Machine, ... 389–410

CONTENTS.

SECTION XVI.

METHODS, DESIGNS, WAYS, KINKS, FORMULAS, AND TOOLS FOR SPECIAL WORK; TOGETHER WITH MISCELLANEOUS INFORMATION OF VALUE TO TOOL AND DIE MAKERS AND SHEET-METAL GOODS MANUFACTURERS.

PAGES

An Improved Washer Punch—Blanking and Piercing a Felt Washer—Device for Controlling Screw-Blanks in a Thread-Rolling Machine—Experience with Thread-Rolling Dies—Making Thread-Rolling Dies—Finding Diameters of Shell Blanks—Finding the Size of Blank for a Cup—Sizing Blanks for Drawing into Shells—Lubricants Used in Re-Drawing Cylindrical Shells—Lubricant for First Cutting and Drawing Operation—Lubricant for Drawing Cold-Rolled Stock and Sheet Steel—Copper Coloring Brass for Laying out Punchings—Hardening Blanking and Cutting Dies—Acid Bath for Hardening Dies at Low Heats—Bolster for Holding Round Dies of Different Diameters—Templet Turning of Punches and Dies—Fastening Punches and Dies in Holders—Cast-Iron Bolt Dies—A Stripping Kink in a Drawing Die—Squaring up a Punch Press—Improved Method of Bending Tubing—The Flattening of Punched Disks—Table of Punch-and-Die Allowances for Accurate Work in Blanking and Perforating Tools, ... 411–439

SECTION XVII.

SPECIAL AND NOVEL PROCESSES, PRESSES, AND FEEDS FOR WORKING SHEET METAL IN DIES.

Special Compound Punching and Forming Die for Sheet-Steel Sheave-Pulley Sides—A Dividing Head for Punch Press—Hydraulic Drawing Press for Heavy Shells and Flanges—Simple Hydraulic Press for Hobbing Jewelry Dies—An Automatic Feed for Medal Embossing Press and Dies—Ratchet Feeds for Power Presses—The Ratchet Pawl for Roll Feeds—Automatic Throwout for Misplaced Shells—Oscillating Die-Slotting Machine, . 440–467

PUNCHES, DIES, AND TOOLS.

SECTION I.

Simple Bending and Forming Dies, Their Construction, Use, and Operation.

CONSTRUCTION OF BENDING DIES.

TOOL and die-makers who have to do with the construction of tools and dies for the duplicating of shapes, and who have experienced trouble in attaining satisfactory results economically, will find much to aid them in the designs and constructions illustrated and described in this section.

Fig. 1 illustrates the simplest form of a bend, a right-angled bend. As far as the bending of the piece is concerned, it is very simple; but in making this bend it is necessary that the bend be made exactly in the centre of the piece, and that every piece be a duplicate of the first. The die illustrated in Fig. 2 will answer the purpose as long as the point of the punch is sharp and the drawing edges of the die are kept smooth; but after the punch gets dull and the edges of the die get worn, there is no certainty that a die of this form will continue to produce satisfactory results.

Fig. 3 shows a die that has been found to overcome the difficulty to a very large extent. While its first cost is more, it is more than compensated for in the uniformity of the work and the life of the die. A in Fig. 2 represents the punch used in both dies. C represents the gauges, and D the piece to be bent. Fig. 1 represents the finished piece. Fig. 3 shows the die slotted out and fitted with a pad E, which works freely in the slot. The pad E has a round shank L, which passes through the die. At the lower extremity of the shank L are placed two jam-nuts K, which regulate the height to which the pad is raised. The die bolster *J*, shown in section in Fig. 3, is tapped out directly under the pad, and a brass or iron tube *F* is screwed in. The spring *G* acts on the pad, forcing it upward until further progress is prevented by the nut *K* coming in contact with the bottom of the die. The top of the pad at the points made by the angle should then be on a level with the surface of the die, as illustrated.

No difficulty will be experienced in understanding the action of this die from the description and the engraving. In the

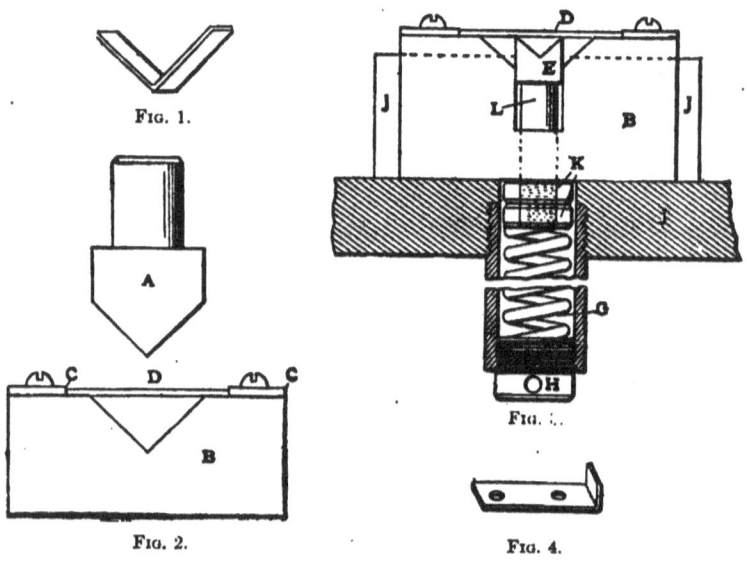

making of the die to secure good and satisfactory results, the die should be slotted and the hole drilled for the pad, and the pad should be driven moderately hard before planing the angles. This principle of working a pad in a bending die is productive of excellent results, and should be used in all cases where the nature of the work will permit. In the piece just described the bend was made in the centre of the piece, which allowed the die to be made with both sides at 45° with the surface or face.

Fig. 4 shows a blank in which the distances of the bend from the ends are in a ratio of about 4 to 1. This necessitates the changing of the angles of the die in relation to the face or surface of same. It is at once obvious that this bend could not be made satisfactorily in the die shown in Fig. 2. While it might be made in the die shown in Fig. 3, the chief difficulty would arise in gauging the piece, and it is manifestly better to change the angles.

Fig. 5 shows us one way to make this die. Fig. 6 shows us a better way, which is to apply the principle to the pad. In both cases, as in the previous example, the same punch may be used.

The relation of the angles to the face of the die is not imperative, but it is regulated to a large extent by the ratio of the distances of the bend from the ends.

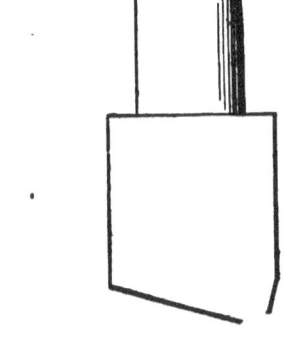
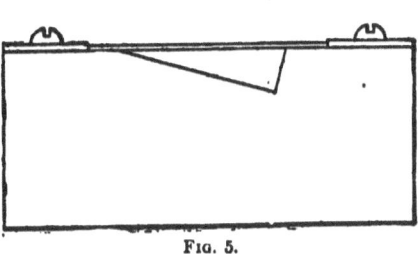

Fig. 5.

In the example the long angle should be 80° and the short angle 10° from the vertical.

Fig. 4 shows the blank as having two holes pierced in it. These holes will be utilized to gauge the bend, as shown in Fig. 6, where the piece lies on the pad before bending. As there are two holes in the piece, we put two pins in the pad to correspond, care being taken to locate the pins correctly. The use of the pins necessitates drilling clearance holes in the punch, as the pins should project through the blank and be pointed so as to

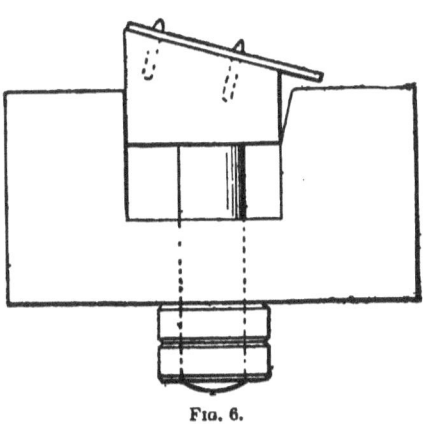

Fig. 6.

16 PUNCHES, DIES, AND TOOLS.

allow of expediting the locating of the work. If ten thousand blanks are bent in this type of die, the last one bent will be a duplicate of the first one.

There is another way in which pieces of the type of Fig. 4 may be produced advantageously. The punching shown in Fig.

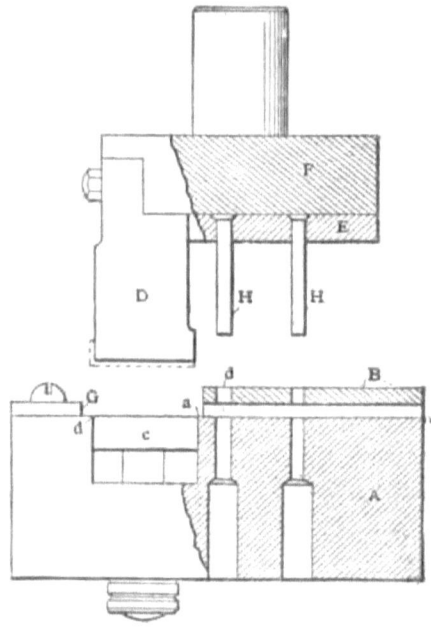

FIG. 7.

4 is a piece of metal whose sides are parallel, is bent at right angles, and has two holes pierced in it. In work of this type it is not always necessary to make a blanking die. Fig. 7 is a sectional view of a combination shearing, piercing, and bending die to produce this class of work in one operation. In this illustration A is the die, having a stripper gauge B, which must be slotted sufficiently wide to allow the material of which the piece is to be made to slide through easily and still not have any side motion. The strip of material is entered in the slot of the stripper gauge B, and the end is brought to the cutting edge of the die at a. The punch-holder F, having the shearing and bending punch D, and the punch-plate E, carrying the piercing punches H, is allowed to enter the die, piercing the two holes.

When the punch rises the strip is moved along until it brings up against the stop G, and after the holes are punched in the end of the strip the movement is continuous, producing a complete piece at every revolution of the press, of the shape shown by the dotted lines on the bottom of punch D.

The principle of the coiled spring may be applied if the metal to be bent is not too heavy. If the material runs over No. 16 B and S gauge, soft rubber should be used under the pad C for a spring. Of course on the temper of the metal depends the amount of tension which will be needed between the pad C and the punch D. There must be sufficient tension at this point to prevent slipping while the blank is being drawn over the edge b. In ordering the material to work in a die of this kind care should be taken to specify that it must be straight and to the exact width. As far as possible the holes, which are in most cases punched in blanks, should be utilized in gauging for the subsequent operations to which they may be subjected. It helps to keep the work uniform. In the class of dies which are here described it is necessary that the length of the blank and the distance between the cutting edge of the die a and the small hole d be known to a certainty before locating either the small holes or the stop G. After the small holes are punched, while the length of the blank may be changed by altering the stop G, it is done at the expense of throwing the holes in the finished piece out of position. In making the die, therefore, proceed to make the die and the punch, with the exception of the small holes in the stripper gauge on the die and the small punches. All other parts may be finished with the exception of the hardening and tempering.

Fig. 8 is a sectional view, enlarged, of the finished piece produced by the die shown in Fig. 7. The distance m is fixed, and is the measure for the shearing and bending punch D. The distance n is m plus the thickness of the metal, and is the measure for the

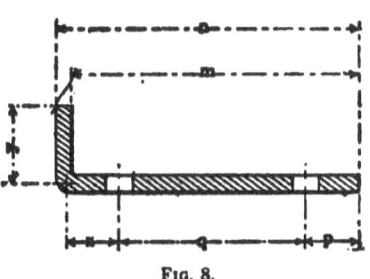

FIG. 8.

die between the shearing and bending edge. The distance *xy* is the one by which to locate the small hole *d*, and is the measure which is problematical. To find this distance proceed as follows: After finishing the die with the exception noted in the previous paragraph, put it, together with the punch, into the press with a blank of any reasonable known length, having holes drilled in proper position, placed between the punch *D* and the pad *C*, and bend into shape. The proper position for the holes is the distance *pq*, Fig. 8, and is shown by the finished sample of work or drawing which the mechanic may be working from. As the length of the piece before bending is known, all that is over the measure at *y* is to be deducted from the length which was started with, which will give the true length of the blank. After deducting the distance *pq* from the true length of the blank the remainder is the measure of the small hole *d* from the shearing edge *a*. This method of finding the length of a

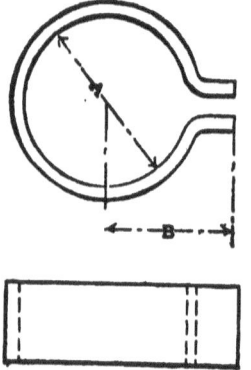

Fig. 9.

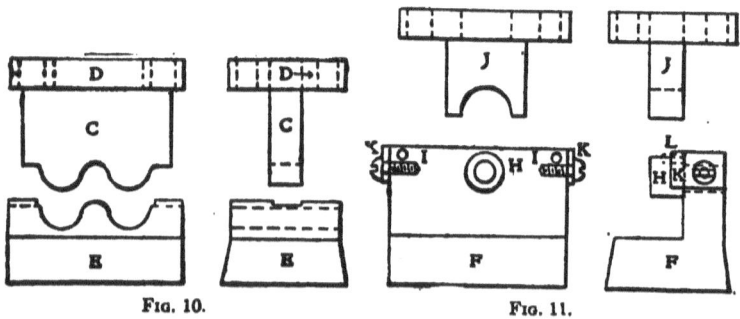

Fig. 10.   Fig. 11.

blank which is simply sheared from a strip and bent into shape, can be applied in a great many cases, and gives the correct solution in one trial. After finishing the small holes in the die, drill and tap the screw holes for the stripper, also drill the dowel-pin holes, and temper before finishing the stripper or transferring the holes for the small punches.

COPPER CLIP BENDING DIES.

Figs. 10 and 11 are illustrations of a method for forming copper clips of the shape shown in Fig. 9. With clips of this kind it is very important that they be true circles and correct to dimensions at A and B, so as to fit perfectly around whatever post they may be used for. The material coming to the right width it is sheared to length and then placed in the groove in die B, Fig. 10, and the punch C descending forms it into the first shape. For the second operation one end of the formed piece is placed under pin I, Fig. 11, and up to stop K, which brings the centre of the piece central with the forming stud H. The punch J descending, the work is formed to the shape shown in Fig. 9. All parts of the tools should be made from mild steel; and the punches C, J, Fig. 10, die E, Fig. 10, and stud H, Fig. 11, should be case-hardened.

MAKING A SET OF HINGE DIES.

From the countless varieties of articles which have to be produced by bending dies, there is always some varied form and unique feature or "kink" that taxes the ingenuity and skill of the die-maker to turn out work satisfactorily.

In the following is contained a method for the constructing of a set of dies for the production of plain hinges.

There are two distinct forms of hinges in use, and numerous modifications, probably all derived from these two styles. The first is where the joint stands out equally on both sides, and the other is where it is placed on one side only. It is of the latter style that the following treats.

The two parts of the hinge were perforated and blanked out in one piece, as shown in Fig. 12, this being done in order to save stock and also reduce the operating time. There is no waste by this method, as will be seen.

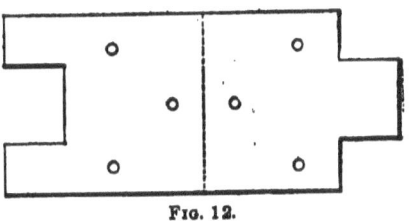

FIG. 12.

In the next operation the two parts were separated and the end formed. Figs. 13 and 14 are the parting and forming punch and die, which was made of cast iron; *a, a* are inserted pieces of tool steel for cutting; *B* is a pressure plate for removing the

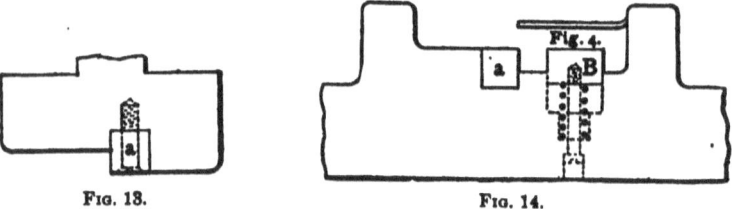

Fig. 13.   Fig. 14.

lower half from the die. For making hinges or butts, a parting die of this construction should always be used, as there is no loss of stock with it.

The next operation is the placing of the work in a vertical

Fig. 15.   Fig. 16.

position in the curling die, Figs. 15 and 16. This die is made of cast iron; *a* being tool steel, hardened and lapped in order to make the hole smooth for rolling. Pin *B* holds *a* in position and prevents its turning, thereby making the alinement perma-

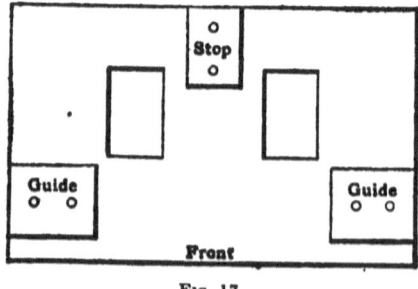

Fig. 17.

nent. Fig. 15 is the punch, which requires no explanation to be understood. The work is removed from a die of this construction by a hand lever which forces a sliding pin through the die.

PUNCHES, DIES, AND TOOLS. 21

An attachment operated by the up stroke of the press can also be used where large quantities of the parts are produced.

As the hinge parts, as produced in dies of the aforementioned design, come through the operation they require to be trimmed so as to allow of assembling easily. For this purpose a trimming die of the style shown in Fig. 17 should be used and a punch to match it.

DIE FOR PUNCHING, FORMING, AND CUTTING OFF PIPE STRAPS.

The die illustrated in Figs. 19 and 20 was used for the purpose of forming, cutting off, and punching the holes for screws in pipe straps of the shape shown in Fig. 18, making a complete strap at each stroke of the press. The tool was made adjustable

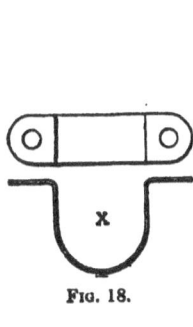

Fig. 18.

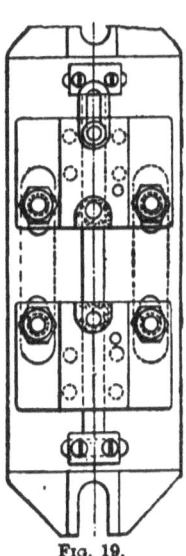

Fig. 19.

to take punches for four different sizes of pipe, the feet of the strap being the same in all four sizes, and having the same size holes. The punch shown in Fig. 20 is for 1½-inch pipe, and the other sizes for which punches were made were 1¼, 1, and ½-inch pipe.

The blocks F, F are moved toward each other and shims of

the proper thickness are placed between them, and the screws
M are for the adjustment of the dies to suit the various sized
punches. They are then locked to the bed of the die by the
bolts shown in the plan view. The alinement of the die is fixed
by the keys or tongues *K*, which are dowelled to the blocks, and
which slide in the groove *G* which is milled in the bed. The
bed is of cast iron and has grooves milled in the bottom to receive the heads of the bolts, and has two stay-blocks keyed to it

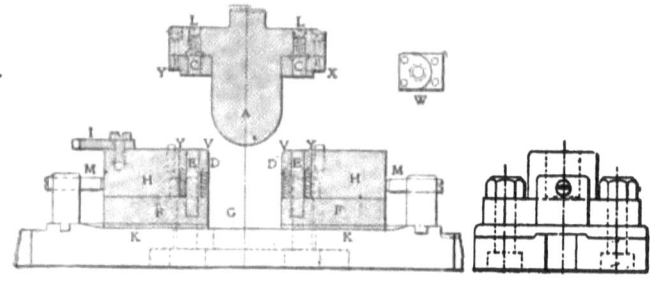

FIG. 20.

and fastened by screws. The screws *M M* pass through the
stay-bolts and butt against the die or shim, as the case may be,
and overcome any tendency of the dies to spread while the work
is being forced between them.

The stop *I* is used for three sizes only, and is removed for
the punch represented, which is the largest size, and another
stop substituted.

The drawing blocks *D* are of tool steel and hardened, and are
held up against the drawing strain by twelve springs in each,
holes being drilled in the bottoms of the blocks to receive them.
These springs are wound on an arbor the diameter of the spring
wire; they are made of drill rod and are tempered after wind-
ing. They work very nicely in the die, since the motion they
must overcome is only $\frac{1}{16}$ inch.

The drawing blocks are held in position by the pins *E E*,
which also serve as punches to punch out the blanks for the
holes in the straps. The body of the punch *A* is of machinery
steel and the cutting punches *B B* are fastened to it with screws
and pins. These punches also act as dies to receive the punches

PUNCHES, DIES, AND TOOLS. 23

E E. The blanks made by *E E* are immediately ejected by plungers *C C*, acted on by powerful springs above them, which have their tension adjusted by the screws *L L*. The extensions *X X* are intended to prevent the metal operated upon from scraping over the cutting edges while drawing into the die. As a matter of fact, however, the scraping does occur to a small extent, but not enough to hurt the work or the tool. The extensions are also convenient places to fasten the punches to the body. At *W* is shown the bottom view of one of the punches. The material used for the straps was soft brass in ribbon form, and of the right width. The thickness of the stock was No. 14 *B* and *S* gauge.

DIES FOR MAKING A SPRING LATCH.

Figs. 22 to 26 illustrate tools used for making a spring latch. The latch is shown in Fig. 21. Fig. 22 shows the blanking punch and die, which is of the ordinary "follow" type, the only special feature being that the punch is rounded at *r* to start the two wings for curling. *D* is a cross view of the die, while *B* is a plan view. *A*, Fig. 21, is the blank as it is produced by the first operation. Fig. 23 is the splitting die for turning up the

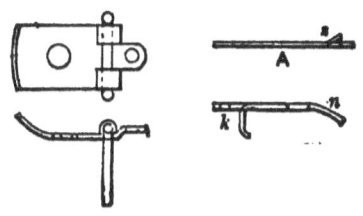

Fig. 21.

two ends as shown at *k*, Fig. 21. The blank *A* is laid on the die *D* with the bent ends *s* pointing down over the bending forms *I*. The punch *L* descends, bending the ends down, as shown at *k* in the lower view of Fig. 21, and curving the back of the latch, as shown at *n*.

The staple die, Fig. 24, is the next to be considered. The wire being laid in the die as shown, the punch descends and the knife cuts the wire off at *C*. The punch, which is ⅛-inch shorter than the knife, pushes the wire, which is left lying in the groove against stop-pin *G*, down into the forming grooves *O O*. Then the punch rises and with it the spring stripper *R*, thus ejecting

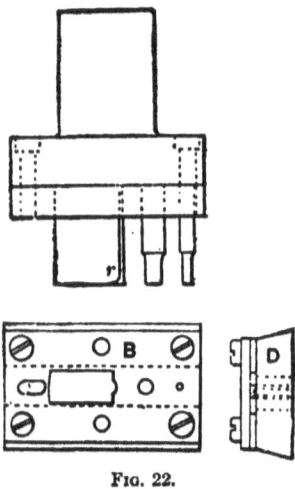

FIG. 22.

the staple when the wire is inserted in the die again and the operation repeated. The die is used in an inclined press, which causes the staples to fall by gravity after being formed. End

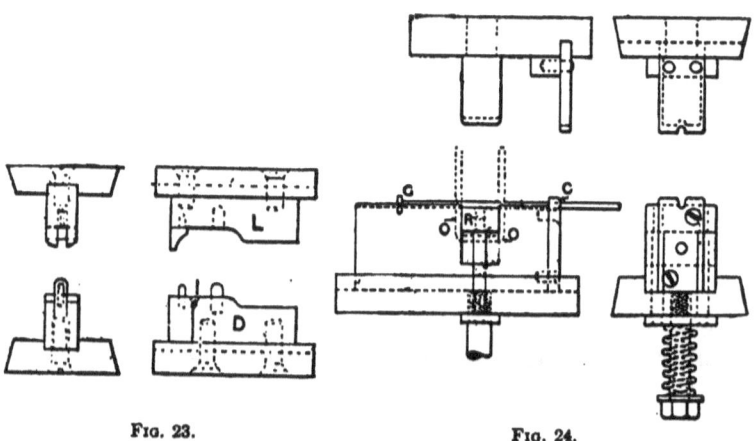

FIG. 23.   FIG. 24.

views of the punch and die are shown at the right of Fig. 24.

The curlings of the two wings, Fig. 21, are now in order. These wings are to be curled over the staple, and for this purpose the punch and die, Figs. 25 and 26, were made. The punching A, Fig. 21, is laid on the die as shown with the ends resting on the holders on either side of the die. The punch in descending strikes the blank with the spring foot, which holds it in position while the curling punch coming on down forms

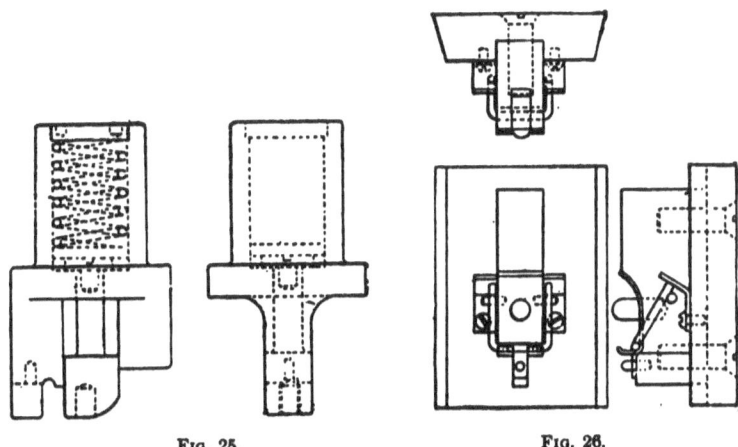

FIG. 25.   FIG. 26.

over the ends k, Fig. 21, around the staple. At the end of the stroke the tongue is formed over the raised die by the punch, thus completing the latch. It is not necessary to explain in detail the construction of the tools, as this can be clearly understood from the illustrations.

DIES FOR MAKING JOINT FERRULES.

Die makers, when making dies for forming joint ferrules, of the style shown in Figs. 27 and 28, out of sheet steel, often ex-

FIG. 27.   FIG. 28.

perience difficulty in determining the length in that part of the break-down die which forms the main part of the ferrule. In order to make this clear, the following contains a simple rule that may be always used when laying out a set of these dies. The illustrations of the different operations are sufficiently plain to enable their being understood clearly.

Fig. 29 is a section of the break-down die; Fig. 30 is a plan, and Fig. 31 an end elevation. In Fig. 29 the distance *a a* is obtained as follows: Multiply the diameter of the hole in the sam-

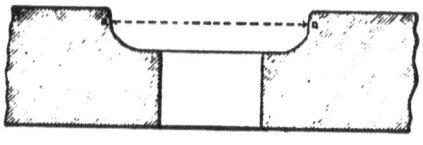

FIG. 29.

ple by 3.1416, divide the result by 2, add once the diameter of the hole and twice the thickness of the stock to be used; the final result will be the distance to work out the die.

A block of tool steel, which has been previously planed true, should be fastened in the swivel vise of the milling machine, and the different circular portions bored by an ordinary cutter held in a boring bar. The rest of the work can be done on a shaper. Any angle can be obtained by the use of the graduated swivel

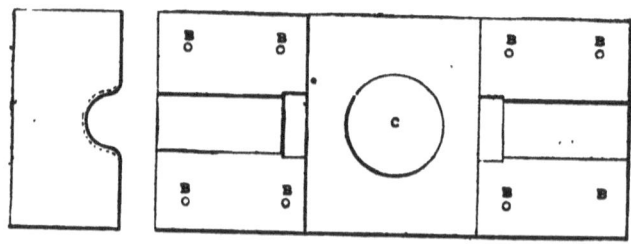

FIG. 30.

vise. Ferrules with acute angles generally require two break-down dies, the first with corners, where the branch joins the main part, extensively rounded.

B B, Fig. 30, are holes for gauge screws. *C* is a hole for knockout, but operated in a press with a cam attachment. Care should be taken to produce a perfect blank before making the cutting die, and also to properly locate the work; for then trimming or filing the edges before passing to the next operation is unnecessary.

Fig. 31 is a view of the punch, which is bored somewhat like the die, the circular bearings being only about ⅛ inch deep in the

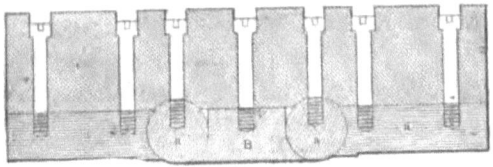

FIG. 31.

large ones and correspondingly so in the others. The round pieces *a* are turned to size, and held in place by two screws each. The block *B* filling in between the round pieces must fit

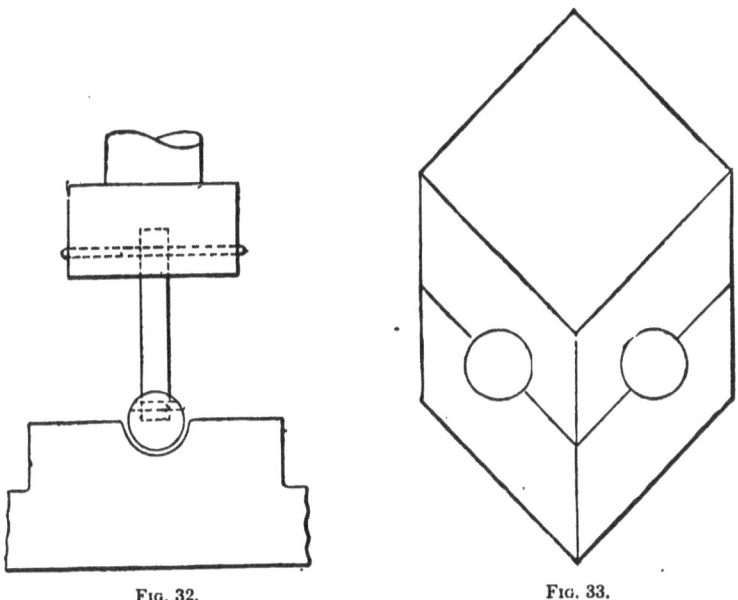

FIG. 32.   FIG. 33.

perfectly, as it keeps the two pieces from being forced out of position and closed together.

The second forming operation, Fig. 32, is to strike the material in the middle after it has passed through the break-down die, and bend it to as near a closing position as possible. The

28 PUNCHES, DIES, AND TOOLS.

essential features of the die and the method of operating will be readily understood from the engravings.

Fig. 33 is an exterior view of the closing die. The two parts, the punch and the die, respectively, are dowelled together by two pins that are not shown in the engraving, and bored to the required dimensions. About $\frac{1}{32}$ inch should be planed from the faces of the two parts, an equal amount being removed from each. This will allow the punch to come down hard on the work without coming in contact with the face of the die. Arbors must be used in the closing operation if perfect work is to be obtained.

PIERCING, SHEARING, AND BENDING DIE.

Fig. 39 shows an elevation of a combination piercing, shearing, and bending die, for producing in one operation an article like Fig. 34, and it is adapted to a large variety of work. Figs. 35, 36, 37, and 38 show some of the varieties which have been produced in dies of similar construction. The die-bed, A, is a flat plate, preferably a steel casting, slotted to admit the form-

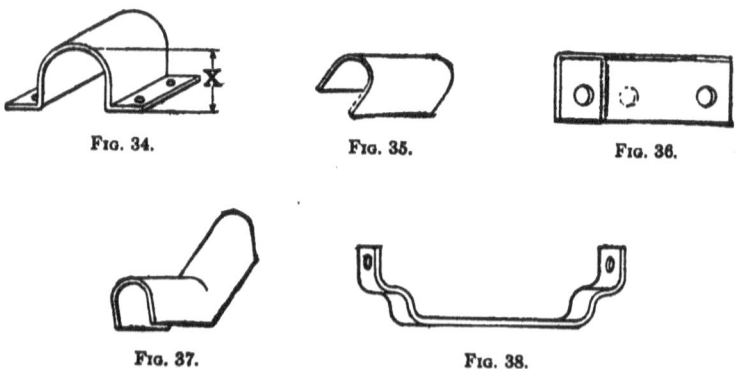

FIG. 34. FIG. 35. FIG. 36.

FIG. 37. FIG. 38.

ing die D and the piercing and shearing die B to be screwed and dowelled thereto. The die-bed A is tapped out directly under the forming die D, and the stud L is screwed in.

The pads E have four pins M riveted into them; these pins work through the die-bed on to the plate P, which is forced upward by the action of the spiral spring Q. In making a shape like Fig. 34 it is necessary that the thickness of the pads be such that they will bottom on the die-bed when the distance from their upper surfaces to the top of the forming die D equals the distance X, Fig. 39. The length of the pins M must be such that when the plate P is forced against the bottom of the die-

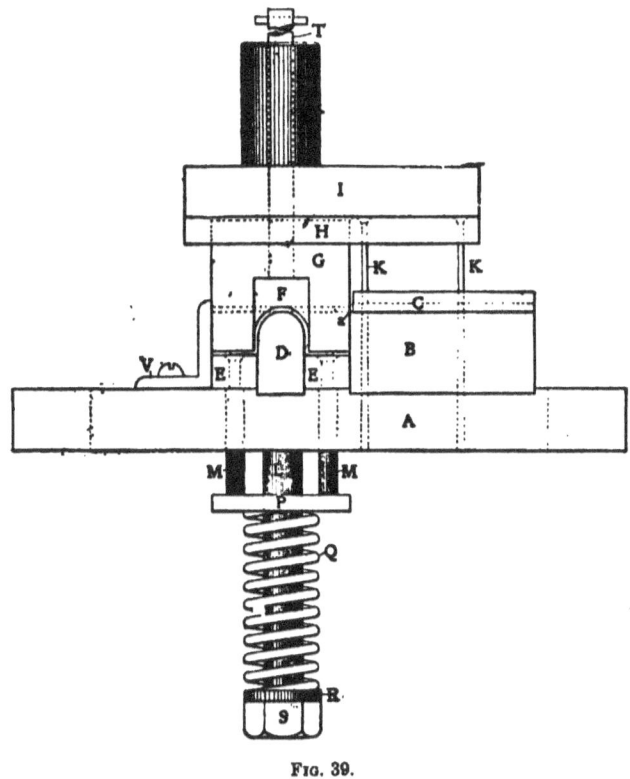

Fig. 39.

bed, the upper faces of the pads E will be on a level with the forming die D and the piercing and shearing die B.

The shearing and forming punch G should come directly under the shank of the punch-holder, to allow the working of the knock-out F which is actuated by the lever of the press on the top of rod T, on the upward stroke, thus freeing the punch G of the finished article. The press being set on an incline, the finished piece falls off and out of the way by gravity, thus allowing of the continuous working of the die.

The top of the forming die *D* should be the same height as the piercing and shearing die *B*, so that the pressure of the shearing punch *G* will be upon the blank before the act of shearing takes place. The blank is represented by dotted lines at the top of the forming die *D*. The shearing edge of the die *B* is represented at *a*.

The production from a well-made die of the foregoing description is very large, the work uniform and satisfactory, and there is no waste of material.

SPECIAL BENDING DIE AND ITS WORK.

Of the different classes of dies in use for the press working of sheet metals probably there is none that taxes the skill and inventive powers of the die-maker more than bending dies. There are rarely two alike; every new one usually differs in some respects from all previous designs; each punch and die is a law unto itself, and after completion it is dismissed from the maker's mind, to be followed by something new. The peculiar

FIGS. 40, 41, and 42.

or special features of each job, however, should be recorded, as it will surely prove of value some time in the future when constructing a similar tool.

Figs. 43 and 44 show two punches and dies that were made and used to bend a piece of flat copper 2⅛ inches long, ½ inch wide, and .025 inch thick, into a circular form with an irregular contour and overlapping ends, the outside one being flared out-

PUNCHES, DIES, AND TOOLS.   31

ward at the extremity. A blank and two stages of the bending operation are shown in Figs. 40, 41, and 42. On one side the blank, almost in the middle, will be seen a projecting part, that,

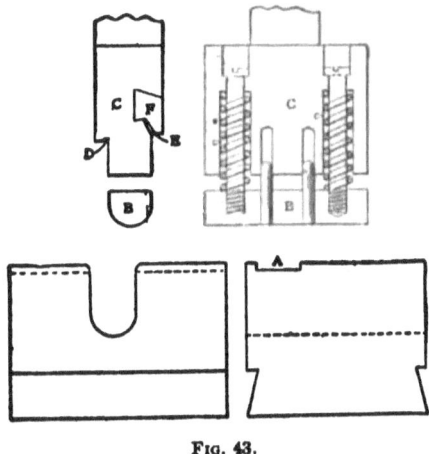

FIG. 43.

according to the specifications, must be flat and unaffected by the bending operation, and also central with the loop.

Following the regular experiments to ascertain the size and shape of blank required, the piercing and cutting die is made, and requires no description here.

Fig. 43 is a sectional and front view of the punch and die

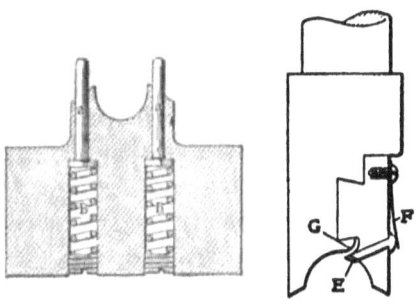

FIG. 44.

for performing the first bending operation, which comprises the bending of the ends up not unlike a letter U, with the highest end turned inward, the bend being $\frac{1}{16}$ inch from the top. The blank is placed in the recess A, Fig. 43, which acts merely as a gauge for locating it sideways; stops for the ends are ⅛ inch pins, not shown. The work is brought well forward to allow the projecting part to stand out over the front of the die to avoid bending when the other part passes down into the die. The downwardly-extending portion B of the punch is held stationary by the coil springs $C\ C$ until it comes in contact with the bottom of the die. As the ram of the press continues on the downward stroke, the notch in the shoulder at D strikes the end of the work on that side and forces it down to its proper height, thus securing uniformity. At the same time the opposite end of the blank is passing into the slot E and bending inwardly. F is a dovetailed piece to facilitate the making of the slot E in the punch.

It is a good plan in this case to locate in the bottom of the die, in rear of where the work comes, a strip of sheet metal or a rigid pin, the same height as the thickness of the stamping, to permit the punch to rest on. This will prevent a hard pressure on the work and allow it to slide around under the punch and avoid buckling, when the shoulder at D presses on that side, to correct the height.

Fig. 44 shows the closing punch and die. Following the first bending operation, the work is located on the die between the two pins $A\ A$ that hold it in its proper position for closing. The pins are supported by slender coil springs $B\ B$, which are regulated by the headless screws below them. Screwed and dowelled on to the middle front of the die is a small block of cast iron, not shown, with two pins, one on each side, to guide the pivoting part on the work during the closing of the circle. On the descending movement of the punch the pin E gives a slight pressure on the work, and causes the end, that has been previously bent in at its extremity for this purpose, to pass on the inside. The pin is retained in position by the flat spring F. As one end of the stamping goes down on the inside, the opposite follows along the punch until it comes in contact with the upper surface of the pin E, when it is forced to pass up into the groove G and bent outwardly. The use of a round arbor is necessary in this operation to size the circle in the work.

SECTION II.

Intricate Combination, Bending and Forming Dies, for Accurate and Rapid Production.

NOVEL AND INGENIOUS CUTTING AND BENDING DIES.

IN the making of dies for bending and forming sheet-metal parts there are many points to be considered besides the mechanical operations. The requirements may be for a limited number, in which case the dies would necessarily be cheap and simple. But for regular, extensive, and continued production there is often much inventive ability required to get the product out as

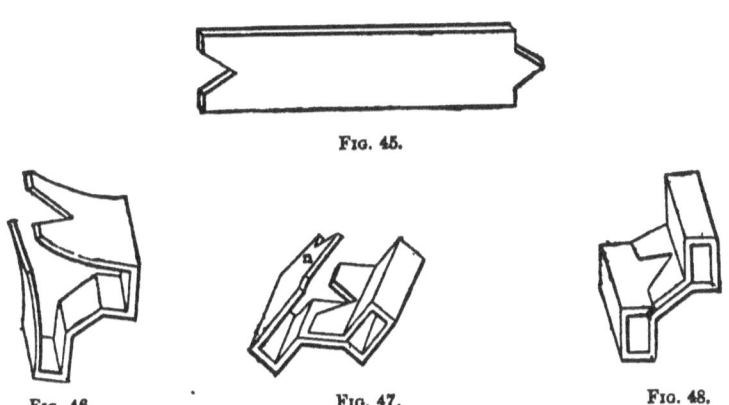

FIG. 45.

FIG. 46.   FIG. 47.   FIG. 48.

quickly and cheaply as possible. Stock must be saved in getting out the blanks, the bending must be done in as few operations as possible, and the die must be entirely practicable to make and to operate.

Fig. 45 shows a blank, which is cut from a continuous strip of soft copper, $\tfrac{1}{16} \times \tfrac{5}{8}$ inches. Figs. 46 and 47 show succeeding bending operations, and Fig. 48 is the completed article. Fig. 49 is the cutting-off die, which continuously gives the correct shape to the ends of the blanks. The stock lies in the channel *a*, and at each ascent of the punch it is fed along to strike the gauge *b*. The back of the punch, when cutting, slides against the projections *c*, so that it will not spring away from the cut-

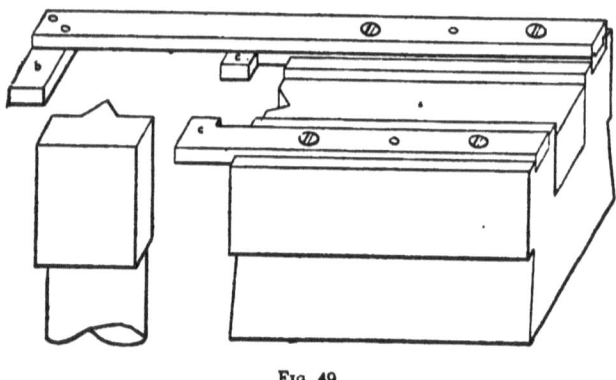

Fig. 49.

ting edge of the die. It will be seen that after the first end of the strip is trimmed and notched, no stock is wasted.

The bending of the blank to the completed shape shown in Fig. 48 is done in three operations. The die for the first of

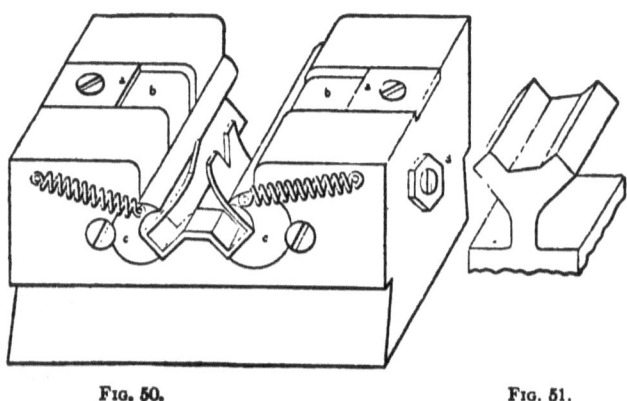

Fig. 50.   Fig. 51.

these is shown in Fig. 50 and its punch in Fig. 51, the resulting shape of its product being that shown in Fig. 46. The punch used in connection with the die Fig. 50 is shown inverted in Fig. 51. This die employs a device which is very novel—that of the rolls *c c* as a part of the bending die. These rolls are fitted to turn easily in channels in the die, their length being equal to the width of the die; they are held in position endwise by the round-head screws as shown. At each end of the die is a screw *d*, with a check nut. The points of these screws project into grooves in the rolls and determine the limits of the rotation. When the dies are free from the punch and the work, the springs

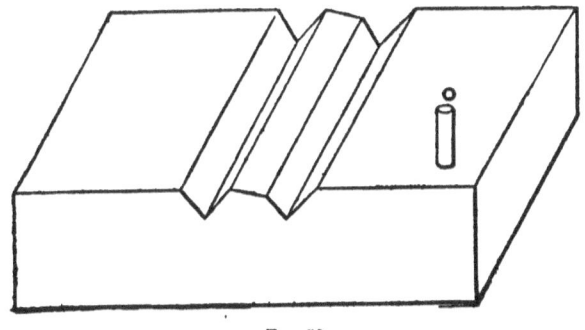

FIG. 52.

pull them to their natural position, which brings their upper lips nearly coincident with the vertical sides of the opening in the punch. The position in which they are shown in the cut is that which they assume when the punch has reached the bottom of its stroke and the work has been bent over it. When the punch rises the work rises with it and is slid off endwise by

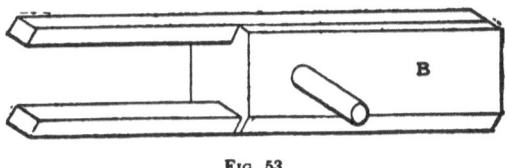

FIG. 53.

hand, the rollers, of course, moving backward to release the piece. The work is not shown in the cut in its correct position, as it has been brought forward to the edge of the die in order to show it more satisfactorily. The work, of course, is done in the centre of the die, the blank being laid upon the surfaces *b b*, between the adjustable gauges *a a*. The use of the rolls in this die saves at least one operation upon the work, as by their action

36　PUNCHES, DIES, AND TOOLS.

the material is bent considerably farther than it would otherwise be possible to do.

In the next operation of bending, which leaves the work as shown in Fig. 47, the die shown in Figs. 52 to 55 is used, in connection with the punch Fig. 55 and former Fig. 53. The

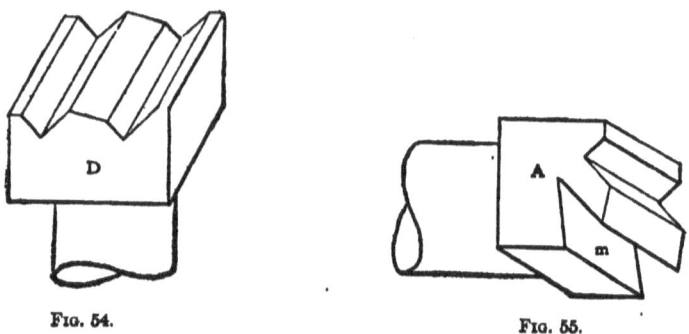

Fig. 54.　　　　　　Fig. 55.

surface marked m on the punch is for bending as far as possible the end n, Fig. 47. In the last operation the same die is used in connection with punch Fig. 54, the former B being again used. The little pin o on the corner of the die is used for pulling the work off the former, the last operation setting the ends down close.

DIES FOR MAKING SHUTTLE-CARRIERS FOR SEWING MACHINES.

The method employed in the making of shuttle-carriers for sewing machines varies considerably, as in the making of other parts. It is governed and depends on the following conditions:

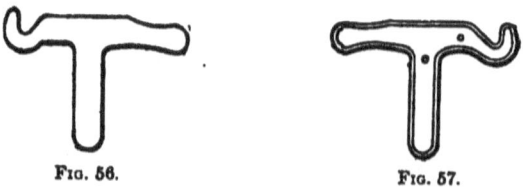

Fig. 56.　　　　　　Fig. 57.

1. The shape of the shuttle-carrier, which varies more or less with every different type or make of sewing machine; 2. The

PUNCHES, DIES, AND TOOLS. 37

quantity to be made, consequently the justifiable sum to be expended on tools, as well as their durability; 3. The machinery in the factory, where the tools are to be made, must be worked in; 4. Tool-room facilities where the tools are to be made; 5. The amount of time the "management" department allows for designing, making of drawings, and building them; and finally, 6. The intelligence or the lack of it as well as the prejudices of

FIG. 58.   FIG. 59.

the factory help, who are to handle the tools, as well as their "boss" who always knows or thinks he knows more than the "man behind the pencil" in the drawing-room.

In the accompanying illustrations, Figs. 56 to 59 inclusive are intended to show the method employed in the making of a shuttle-carrier for a certain type of sewing machine newly put on

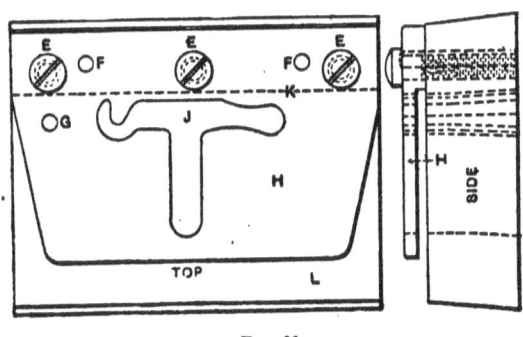

FIG. 60.

the market. A finished shuttle-carrier is shown in Fig. 59. It is made of cold-rolled steel $\frac{1}{16}$ inch thick, the outer edge rounded, the end A bent at right angles, and the opposite end B bent double as shown. It also has two small holes $C\,C$ for riveting on a flat spring and one tapped hole D for the connection with the shuttle-carrier lever. With the tools herein described it

38   PUNCHES, DIES, AND TOOLS.

takes six machine operations to bring the carrier into the final shape.

The first operation is cutting out the body, as shown in Fig. 56, and is done in a cutting press with a punch and die as shown in Figs. 60 and 61, which blanks the bodies of the carriers out of a strip of cold-rolled steel. This, of course, is done in a way to insure the greatest economy in the material, as in Fig. 62,

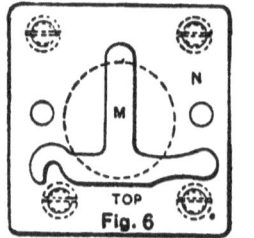
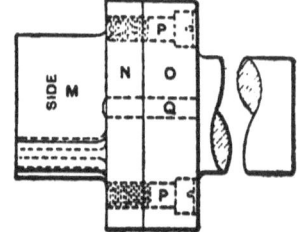

FIG. 61.

which shows a piece or strip of steel (cold-rolled) after the blanks have been punched out of it. The body of the die L, Fig. 60, is a block of tool steel of sufficient thickness to prevent its springing while in operation. The cutting sides of the opening J taper inwardly about 0.001 inch to the depth of about $\frac{3}{8}$ inch to insure the die keeping its size, approximately, after repeated grinding, while the rest of the depth of the opening is

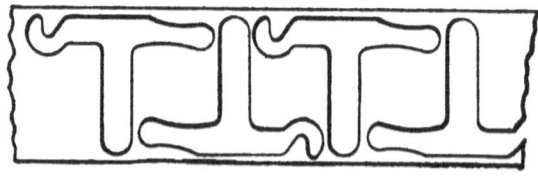

FIG. 62.

tapered considerably more, to facilitate the egress of the punchings from the die. The stripper H is planed out at the bottom to act as a guide K for the stock, and the top is part heavy enough to strip waste from the punch without bending. It is fastened to the die block with screws and dowels $E\ F$, only on one side, the other side being left open to facilitate its operation.

The gauge pin G is sufficiently removed from the extreme cutting edge to permit the arrangement of the punchings, as shown in Fig. 60, which is done by punching one row first, then turning the strip over and punching the other row between.

The punch, M, Fig. 61, is milled out from a solid block of steel to the size of the flange. This is better practice in *some*

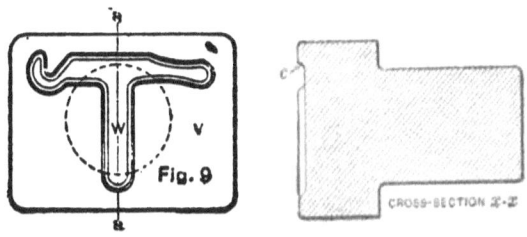

FIG. 63.

cases than planing the punch across and fitting it in a flange or pad and riveting it slightly on the back, through a mistaken notion of economy in making. The punch is fastened to the holder O with screws and dowels P Q.

The next operation in the production of the desired result is the rounding of the corners on the outer side of the shuttle-

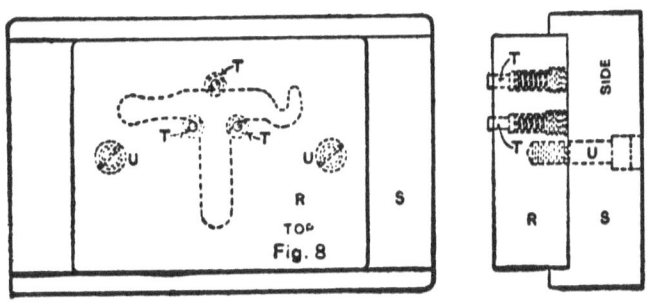

FIG. 64.

carrier, as shown in Figs. 63 and 64. The operation may be best designated as a squeezing or a squashing operation; the bottom part of the die, Fig. 64, is a tool-steel block, hardened and ground straight and smooth on the surface. It is fastened with screws U to the holder, S. The shuttle-carrier blank, of the form shown in Fig. 56, is laid on the surface of the block

40 PUNCHES, DIES, AND TOOLS.

R, between the gauge pins *T*. The punch, Fig. 63, is of hardened tool steel, and has an impression *W* in its face *V*, the same form as the blank, Fig. 57; its depth is equal to the thickness of the blank when the punch fastened in the ram of the press comes down on the blank. Lying on the surface of the die it presses the gauge pins inwardly, out of the way, and the round corners *c* in the depression in the face of the punch, strike, squeeze, and

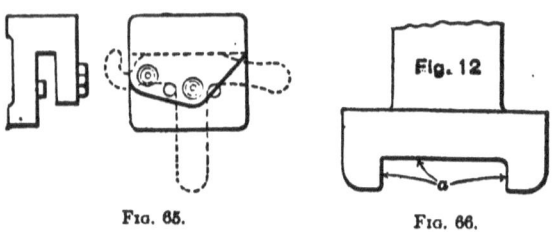

FIG. 65.   FIG. 66.

round over the sharp corners of the blank, making it in the form of Fig. 57. The blanks are then "tumbled" for some time, to remove all sharp edges and to brighten them up a bit. When taken out of the tumbler the two rivet holes *C C*, Fig. 59, are drilled in a jig, Fig. 65, and the hole *D* is punched in a small piercing die, and is afterward tapped.

The two ends of the blank *A B* are then bent over in the bending punch and die, Figs. 66 and 67. It will be noted that

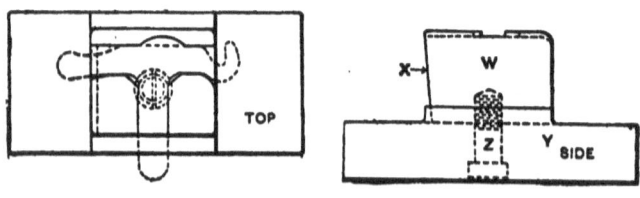

FIG. 67.

the side of the bending block *X*, Fig. 67, is slightly tapering inwardly, in order that the end of the blank bent over that side may, when taken out, remain at right angles, which would not be the case if the slide in the side of the block was made square, as the material bent always springs back more or less. The block *W* is of of tool steel, hardened and fastened with the screw *Z* in holder *Y*. The faces of the bending punch, Fig. 66, are hard-

PUNCHES, DIES, AHD TOOLS. 41

ened and ground out to the thickness of the bending block, plus two thicknesses of metal blank; its operation is obvious from the illustrations. After the blank is operated on in this tool, it is in the form shown in Fig. 58.

We now come to the last operation on the blank, that of forming the second bend on the side B of the point E, as shown in Fig. 59, which has to be brought to an exact position, without in doing so distorting the shape and form of the remainder of the body. And while it is done in the ordinary punch press, having the requisite stroke, the forming tools, which are in the form of a punch and die, are of a peculiar design and make, as will be noted from the illustrations, Figs. 68 and 69. Like a good

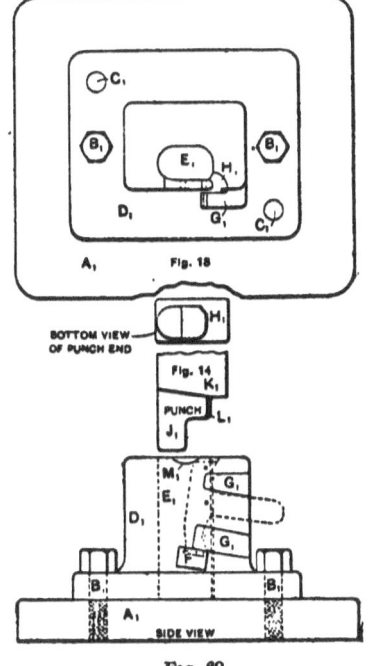

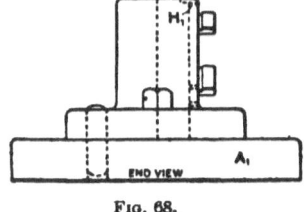

Fig. 68.   Fig. 69.

many other tools that, when made, are of simple manipulation, it may not impress the superficial observer. The difficulty is the original conception and the determination of its shape and form, to meet in an inexpensive and simple manner the requirements for the successful performing of its operations. In Fig. 68 the base plate A is cast iron. Fastened to it with screws and dowels B C is a tool-steel block D, through which there is a perpendicular hole E; on the side near the bottom there is a square hole F, leading into the opening E; two "jaws" or lugs G; a depression H, and cut-out port M, on the top. The shuttle-carrier as shown, with dot and dash lines, lies up against the side of the block, its bent end B fitting in the depression H, with the point E overhanging in the opening E, while its other end A as in opening F. The two jaws G keep it from "tipping" when the point E is getting bent by the punch Fig. 69, which, it will be noted, has a longer end J on one side that enters the opening E first, and when the punch port L comes up against the point of the carrier E and bends it, it keeps it from getting away. The part of the punch marked K flattens and sets the side of the carrier, after the point E is bent over; when that is done the punch ascends. To take the completed shuttle-carrier out, it is necessary to raise it up slightly to move it forward out of the jaws G; when the top of the carrier gets in the middle of the recess M, it is taken out completed, as shown in Fig. 59.

DIE FOR CRIMPED BOX-CORNER FASTENERS.

The reader is no doubt familiar with the crinkled metallic fasteners that are used for strengthening the sides of boxes after they have been put together, as is shown in Fig. 70, which represents a corner of a box into which one of these fasteners has been driven. These articles are used in large quantities and are sold very cheaply, as they are produced in one operation by

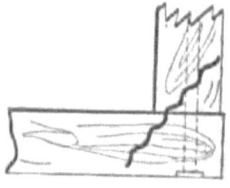

FIG. 70.

means of dies to which the stock is fed automatically. As the manner in which they are made is suggestive for other work of a similar nature a description of the tools used will be of' interest.

In Fig. 71 is shown a front view of the punch and die as they appear when in the press. As will be seen the tools are constructed so as to allow of feeding the metal in strips and producing the finished article in a single operation. In the die A

PUNCHES, DIES, AND TOOLS. 43

is the die-block, which contains the spring pad *E* along which the metal is fed, and the combined cutting-off and forming die

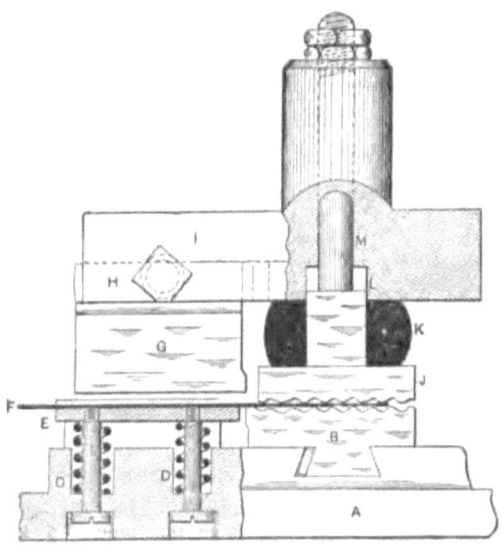

Fig. 71.

B. Fig. 72 shows a plan of the punch, looking at it from below. The punch consists of the usual cast-iron holder *I*, the cutting-

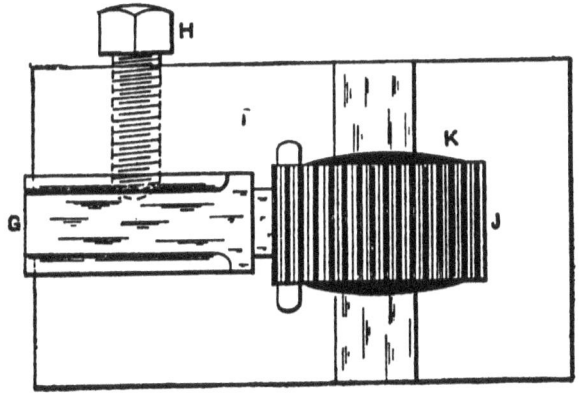

Fig. 72.

off punch *C*, and the forming punch *J*. It will be noticed that the cutting-off punch is let into a dove-tailed channel in the face of the punch-holder and is fastened in position by the large set-screw *H*. This allows for adjustment of the cutting-off punch when its face and that of the die *B* are ground. The forming punch *J* is furnished with a stem *M*, which fits a reamed hole running through the stem of the punch-holder, while the punch proper locates in the channel *L* cut in the face of the holder. *K* is a piece of hard spring rubber placed between the face of the holder and the back of the punch. The punches *G* and *J* were hardened and drawn to a dark straw temper, while the die *B* was drawn to a light straw temper.

When the tools are in use they occupy the position shown in Fig. 71. The strip of metal is fed along the face of the spring guideway *E*, the feed being automatically adjusted to project a certain length of strip across the face of the die at each stroke. As the punch descends, this portion of the strip that is to be formed is cut off by the punch *G*, and then the section is held on the face of the die by the punch *J*, which remains stationary while the cutting-off punch and the spring guide may continue to descend, and the rubber compressing meanwhile, until the punch *J* bottoms in the punch-holder, when it presses the strip into the desired form. As the press is inclined the finished work slides off from the face of the forming die as soon as the punch has risen sufficiently to release it. By the use of these tools the articles are produced very rapidly and cheaply.

BLANKING, DRAWING, AND BENDING IN ONE DIE.

The tool described in the following performs three distinct operations: blanking, drawing, and bending. It may be termed a built-up die, as there are nine parts to it without the screw, the plunger also being made in sections and containing four parts.

This tool was made to manufacture what are termed "boxes." These "boxes" were made of No. 30 B. and S. brass stock. Fig. 73 is a side view and section of the "box." The first work was the making of the bottom plate *A*, Fig. 75, of tool steel $2\frac{1}{2} \times 2 \times \frac{3}{8}$ inches, the sides being planed with a 5° angle to fit the die-block. The hole *B* was then cut through the die, the width

PUNCHES, DIES, AND TOOLS. 45

being the width of the blank to be cut and long enough to allow a surplus on the plunger outside of the drawing part. The hole was then recessed back to the dotted lines C, this recess being the thickness of the tips on the drawing plates.

The plate was then returned to the shaper for the cutting slots, D for the guide-plate, E for the cutting-plate of the die and slot, F for the drawing-plates, the first two slots being $\frac{1}{16}$ inch deep and the last $\frac{1}{16}$ inch deep. The drawing-plates were planed to fit slot F, with a tongue $\frac{1}{4}$ inch deep and $\frac{1}{16}$ inch thick, extending down through the die to dotted line G. The top edge of this tongue was then curved to dotted line C, the same radius as the required form of the "box," this being shown by dotted line H. These pieces were then hardened and tempered. The

Fig. 73.

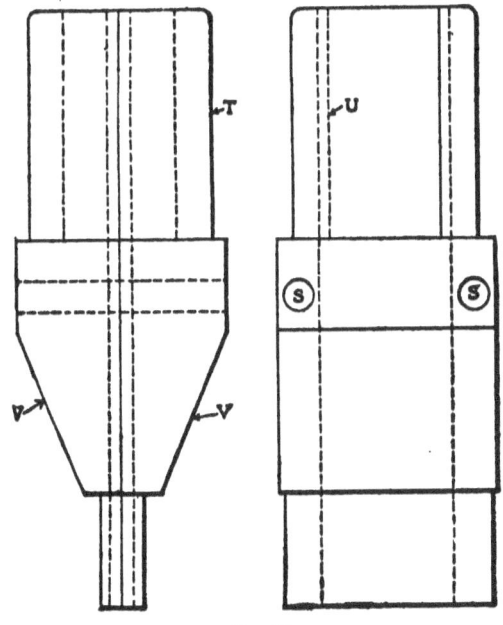

Fig. 74.

cutting blade was fitted into groove E, being $\frac{1}{16}$ inch thick, the centre part on one side extending to the edge of the hole in the die for the cutting plunger, the other side extending to the end of the die. The guide J was fitted into groove D, the same as the cutting blade except that the thickness was $\frac{3}{16}$ inch, or equal to the depth of the slot. This guide served as the gauge to which the stock was pushed into the die.

The two guides K and L were fitted into slot E and fastened on top of cutting blade, guide K being made in one piece with a projection on each side, one side extending nearly to the edge of the cutting blade, the opposite side to the edge of the die. The opposite guide L was fitted into the opposite side of the slot. This guide was formed on a different plan. A piece was planed

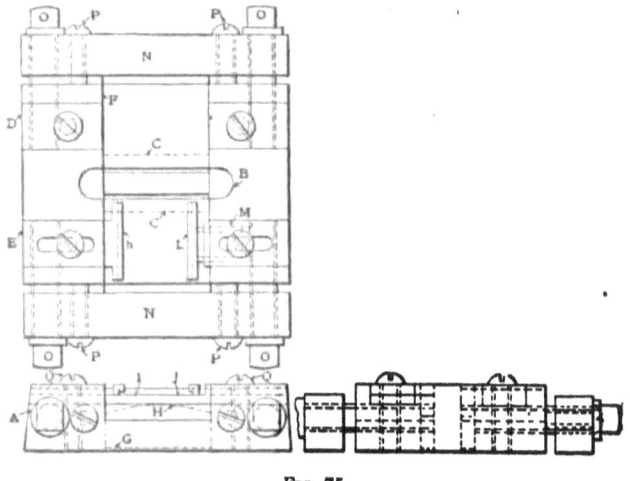

FIG. 75.

to fit the slot, and two holes for pins M were drilled. A narrow strip was then cut to length of the projecting ends of the left-hand guide; two holes were drilled through the part to correspond with holes in part just described, and into these holes were driven pins. Two small spiral springs, open, were then made and one placed in each hole in the stationary part of the guide. The faces of each of these guides were lipped under $\frac{1}{16}$ inch. This served to hold the stock on the face of the cutting blade. An elongated hole was then cut in each guide for fastening. The end or adjusting blocks N were then planed $\frac{3}{8} \times \frac{1}{2}$ inch and as long as the width of the die, the ends with the same angle as the bottom plate. These blocks were fastened by screws O, the screws P being tapped into the blocks and the ends resting against the ends of the die. These blocks served to adjust the drawing-plates F, the end of the plate projecting beyond the ends of the die. The screws Q fastened the guides and the cutting-plate to the bottom plate. This cutting-plate was then tempered and the die was ready to be assembled, which, however, could not be done until the plunger had been completed.

The cutting plunger, Fig. 74, with end or face view, Fig. 76, was of tool steel, ¾ x 1½ x 8 inches, with grooves planed on one side and in the centre 1⅜ inch wide and .085 inch deep, or one-half the thickness of the drawing plunger. This piece was sawed into two pieces of equal length.

The cutting plunger was now laid aside and the drawing plunger, Fig. 77, was taken in hand. The blade R was planed

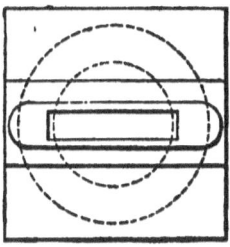

FIG. 76.

.170 inch thick and 1⅜ inch wide. After this blade had been finished and fitted to the pieces just planed for the cutting plunger, it was placed in the slot planed in the cutting plunger, and the two halves were clamped together. The pin holes S S were then drilled and pins fitted, a driving fit, into these holes, this being the only means of the fastening the two parts together, as the screw fastening the shank into collet was relied upon to do the remainder. The shank T was turned to fit collet in the press, the hole U was bored a trifle larger than the shank of the drawing plunger, the sides V were bevelled off, and the end which served to do the cutting was planed as near as possible to size. All that now remained was to finish the drawing plunger. The lower end rounded out the radius of the bottom of the "box," the shank W was turned to fit the drawing plunger in

48 PUNCHES, DIES, AND TOOLS.

the press, and the end grooved for the holder which held it in place. A groove was planed into the opposite end into which the blade was fitted and pinned.

The drawing-plates being placed in the slot F, the guide or stop J in slot D and fastened with the screws Q, the cutting-blade I was placed in slot E. On top of this were placed the two guides K and L, and the end blocks N were fastened on.

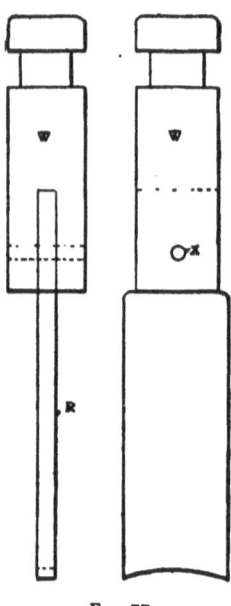

FIG. 77.

The drawing plunger was now brought into use. Taking two pieces of stock to be used, and placing one on each side of the blade of the drawing plunger, then placing the same between the two drawing-plates, these plates are adjusted by unscrewing the screws P and tightening screws O. The guides K and L are then adjusted to the width of the stock. Further description is unnecessary as the operation of the tools can be clearly understood by the reader.

PUNCHES, DIES, AND TOOLS.   49

DIE FOR FINISHING BUTTONS WITH CELLULOID TOPS.

Fig. 78 represents a button, the face of which is a shell of figured metal with a lip, and with a transparent covering of celluloid. A regular button back is used inside the shell, and the shell with the lip holding the celluloid cover is closed against it. This particular operation described here is the assembling or finishing one.

Fig. 78.

The "die," Fig. 80, has a post in the bed with a sleeve, as shown in Fig. 78, over it, actuated by a strong spring which keeps it up in position. The shell is dropped inside this sleeve with a flat celluloid blank upon it, as shown. The punch, Fig. 79, which when at rest stands considerably higher than here shown, descends and pushes the sleeve which it carries down far enough to allow the celluloid blank and the shell to enter it, and

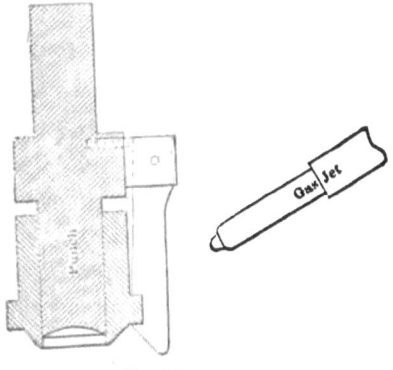

Fig. 79.

they are thus picked up and held for the second stroke. The gas jet, Fig. 79, keeps the upper sleeve and the punch hot enough to form the celluloid without breaking it or wrinkling it. The sleeve on the punch must of course be finished the thickness of the celluloid larger than the shell.

The tube shown in Fig. 81 now comes into use. It is held in the left hand of the operator, and he places it in the back of the button, as shown. It is then placed to fit over the top of the lower sleeve, and is held there until the punch descends. The

4

hook which holds the upper sleeve in position is pushed aside and the sleeve comes up against the shoulder on the punch. The lip on the shell comes between the interior curve of the tube and the top sleeve, throwing it down and against the celluloid cover, and they are pressed together against the back of the button. In this last stroke the hand-held tube is pushed down until the

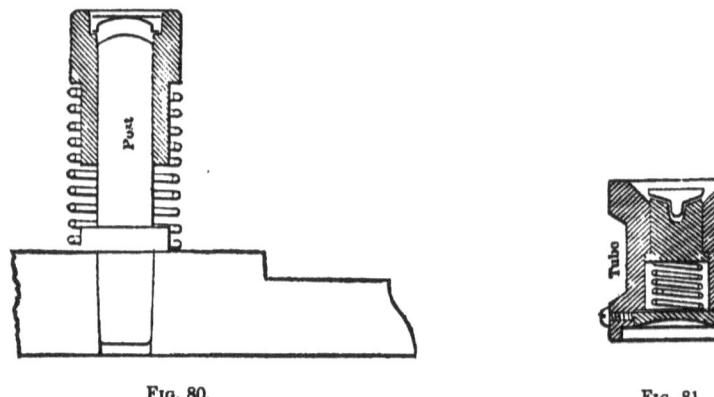

Fig. 80.   Fig. 81.

washer W rests solid on the post, thus giving a firm closing pressure. The back of the button fitting loosely in the shell enables it to enter before closing occurs. It rests on a sliding central plug, which is held up by a light spring resting upon the washer, thus allowing it to back down during the closing operation. As the punch ascends the completed button is removed, and the sleeve is automatically hooked in place again for the next button, and so on.

STOVE-RIM CURVING DIES.

Figs. 82, 83, and 84 illustrate a pair of dies used for curving stove rims. The strip for the rim is first bent around the corners of the lower die and then put in the slit, its lower edge resting on the steel ledge, while its upper edge comes a little above where the lower die commences to curve over. The upper die comes down to within the thickness of the plate of the lower die; then both dies descend together, forcing the strip to slide

PUNCHES, DIES, AND TOOLS. 51

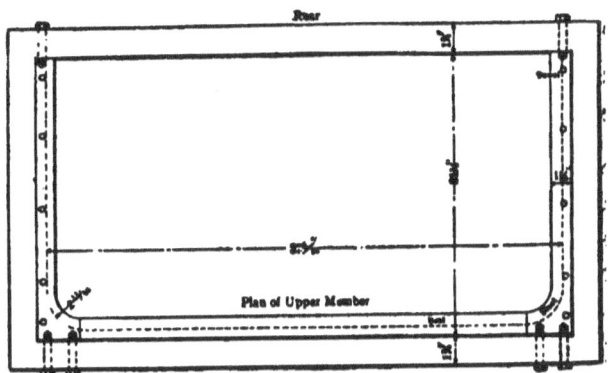

Fig. 82.

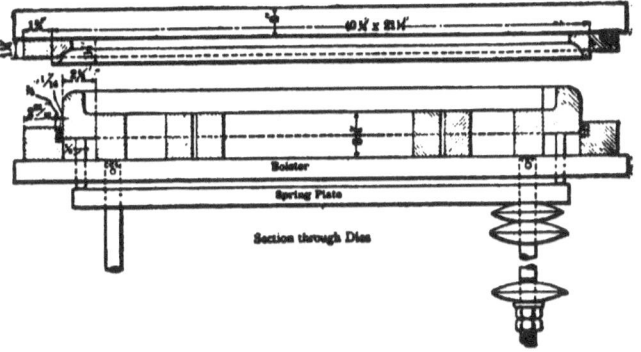

Fig. 83.

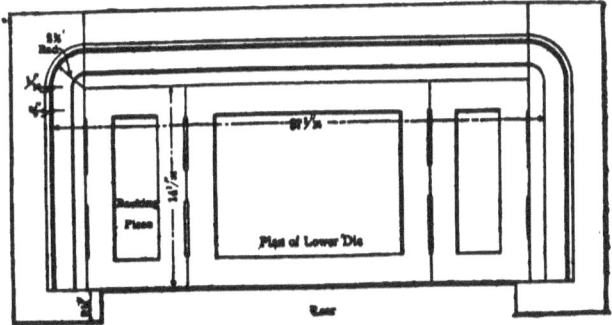

Fig. 84.

into the curved passage between the two dies, and leaving no opportunity for the metal wrinkling.

PRESS TOOLS FOR FORMING STEEL RANGE BASES.

The die and punch illustrated and described in the following were designed for the forming of steel range bases.

This die performs the second operation on the base which has been previously formed to the shape shown in reduced scale in Fig. 97, *J*. Fig. 88 is the die-block upon which the steel sheet *J* rests during the operation. It is of composite structure, the four pieces *G*, Fig. 94, being placed at the sides and the ends as indicated at *F*, Fig. 93, which is a sectional view of the bottom die with the side pieces *G* in position. The top part or punch is shown at *B*, Fig. 89. It is inlaid with steel strips of the shape shown at *H*, Fig. 95, which form end and side pieces to match the pieces *G* in the bottom die. The pieces *C*, Fig. 90, are fitted across the channels in Fig. 89 to hold the steel end pieces and complete the shape of the punch. The cushion plate, Fig. 91, is fitted with side strips which project about ¼ inch below its base. This is indicated in the cross section of the

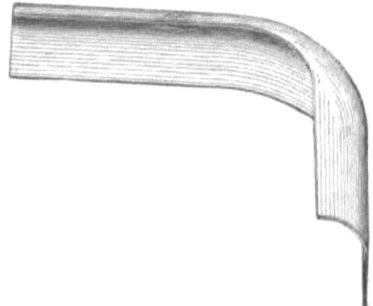

FIG. 85.

cushion-plate at Fig. 92. The reason for having these strips on the cushion-plate is that they drop into the previously formed groove in the sheet and over the edges of the die-block and hold the piece from jumping up when the side-forming pieces move inward.

The lower die-plate, Fig. 88, has dovetailed grooves planed

PUNCHES, DIES, AND TOOLS. 53

at right angles for the reception of the similar shaped projections as on the bottoms of the steel strips, Fig. 94. The holes shown in the elevated portion of Fig. 88 are for springs which shove the side pieces back when the punch ascends after the forming operation is completed. A section of a completed base is indicated in reduced scale at *J*, Fig. 98, and a complete base is

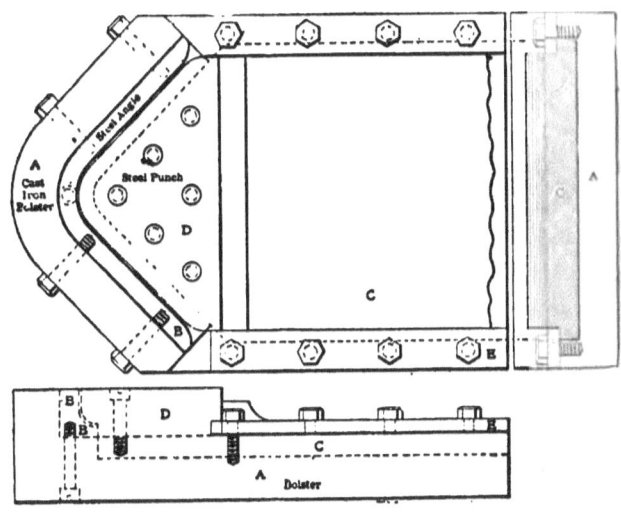

FIG. 86.

shown in the engraving, Fig. 99. At *I*, Fig. 96, is shown a sectional view of the punch and die when closed.

This die was made for three sizes of stove bases; the bases measuring 29 x 20 inches, 31 x 20 inches, and 33 x 20 inches, respectively.

DIE FOR FORMING CORNERS FOR STOVES AND RANGES.

The die shown and described in the following was used to form corners like that shown in Fig. 85, without marring or wrinkling the stock. When finished they were nickel-plated and used on up-to-date stoves and ranges.

In Fig. 86 *A* is a bolster made of cast iron and heavy enough to stand considerable pressure in shoving the stock under. *B* is a steel angle piece screwed to the bolster form beneath and through the wall. It is made of tool steel and is hardened and ground. C is the ram, also of cast iron, to which is fastened the former D of tool steel, also hardened and ground and fastened to the ram with fillister-head screws. The ram was operated by hand by what is called the link-lever motion, not shown in the engravings.

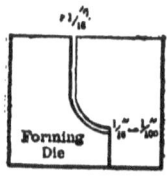

Fig. 87.

The bolster after having the bottom planed was milled out to take the ram. The wall and base were also milled, and to them was fastened the forming angle, the centre of which was just in line with the centre of the ram. The ram was planed to slide in the bolster and was kept in place by plates E on each side. The steel forging for the angle was of tool steel, and after being roughed out a special milling cutter was used, of the same shape and diameter as the corner to be formed. Instead of milling the same to an angle of 90°, it was made to only 87°, so that when hardening it would spring outward, which it did in this case to 90°. All this milling was done in a vise. After hardening this angle was ground all over and fastened in posi-

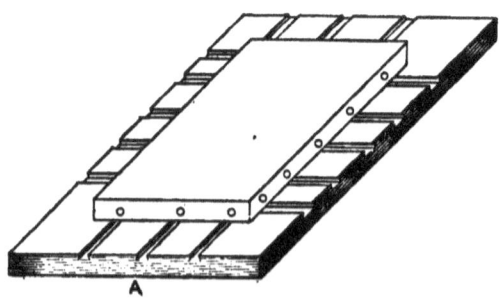

Fig. 88.

tion in the bolster as shown. The punch or nose of the ram was a solid piece of tool steel, thick and large enough to fill the angle all over. After roughing this piece to near the shape, a special face-plate was made to fit the dividing head of the milling machine, and a special cutter was used, just the opposite of the cutter made for the angle minus the thickness of the stock to

PUNCHES, DIES, AND TOOLS.

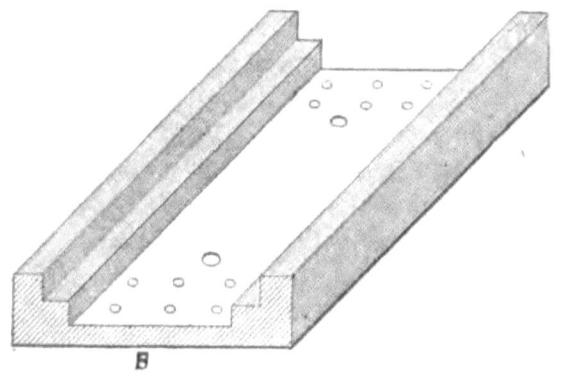

Fig. 89.

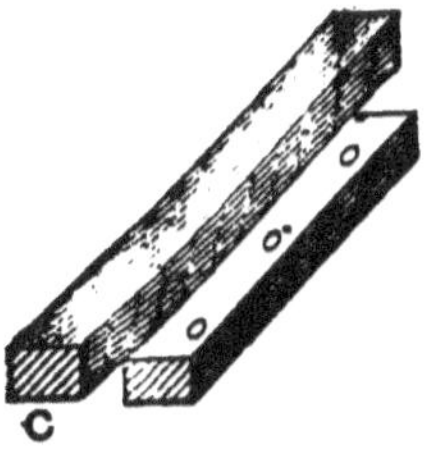

Fig. 90.

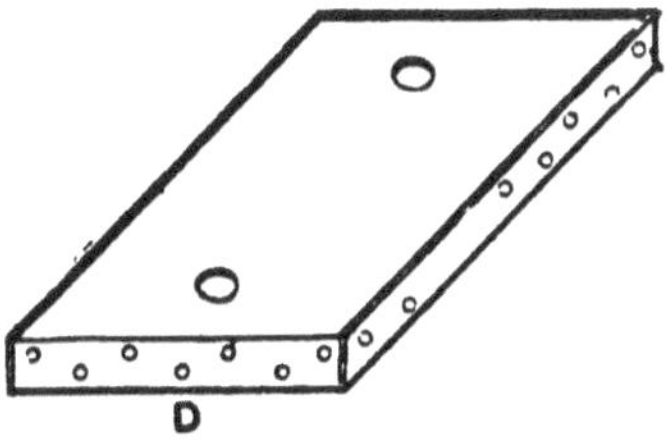

Fig. 91.

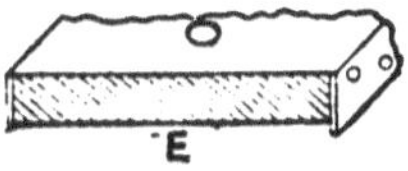

Fig. 92.

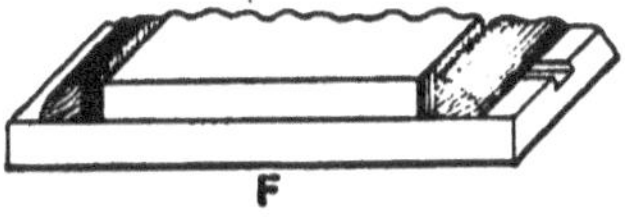

Fig. 93.

Fig. 94.

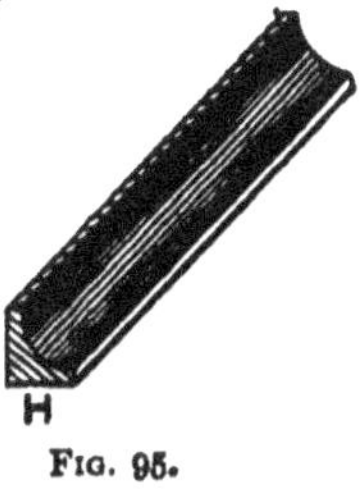

Fig. 95.

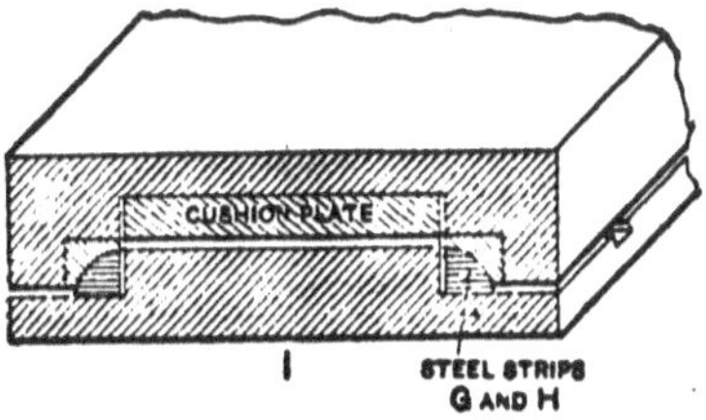

Fig. 96.

Fig. 97.

Fig. 98.

be worked, which was $\frac{1}{16}$ inch. The punch was milled to an angle of 90°, and the point was milled the proper radius by having the milling cutter directly over the centre of the dividing head and swinging the same ten turns, or 90°, and then allowing the cutter to follow along the straight wall. This also was hardened and ground. In grinding the base of this punch when in position, the space at the bottom was $\frac{1}{100}$ inch less than the thickness of the stock (see Fig. 85), as the stock in pushing under has a tendency to lift the ram. Both angle and punch were rounded out at the ends, so that the shape gradually kept getting less and less, and so that after forming the corner at each

FIG. 99.

end of the strip there would be no joint after bending and forming the intervening space in another die.

The stock to be formed came in strips $1\frac{1}{16}$ inches wide and $\frac{1}{16}$ inch thick. This was cut off to proper lengths, so that one of these corners could be formed on each end with a long straight part between, as often seen on the front of a stove. These pieces were first bent flat-wise at each end and the right distance apart before they came to this machine. A piece was then clamped in position against the forming angle by the steel nose of the ram, the ram then moving forward by pushing the lever upward, which straigthened out the link and, tripping the press, pushed the stock downward to the shoulder of the form, after which the same was released by pushing or pulling the lever downward, and the same process repeated for the other end. The intervening straight part between the corners, as will be seen, is still flat. This is then formed to the same shape as the

PUNCHES, DIES, AND TOOLS.

two corners in other dies. The punch used in forming the corner is perfectly flat, as the same, by downward movement of the press, has merely to act so as to push against the edge of the strip of metal being worked.

DIES FOR CUTTING OFF, BENDING, FORMING, AND DRIVING DOUBLE-POINTED TACKS.

The dies illustrated in Figs. 100 to 108 were used for making and driving at one stroke twenty-seven double-pointed tacks in a row. The tacks were to be made from a coil of flattened wire and driven through a canvas belt into an oak slat, the points coming through the slat and clinching smoothly. These canvas belts with the oak slats are used on harvesting machines. The vital part of the machine is the die for making and driving the tack. Two different widths of canvas and lengths of slat were used and the dies at the ends of the row each made and drove two tacks about $\frac{1}{16}$ inch apart, while the other twenty-three dies made and drove but one tack each. Before making the full set of twenty-five dies one experimental double die was made with a cheap device for running it, and it was tested for durability and satisfactory operation by making 3,000,000 tacks, several thousands of which at different times were driven through oak slats and canvas. At the close of the test the dies appeared to be good for at least 3,000,000 more tacks before any sharpening would be necessary. No defective tacks were made, except a very few which were caused by short kinks in the wire.

The complete apparatus consists essentially of the rear and front dies, the sliding shear and bender, the sliding bender, the swinging bender, and the driver, all as here shown in general elevation and in details in Figs. 100 to 108. The two flattened wires, which are fed into the die from the rear side, lie over the swinging benders, so that when the sliding shear, descending, cuts them off, the ends are bent over the swinging bender by the sliding bender and the shear and bender' which are forced down together by the disengaging hook on the cross-head of the machine. When these two benders have completed their stroke, which brings their lower ends flush with the bottom faces of the

58   PUNCHES, DIES, AND TOOLS.

dies, the tripping-pin, striking the bevelled face of the disengaging hook, disconnects it from the caps, which are not further affected by its continued downward travel.

About this time the lower end of the driver, whose upper end is positively clamped to the cross-head, is pushing the com-

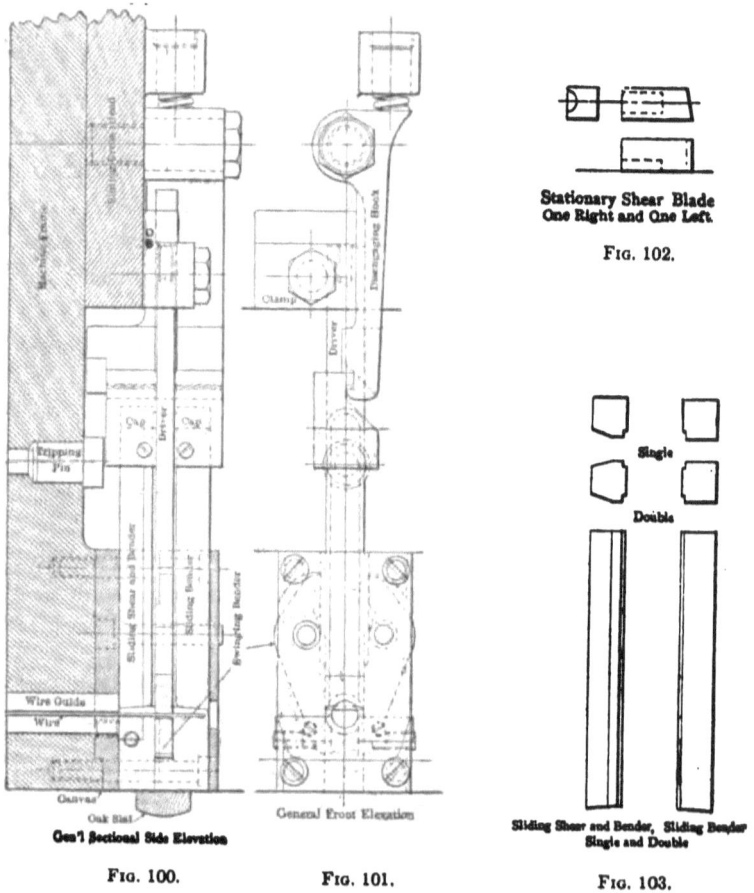

Fig. 100.   Fig. 101.   Fig. 102.   Fig. 103.

pleted tacks downward. The enlarged portion of the driver when in its upper position, as shown in the front elevation, holds the upper ends of the swinging benders outward and the lower ends in position for bending the tacks. In descending after the shear and benders have completed the tack, this enlarged part of the driver forces the lower ends of the swinging

PUNCHES, DIES, AND TOOLS. 59

benders out of the tacks and clear of the descending driver. The tack is positively guided in bending and in driving until it is clinched, by the corner grooves in the shear and bender and the sliding bender. It is so well and firmly guided that it will drive

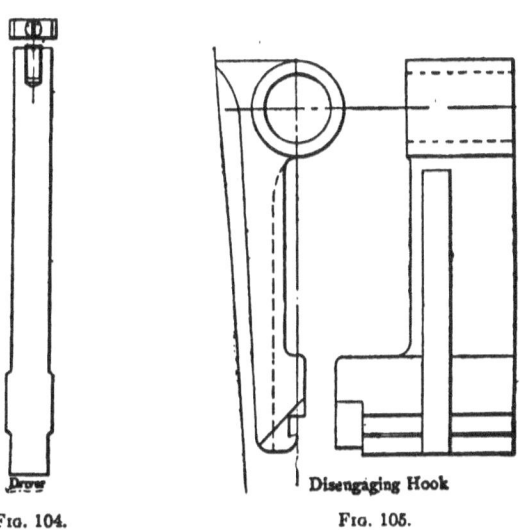

FIG. 104. FIG. 105.

into oak knots without any difficulty, and without breaking or doubling the tack. On the return stroke of the cross-head the disengaging hook takes hold of the caps and draws back both sliding benders so as to clear the entering wire, while the re-

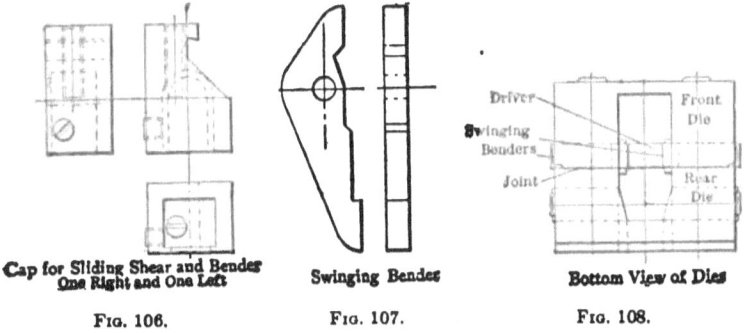

FIG. 106. FIG. 107. FIG. 108.

turning driver throws out the upper ends of the swinging benders, bringing the lower ends back into position again for bending the tacks. The stationary shear blades are held in position

60 PUNCHES, DIES, AND TOOLS.

endwise in the rear die by a set-screw in the half-round groove. The rear die is threaded to fit this set-screw, but the shear blade is grooved to fit the outside on the screw. The stationary shear

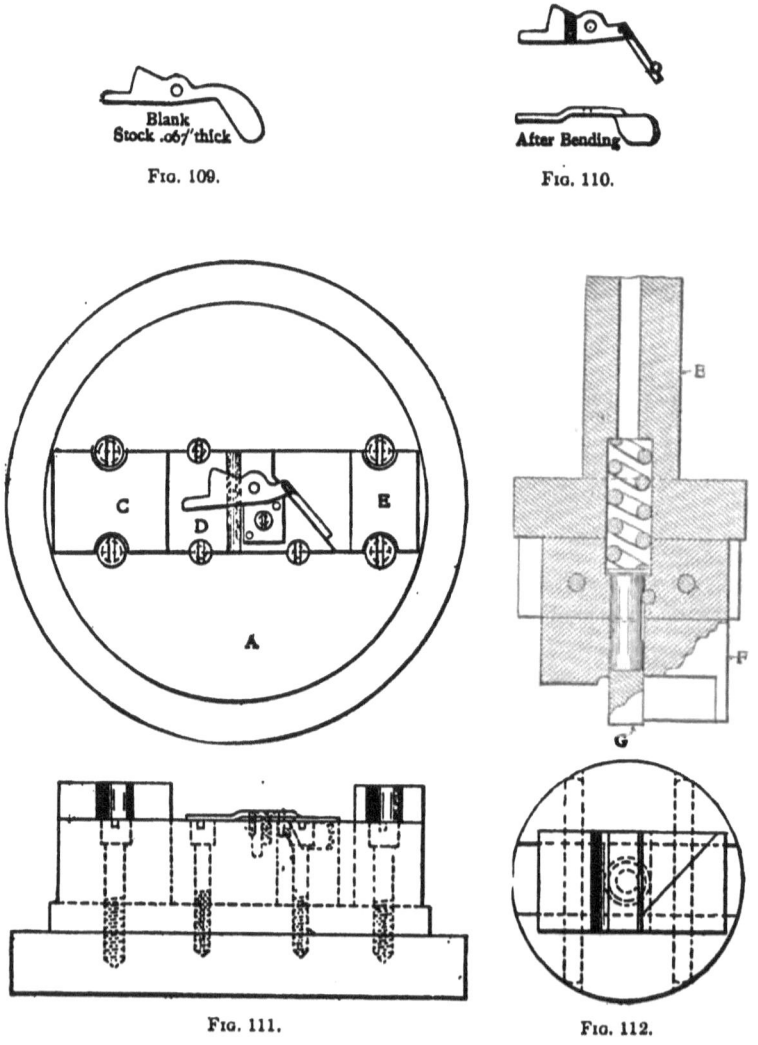

FIG. 109.

FIG. 110.

FIG. 111.

FIG. 112.

blade is held sidewise by the set-screw in the back of the rear die.

The single dies for making one tack were made with but one

PUNCHES, DIES, AND TOOLS. 61

shear blade and one swinging bender, and the driver had the projection for operating the swinging bender on one edge only. The rear die is shaped, of course, to fit the single shear and bender. The machine makes one revolution at a time, like the ordinary power press, stopping always with the cross-head at the upper end of its stroke and the parts in position to allow another length of wire to be fed in. The slat and canvas are held firmly against the bottom face of the dies by a tempered steel bar curved to fit the slat. All important wearing parts were made of Styrian tool steel carefully hardened and tempered.

BENDING AN ODD-SHAPED PIECE.

The engravings, Figs. 109 and 110, show a rather odd-shaped piece to be bent and the punch and die used for bending same.

A, Fig. 111, is the usual style of round die-bed made of cast iron. This was milled out for the parts C, D, and E, which, after being machined and made a good fit in the bed, were then secured in place.

The piece is shown in position in the die after the bending operation is completed in Fig. 111. It is located by the pierced hole in the blank and also by the "nest" piece in front.

The punch-holder, B, Fig. 112, was of the round, cast-iron type; it was milled out for the punch piece F, which was then machined to a nice tapping fit in the holder, and afterward slotted and bored out for the steady plunger G. This steady plunger came in contact with the blank first and kept it from tipping in the die while the angle bend was being accomplished.

Pieces C and E were made higher than the die piece D for the purpose of locating and backing up the punch.

PRESS TOOLS FOR FORMING A BRASS STUD.

The blank to be formed is shown at A in Fig. 114, and it has to be formed to the shape shown at B in the same figure. The blank is placed on the former, Fig. 113, and is held in position by the disappearing stops 1, 2, and 3, and by blocks 4 and 5.

62   PUNCHES, DIES, AHD TOOLS.

The punch, *A*, Fig. 112, is slightly in advance of punch *B*, which allows *A* to partly form the stud before *B* starts to work at all. When the punch *B* comes down it strikes the slide *C* and pushes it over against the body of punch *A*, which has a recess

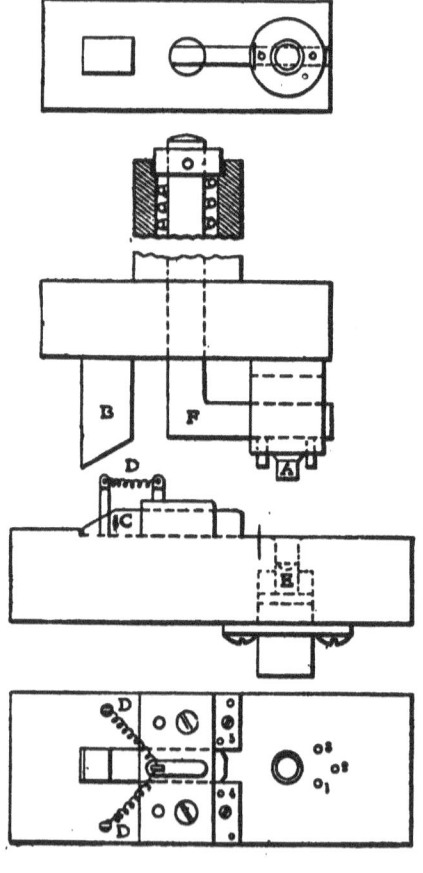

FIG. 113.

cut in it to suit the shape of *B*. When the punch returns *C* is brought back by springs *D*. The die has a stripper to push the stud out, shown by the dotted lines at *E*, and *F* is a knockout held by spring shown in section, *F* being shown in the sketch as if a piece had just been knocked off.

A CUTTING, BENDING, AND FORMING DIE.

The die shown in Figs. 117 and 118 was made to produce in one operation the part shown in side and end views in Figs. 115 and 116. This part was made of tin or light iron. It was 2½ inches long and 1¾ inches wide (inside measurement) with a ⅛-inch lip bent down on either edge and a groove (double) cut

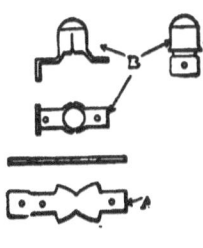

FIG. 114.

and formed through the centre, so that the article when finished might be assembled to turn on a $\tfrac{1}{16}$-inch wire.

In making the punch and die, the head-block g and the die-holder f were cast iron. Part g was first finished top and bottom and planed to receive $a\ a$, and also planed at angles of 15° to hold $c\ c\ j$. Two $\tfrac{1}{16}$-inch holes were drilled and reamed to receive studs $h\ h$, which were drive fits. Parts $c\ c\ j$ were then

FIG. 115.   FIG. 116.

planed to drive into g and were held in place by set-screws through $h\ h$ and also by set-screws in g.

The rib on the face of $c\ c$ was now produced as follows: The surfaces each side of the centre were planed to make the piece of the desired thickness, leaving sufficient stock in the centre for the rib. A piece of tool steel, ½ x 1 inch, to fit the tool post in the shaper, was finished on one side (the ½-inch surface) and a hole $\tfrac{1}{16}$ inch was drilled through it and reamed from the back with a taper reamer until it was .003 inch over size. This tool was then cut off and finished the thickness of the stock (about

.020 inch) below the centre of the hole, and the sides of the hole were filed parallel at the largest point. The tool was backed off slightly, and, after hardening, the ribs were planed with it. On part j the piece was planed to the same thickness as $c\ c\ m$ and a semicircular groove, $\frac{1}{16}$ inch plus .005 inch, was cut through the centre. This part was then drilled and countersunk for the four spring-pins, as shown. Parts $c\ c\ j$ were now hardened and drawn to a light straw color, the ends were ground square, and they were driven to position in g. The blocks $H\ h$ were driven in place and set-screws tightened. Parts $a\ a$ were made $\frac{1}{16}$ inch thick and projected below $c\ c\ j$ $\frac{1}{4}$ inch, being

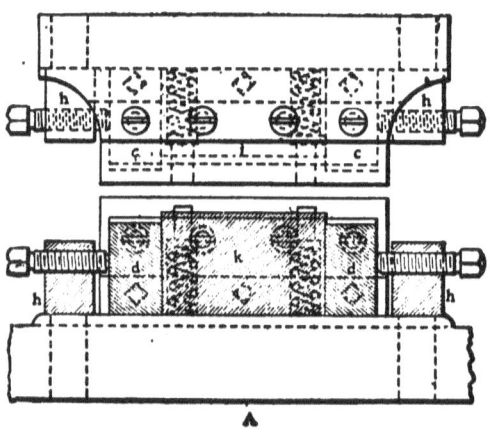

FIG. 117.

rounded on the inner edges to avoid marking the stock in bending. They were held by four fillister-head screws and two dowels each.

In making the die much the same plan was followed, except that parts $B\ b$ were not made until the width of the blank was determined. In this case the semicircular grooves were of course in $d\ d$ and the rib was on k. The guide e was a piece of flat steel, 8 inches long by $\frac{1}{4}$ inch thick, and 1 inch wide, with a groove cut in one edge $\frac{1}{4}$ inch deep and wide enough to allow the stock to slide freely through it. This piece was fastened at one end of the die and at right angles to it, and the operator

PUNCHES, DIES, AND TOOLS. 65

kept the stock against the bottom (horizontally) of the groove in making the punchings.

A trial blank is placed on the die and the punch is forced down over it. The four spring pins, of which the upper were $\frac{1}{4}$ inch diameter and the lower four $\frac{1}{16}$ inch, hold the stock and prevent twisting. As the punch continues downward the blades $a\,a$ bend the stock at the sides over $d\,d\,k$, and the ribs and grooves $c\,c\,j$ and $d\,d\,k$ cut and form the centre.

Having determined the required width of the blank, parts $a\,a$ were ground on the outside to size, care being taken to make them of equal thickness. Parts $B\,b$ were next made, part B

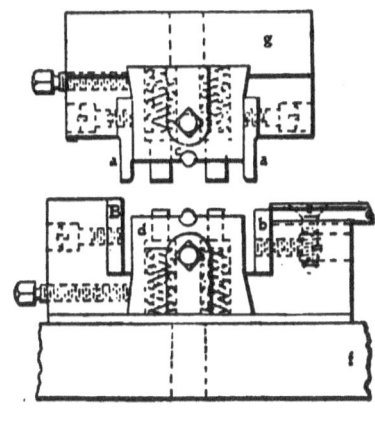

FIG. 118.

being $\frac{1}{8}$ inch higher than part b. These parts were ground to be equal distances from the centre grooves and a snug fit over parts $a\,a$. These parts were also held in the same manner as $a\,a$, and were also hardened and drawn to a light straw color. The guide was then fastened to g, and the die was ready for operating.

In operating this die the pressman slides the strip, which has been cut off $2\frac{1}{4}$ inches wide, along the groove in e until it rests against B, when he starts the press. The stock is cut off by $a\,b$ and bent and grooved as before described. On the return stroke the upper spring-pins release the stock from the punch, and the lower pins raise it, so that it rests freely upon them, and the

5

operator pushes it out of the way and is then ready for the next punching.

When cutters $c\ c$ and k become dull they are ground on their faces, k being ground on both faces and j being ground to the same thickness. Parts $c\ c$ were ground on cutting ends only and parts were reassembled. Parts $a\ a$ and b are ground on a surface grinder without removing them from the die, and the inner edges of $a\ a$ are rounded on a small wheel and smoothed with emery cloth. This die worked very well, and as the distance between $c\ c$ was not an important point, the grinding did not impair its usefulness.

SECTION III.

Automatic Forming, Bending, and Twisting Dies and Punches, for Difficult and Novel Shaping.

AN AUTOMATIC BENDING DIE.

COMBINATIONS of ingeniously constructed parts, operated so as to accomplish difficult bends, forms, and twists in sheet-metal parts, are frequently necessary of adoption in the production of the wonderful variety of shapes and articles in sheet metal, and although in the construction of these tools it is very rarely that anything distinctly new in movements is encountered, the combining of the various parts, the timing of their working, and the rapidity with which they may be operated so as to produce the maximum output, require a wide knowledge of die construction and familiarity with the most approved and up-to-date designs in such tools.

In this section are described and illustrated a sufficient number of automatic tools, which have been in actual use for the forming of sheet metal, to enable the reader to clearly understand what is required to attain any desired result rapidly.

In Figs. 120 and 121 we have a punch and die for bending the blank shown at A, Fig. 119, into the shape shown at B, making a "duck's foot," such as is used in spacing the ventilating openings in armatures. In Fig. 120 is shown an end view (partly sectional) of the dies part way on the down stroke of the press, and with the "starters" just about to take hold of the blank. They consist of the tool-steel die E and punch D, which is screwed and dowelled from above (not shown) to the cast-iron holder C. The die E has a groove milled through it lengthwise, with one side perpendicular and the other side at an angle as shown; to this is fitted the wedge f, leaving room at the perpendicular side for the blank x to be held tightly. Into the wedge f are driven the small pins to which the ends of the springs g

68 PUNCHES, DIES, AND TOOLS.

g are fastened, and then wound in turn around the screws *i i*, and the other ends secured by the small button-head screws as shown. The tendency of these springs should be to lift the wedge up and away from the blank, thus releasing it after bending and permitting a new one to be inserted. The motion of the wedge is limited by the screws *h′ h*, which fit loosely in the die *E*, but which are screwed home in the wedge.

The blank is slid along between the wedge and the die until it reaches a stop, not shown, which locates it in its proper posi-

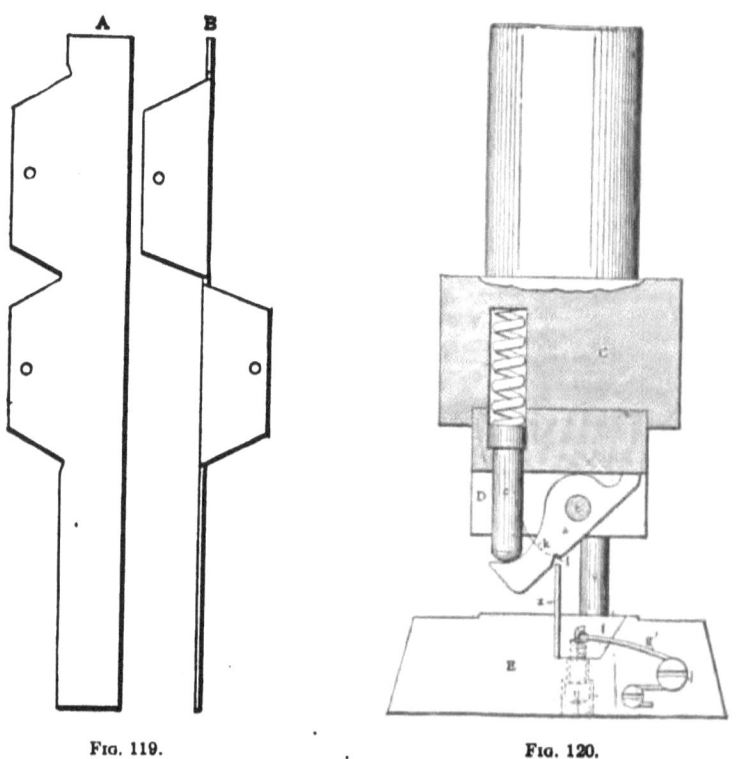

FIG. 119. FIG. 120.

tion; as the press comes down the buffers *e e′* press the wedge down into the die and against the blank, thus securing it and holding it tightly in position. The buffers should be long enough to do this before the starters *a a′* take hold, also the buffer springs *i i′* should be quite stiff. Referring to Fig. 120:

PUNCHES, DIES, AND TOOLS.   69

The starter *a* is pivoted on the pin *b*, which is on the opposite side of the centre line from which the bend is to be made.

The little arm at the upper end serves as a stop to bring the notch, shown at *l*, directly over the edge of the blank, and the projection at the lower end has a cam on its upper side upon

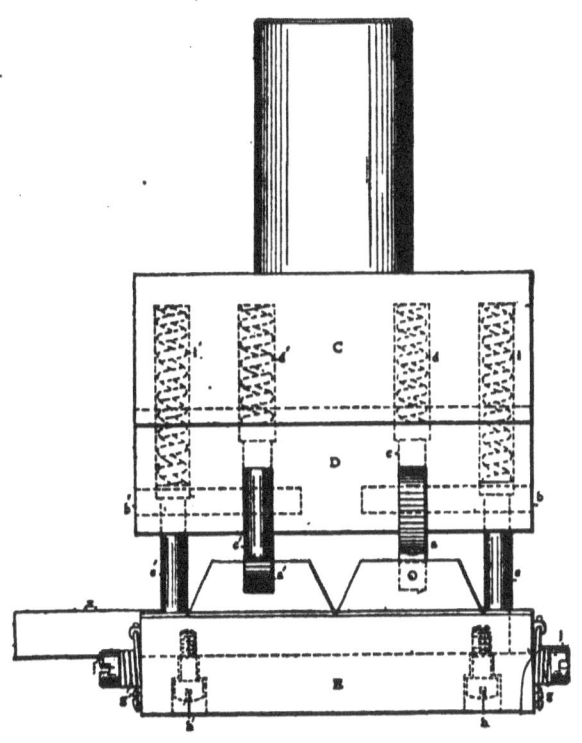

FIG. 121.

which the plunger *c* works, actuated by the slight spring *d*, thus retaining the starter in position shown after each stroke of the press. The top edge of the blank, catching in the notch of the starter during the down stroke, is carried over by it in the direction of the dotted line *k*, until it reaches the face of the punch *D*, which it then follows till the final bump flattens it out at right angles with the part held in the die *E*. As the two "feet" of the blank are to be bent in opposite directions, the two starters *a* and *a'*, must work in opposite directions, the pins *b' b* and the plunger $c\ c'$ must be on opposite sides of the centre line. All parts of this punch are hardened, and the starters have to work very freely on the pins $b'\ b$ and in the slots of the punch D, so that there will be no binding when under pressure. The die is made to bend the blank in one operation, which before had always been handled twice in two sets of dies. It was very successful, bending thousands of blanks in exact duplication of each other.

DIES FOR BENDING EYES OF VARIOUS SHAPES.

Figs. 122, 123, and 124 show an arrangement of dies for the power press for bending and forming small rods of wire to shapes such as are illustrated in Figs. 125 to 130. The shape actually formed by the tools here shown is the "D" shape in Fig. 125. Although the movement of the press plunger is vertical as usual, the bending is done horizontally. Fig. 122 is a

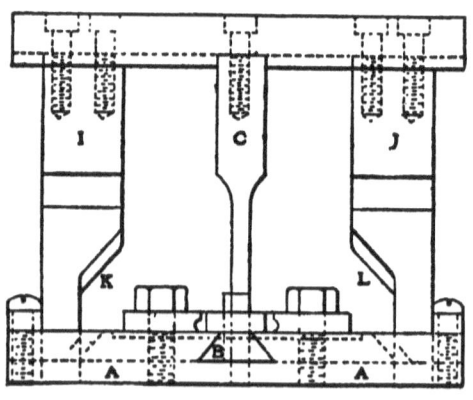

FIG. 122.

front elevation and Fig. 123 a side elevation of the entire arrangement, and Fig. 124 is a plan of the lower part. In the base A there is a dovetail cross-slot in which is the sliding block B, the top of which is flush with or only slightly above the surface of A. The sliding block B is operated by finger C which moves with the plunger. The lower end of this finger, which is narrow and parallel, is long enough to always reach down into

PUNCHES, DIES, AND TOOLS. 71

the base which has a slot which fits and guides it. This finger passes through the slot E in the sliding block B and in its descent the oblique projecting strip a, which works in notch b,

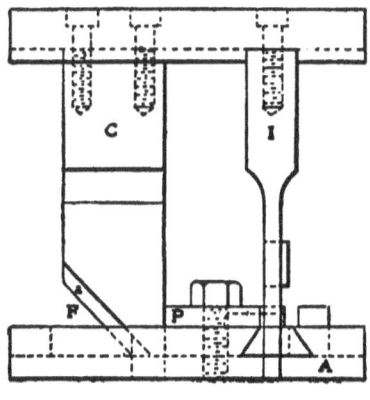

Fig. 123.

acts to draw the sliding block B back to its normal position when the finger rises on the return stroke of the plunger. There are two other sliding blocks, G and H, moving at right angles to

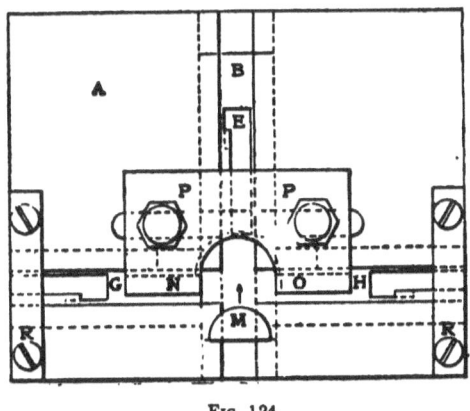

Fig. 124.

block B and operated by fingers I and J. It is only to be noticed that the oblique edges K and L of these fingers are higher, and therefore in the descent of the plunger move the blocks G

72   PUNCHES, DIES, AND TOOLS.

and H later than the movement of block B. On the face of block B is fastened the former M and on the face of G and H are formers N and O.

The wire to be bent, ¼-inch iron, is cut to length and heated, and a piece is centrally placed in front of former M, Fig. 124, and then, when the ram comes down, the inclined edge on finger C draws B forward and the former M, working into the semicircular opening in P, forms the piece into a "U," and then

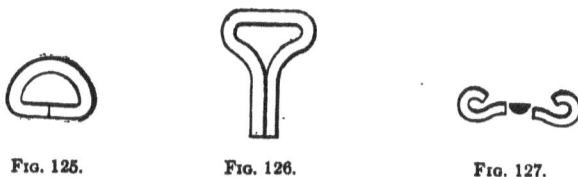

FIG. 125.   FIG. 126.   FIG. 127.

formers N and O advance and bend the ends over, completing the "D" to be formed.

Figs. 128, 129, and 130 were made of No. 9 soft, straightened brass wire, fed to the die from the right, there being a cut-off attached. The forming mandrel for Fig. 129 was caused to slip down out of the way at the end of its stroke, permitting the loop to close tight. For making No. 127, the forming mandrel was stationary, and the forming block is caused to move on account of the mandrel in this case being somewhat weak.

FIG. 128.   FIG. 129.   FIG. 130.

$R R$, Fig. 124, are steel strips fastened by screws on to the bottom plate A and acting as backs for fingers I and J, P acting as a back for C. The piece P has a tongue on the bottom fitting into a slot cut in the bottom plate and fastened on with two capscrews. These tools having been designed before it was known just what press they were to be operated in, lugs for fastening the lower part to the press bolster and a stem for the press ram are not shown.

PUNCHES, DIES, AND TOOLS. 73

AN INGENIOUS BENDING ARRANGEMENT.

Figs. 133, 134, and 135 illustrate a bending die which will no doubt interest many readers. It was required to bend the blank, Fig. 131, of No. 14 gauge sheet brass to the shape shown in Fig. 132, and it was preferred, of course, to accomplish all of the work in one operation.

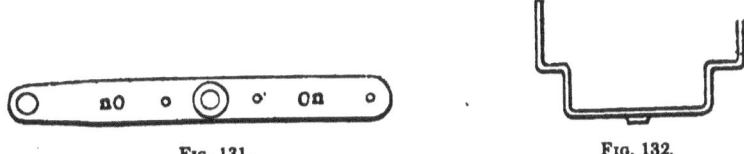

Fig. 131.   Fig. 132.

The blank was got out in the usual way, first perforating, then blanking. The punch for bending is shown in Figs. 133 and 134. The blank is placed on the punch and held by the clips $B\ B$. The two parts of the die $G\ G$ are pivoted, as shown in Fig. 135, and are held in an elevated position by the spring-actuated rod I. As the blank is carried down by the punch the

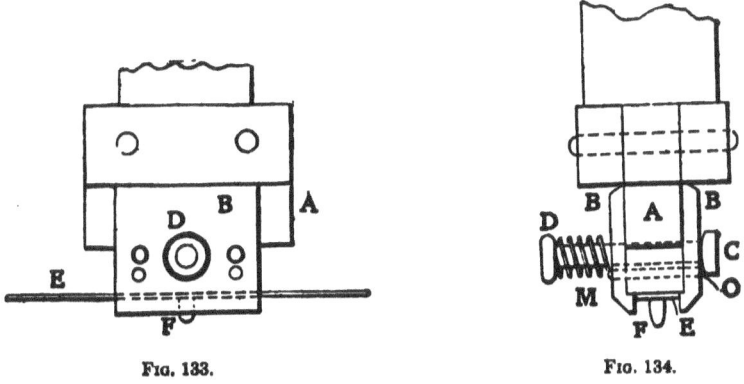

Fig. 133.   Fig. 134.

blanks $G\ G$ assume the position indicated by the dotted lines. This allows the stock to take the desired shape with a minimum amount of stretching. There is no appreciable change in the width of the blank, although it was found necessary to make the holes $n\ n$ in the blank oval, in order to have a round hole in the finished stamping. F is a gauge pin fitting the centre hole in the blank.

The bolt C, with a knurled nut D and spring n, allows the

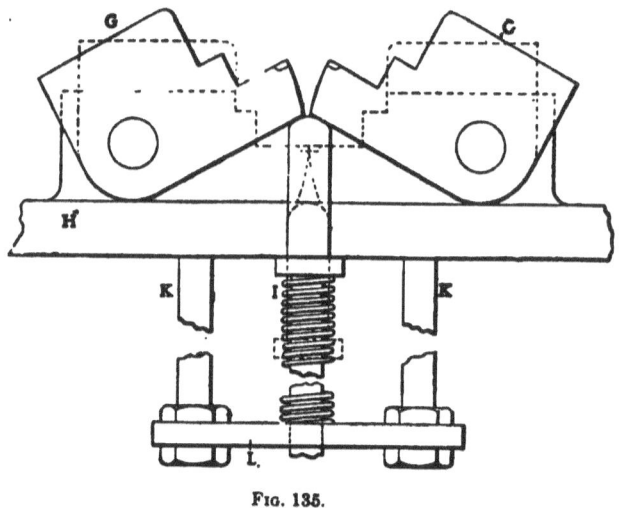

Fig. 135.

clips B B to open enough to clamp the blank securely to the punch. O O are pins fitted to drive through the punch and fit loose in B B.

ACCURATE AUTOMATIC BENDING AND SHAPING DIE.

A type of bending die involving a principle that is extremely useful in the rapid production of work in the press-room is illustrated in Figs. 139, 140, and 141. It consists of bending the

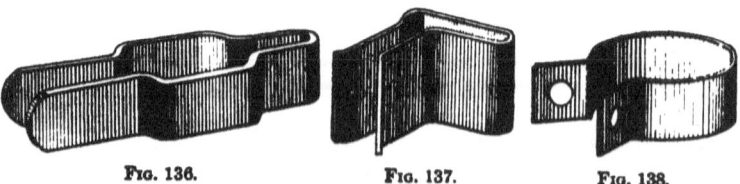

Fig. 136.   Fig. 137.   Fig. 138.

material by pressing it down into the bottom of the die and immediately closing it in on the sides of the punch, which usually corresponds with the inside shape and measurements of the work.

Referring to the engravings, Figs. 136, 137, and 138 show pieces that have been produced by dies of this kind, and the work turned out uniform in size and shape. Fig. 136 is the article formed from $\frac{1}{16}$-inch thick by $\frac{1}{16}$-inch wide hard brass by the

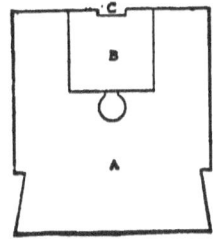

FIG. 139.

punch and die here shown and described. The plan of construction was adopted as giving the best results after several trials with other devices.

Fig. 139 is an end elevation and Fig. 140 a plan of the die and Fig. 141 a longitudinal section of both punch and die. A is a cast-iron die bed, carrying the slides B, which are held down by strips of $\frac{1}{4}$-inch steel screwed to the top of the die bed. $C\,C$ are

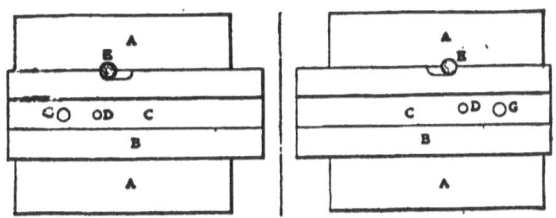

FIG. 140.

grooves planed in the tops of the slides to gauge the work sideways. D represents the pins for the ends. E are pins that fit into the bottom of the die bed and limit the outward movement of the slides, which are grooved to permit them to slide the necessary distance. The slides are forced apart when the punch ascends by the coil springs F. The pins G pass down through the slides and into the grooves and engage with the springs.

The sectional view shows the method of construction. At *H* are lugs with ends turned and fitted to holder *I* and located by taper pins *J*. *K* is the lower or forming end of the punch, dovetailed to upper part *L*, which is machined to slide in a reamed hole in holder, being retained in position by pin *M*. The upper end of the stem at *N* on the punch-holder is slotted to allow the punch to slide when in operation. Springs *O* hold the punch *K* down in proper position until the first bend is made.

The springs *O* vary as to size and number in proportion to

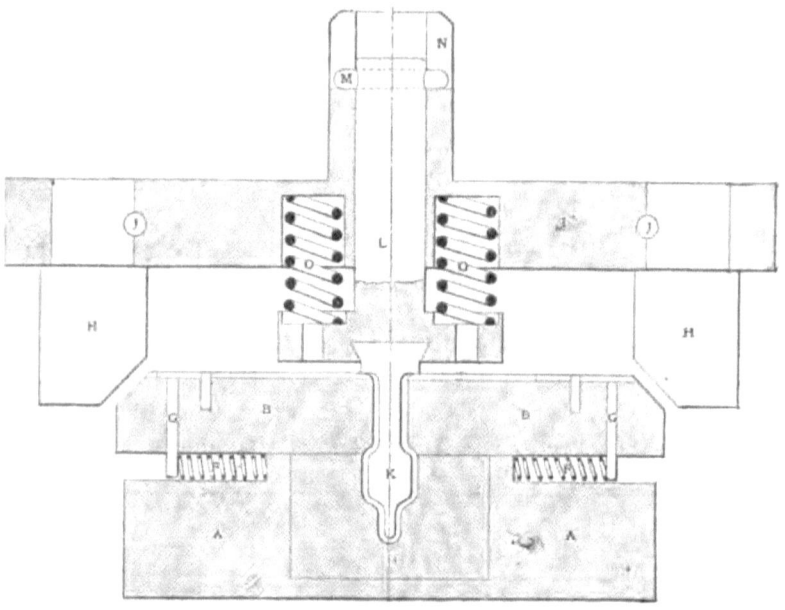

Fig. 141.

the power required to perform the first bend. If necessary, they can be arranged around the punch in large numbers.

In all dies of this kind the springs must be strong enough to perform the first bend by pressing the work down into the die before any perceptible contraction takes place. The punch and die should be so designed that when the lower end of the forming punch *K* rests on the bottom of the die the angles on the sides and lugs should touch. Further descent of the press will cause the slides in the die to be forced in and perform the side bends.

PUNCHES, DIES, AND TOOLS. 77

The parts of both punch and die will differ according to the nature of the work to be produced. When the piece formed has the same style of bend on both sides, the angle of the slides and lugs should be about 45°. For irregular work the angle on one side can differ from that on the other, thus causing one slide to travel a greater distance than the opposite one. An example of irregular bending is shown in Fig. 136. Dies with three and four slides have been built on this plan and operated with success.

PUNCH AND DIE FOR FORMING HINGE-SPRINGS FOR NOVELTY BOXES.

The cut, Fig. 142, illustrates a punch and die bending a sheet-steel piece to the shape of the spring shown in Figs. 143, 144, 145, and 146. The piece was afterward given a spring temper

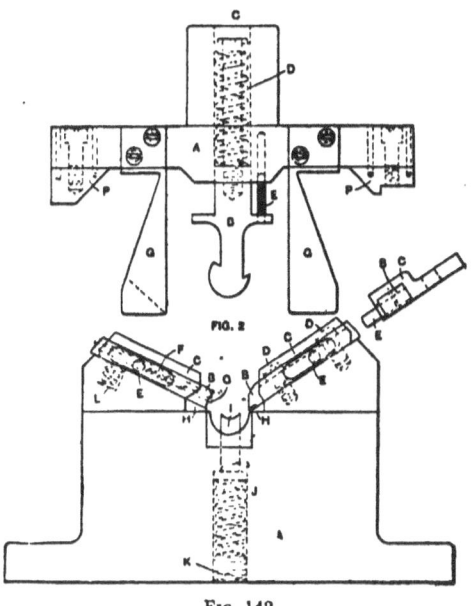

FIG. 142.

and was used as such. The spring in question is used on small novelty boxes—such as jewelry boxes, etc.—it being snapped on the hinge to prevent the cover of the box being thrown back

PUNCHES, DIES, AND TOOLS.

beyond a vertical position, and also to keep it closed with some little tension when down. The demand for these springs was such as to warrant the making of the punch and die for the purpose of adapting it to the rapid handling and bending of the spring complete at one stroke of the press. The spring is made from high carbon steel .012 inch thick, which was first blanked out in a cutting die the required length and outline.

Those who have tried to bend tool steel in a die for the first time have probably not been gratified with the results of their

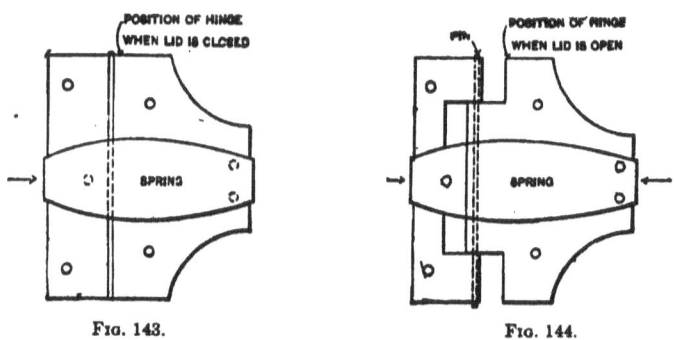

FIG. 143.   FIG. 144.

experiment, and usually not at all satisfied with their success, especially when a high degree of accuracy was required. In this instance, moreover, the difficulty of developing the exact shape and the added difficulties of securing successive lots of steel of the proper carbon and anneal to make continuous production the attainment of uniformity, grow very rapidly in the production of a piece of this type; even with previous experience on this class of work one would be forced to make two or three of the pieces that constitute the forming die, and experiment considerably before the perfect development could be reached. An experienced punch-press hand is essential to operate a die of this kind. It is pretty safe to state that a tool of this design when two blanks, one on top of the other, are put in the die—which is seldom—that something will happen. In the getting out of this job this experience was encountered soon after constructing the first die, and the chances are, as was the case with the constructors, that a new die would be cheaper than attempting the repair of the old one.

PUNCHES, DIES, AND TOOLS. 79

The use of the cams to move the slides outward was found to be better than springs in view of their positive action, which decreases to a minimum the liability of an accident. An accident in this type of die is self-evident should the slides for any reason be in a position with their inner ends over the shoulders of the punch during the upward movement of the press ram. The capacity of these cams for this particular type of die is extensive. Of course, the angle is measured according to the movement of the press ram; it can be varied up to 45°.

The punch and dies shown in the engravings and now to be described consist of the die, Fig. 142, and the punch, at the top of Fig. 142. The base A of the die is of cast iron and carries the rectangular hardened steel slides B B secured thereon by caps C C and four $\frac{7}{16}$-inch filister head-screws D D. The two screws in the left cap are omitted for the sake of clearness in the drawing. The fronts of the caps are provided with elongated slots through which project the hardened $\frac{1}{4}$-inch pins E E that engage with the cams on the punch to throw the slides outward. The pins extending through the slides are retained rigidly in place

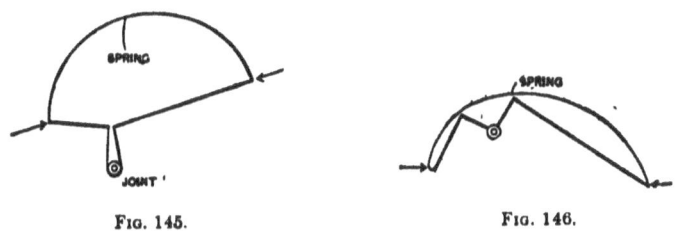

FIG. 145.   FIG. 146.

by small pins driven through them. The pin in the left slide insures the maintenance of the spring F and the sliding pin G in proper position. The base A is cut away at H H to gauge the work. On the downward action of the press the steel blank is clamped between the punch and the square piece I, which is supported by the compression spring J. Directly beneath it, to provide for the adjustment of the tension, the bind screw K is provided. To provide for and control the necessary steady action of the slides, and to prevent a loose movement, which must naturally take place after a short run on account of their inclined position, a spring L is located beneath each slide, the one on the left only being shown in the engraving. The right spring is omitted to avoid obscurity.

Fig. 142, upper, shows the punch in detail: A is the holder, B the punch proper, held in its downward extended position by the screw C. Back of the former is a stiff spring D, essentially of a strength sufficient to bend the blank downward in the die before any perceptible contraction takes place. Pin E, with the upper end fitted freely in the holder, keeps the part B from rotating. The inclined faced pieces $P\,P$, fastened to the holder by screws and dowel pins, on the extreme downward movement of the ram, force the slides in and make the sharp end bends on the work. Seated in the holder to insure true parallelism and angular displacement, and secured thereto by screws are cams $G\,G$ made of $\frac{1}{4}$-inch-thick tool steel, with their working surfaces hardened to obviate abrasion. They project forward just sufficient to pass downward and close to the front of the die, where they engage with the projecting pins in the slides. When the punch is up to its extreme height the lower ends of the cams are in contact with the pins; in fact, they are never entirely clear of the die. When the press carries the punch up beyond the die the lower corners at $G\,G$ are usually machined away at about the same angle, as shown by the dotted line, to avoid striking abruptly on the pin in the slide, if for some accidental reason they should chance to be in when the press is tripped.

All working parts of this punch and die were hardened and drawn, as is consistent with good practice, to insure the life of same. The press was run at a high rate of speed and inclined, as the expansion of the work was sufficient to allow it to fall off by gravity from the punch, and therefore allowed the blanks to be constantly fed to the die, increasing materially the daily production.

FIVE OPERATIONS OF BENDING AND FORMING.

The Figs. 147 to 152 illustrate the work and the means used for a five-operation punch-and-die job of rather unusual character. Fig. 147 shows the blank, and Fig. 148 the finished piece. The piece was required to be very accurate in all dimensions;

PUNCHES, DIES, AND TOOLS. 81

the curve also was important, and it will be noticed in Fig. 148 that there are two distinct curves, one a little lower than the other. The stock used was steel, .040 inch thick, bright drawn. The first operation was to blank and pierce Fig. 147, which calls for no special comment, as the dies were of ordinary construction. Great care had to be taken in sizing the blank, though, so as to save trouble during the following operations.

The second operation was to half-curl the end of the blank,

FIG. 147.

which was done in a small press with the punch and die, as shown in Fig. 149, both doing good work. The third operation was to finish the rounding of the end of the blank. This was done by placing the blank against the gauge-plate, Fig. 150, pushing the small spring lever back in doing so, then raising the lever and clamping the blank to the die face, the punch then descending and rolling the bend, the clamping-plate being released to allow the end to roll over and also to keep the punch

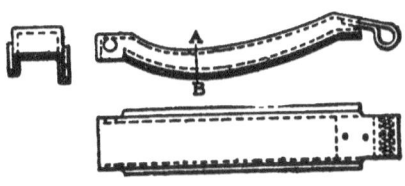

FIG. 148.

from pressing the blank forward; the lever was then pressed down and the trigger forced the blank out, as the bottom of the die was bevelled to allow the blank to fall clear.

The fourth operation was the bending the sides of the blank upward and slightly breaking the corners which were to be forced over, as the corners had to be square and sharp. The gauge-plate was brought across the die, and the blank was steadied by the two springs in the punch. The blank, when pressed into

82   PUNCHES, DIES, AND TOOLS.

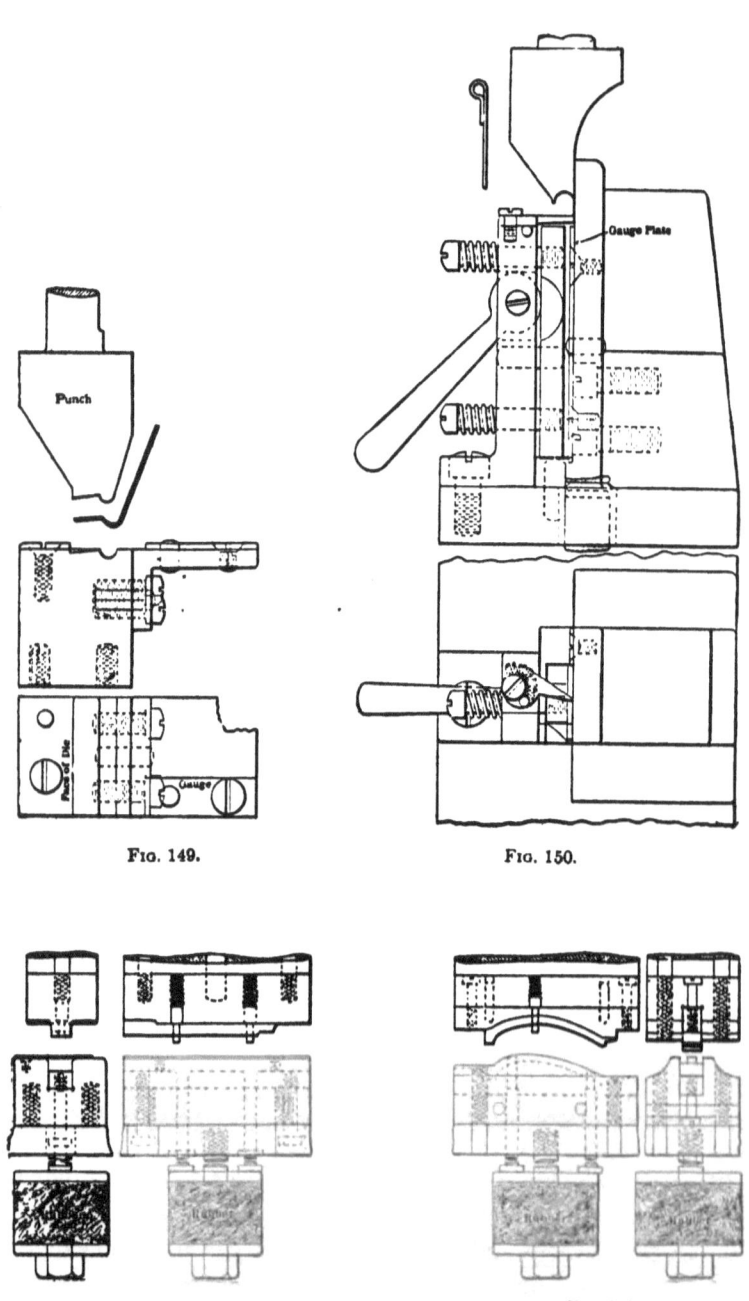

Fig. 149. Fig. 150.

Fig. 151. Fig. 152.

PUNCHES, DIES, AND TOOLS. 83

the die and finished, was stripped by the lifting of the loose piece in the die, the lifting being done by the rubber pad below the bolster-plate.

The fifth operation completed the piece by forming the long curve, closing over the end of the piece, and making a small angle on the tapped bend. This die was made in sections so as to be easily ground and adjusted when wear took place. The blank was stripped by means of the pins through the die, which were worked by the rubber pad as in the fourth operation. These dies did their work remarkably well and overcame all difficulties in what was a rather nice job of punch-and-die work.

HOT-FORMING AND BENDING ODD-SHAPED STEEL SPRINGS.

The bending fixtures shown in Figs. 153 and 154 were designed and used for bending $\frac{7}{16}$-inch by $\frac{7}{16}$-inch spring steel to the shape shown at the black line at x, Fig. 154, in two operations, with only one heating for each operation. The first operation was performed in the fixture shown in Fig. 153, the spring steel piece x being bent into the form indicated by the black line and the dotted lines. It will be noticed that it was a hard, stiff bend, and one that required powerful leverage to enable it to be done by hand. The bending had to be uniform and interchangeable to gauge fit. The fixture was designed so as to accommodate different shapes in bending by exchanging formers.

The frame a was of gray iron, made with a hole in each corner for belting solidly to bench, with the handle at the right by the side of the operator. The frame is made with thin walls at the sides, so as to form a pocket or reservoir for the lard oil which is used on the formers when bending; with a brush in his left hand the operator brushes some oil across the face of the outside jaw before each bending operation. The steel must be heated very low, just hot enough so that the red can be discerned. The formers, b and e, are made of tool steel and are held in position with three screws and two dowel pins. The inside corners of b at $e\ e$ are rounded, so as not to cut the metal when it is forced into the former. Before making the formers it was necessary to determine the correct shape, as they had to be such as to form the desired shape for the finished piece after it was hardened. The shape was calculated very closely by men accustomed to the work; it could have, however, been found by

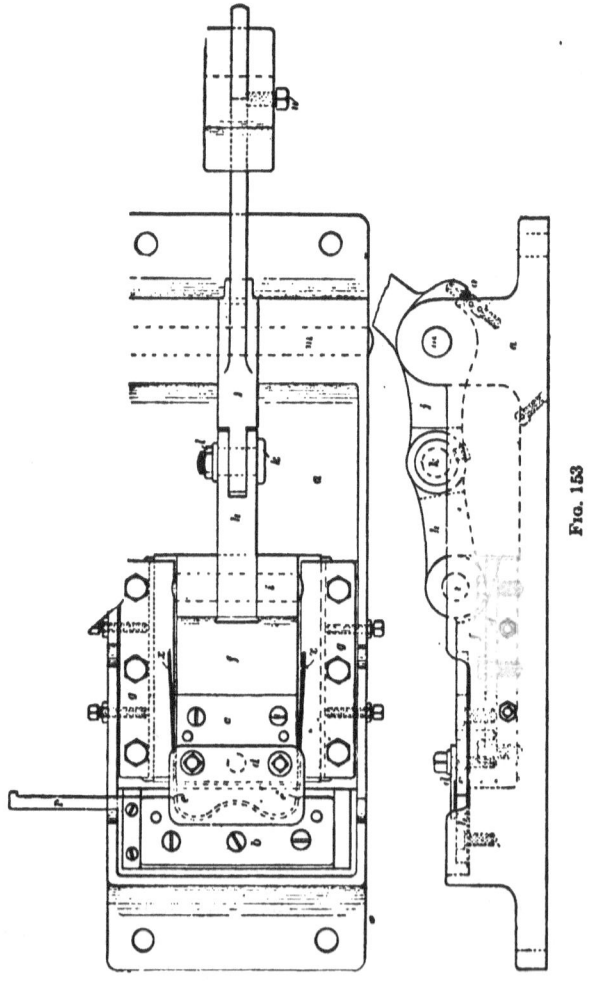

Fig. 153

trial, by using a few pairs of samples bent to the approximate shape, and trying one of each pair until one was found to be right after being hardened. The former was then shaped to the mate. The plate d, held at the top of c by the two screws, was

PUNCHES, DIES, AND TOOLS. 85

for keeping the ends of the piece in alignment during the bending operation; at the same time it prevented the oil from flashing upward and burning the operator. The T-slide f, which carried the former c, was made of cast iron and was finished all over. It was held down by two steel plates g, which in turn were clamped with three screws on each side. The two steel gibs were provided on each side with gib screws and lock nuts to keep the slide in a central position, and to compensate for wear. At the rear end of the T-slide is a toggle joint consisting of the links h and j, and the bearing-pins i, k, and m. The pin

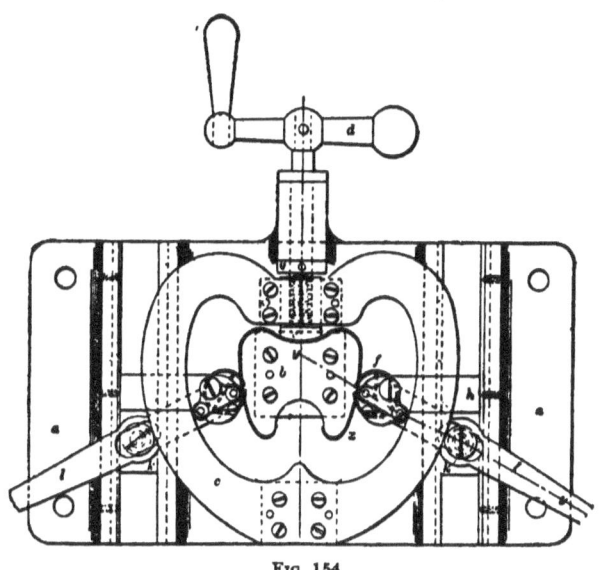

FIG. 154.

i was made of tool steel and was a driving fit in the slide f, and a turning fit in h. The screw k was drawn up solid to a shoulder with a nut l, and was a turning fit in h and j. The handle j swings on the pivot pin m, which is a driving fit in j and a turning fit in the frame casting. On the extreme end of the handle j was fastened a cast-iron weight or block, so as to be adjustable by loosening the screw n. The block on the handle weighed fifty pounds, and that weight, together with the toggle-joint action made, the machine very easy to operate, although a heavy pressure was required in the bending. To prevent drawing the metal too much, screw stops, *o*, were provided which limited the movement of the handle *j* in the forming operation. These screws were made of tool steel and were spring-tempered. Stops were also provided for limiting the withdrawal movement of the formers, as indicated in the engraving. The stop *p* was clamped to the fixture with two screws; and was used to locate the metal accurately before the bending took place, so that each leg of *x* would be of the same length. The walls of the casing were cut away sufficiently to allow the ends to pass over freely. The operator picked up the metal from the fire by tongs held in the left hand, and placed it against the stop *p* and the face of the former. With the right hand he moved the handle *j*, and the first movement of *d* located the metal in a horizontal plane, since there was only just enough room for it to pass over freely. The bending was done very rapidly.

The second operation in the production of the springs required the fixture shown in Fig. 154. This fixture included the frame casting *a*, which had a hole in each corner for bolting it securely to the bench, with the clamping handle *d* on the opposite side from the operator. With a turn of this handle he clamped the piece *x* between *e* and *b*, and then with the right and left hands on the sliding handles *l l*, a simultaneous inward movement bent the legs of *x* around to the shape shown by the black line. In the centre of the casing was the former *b* fastened securely by four screws and four pins. This former required the same attention in design as the one described in connection with Fig. 153; it was made in the same manner, after having ascertained the right shape, of tool steel, tempered to a straw color. The tool steel sliding-block *e* was fitted in a groove planed in the casing, and was provided with a suitable cap for holding it in position. It was tapped out in the centre to fit the screw of the handle *b*. On its face it fitted *b*, and both conformed to the shape produced in the first bending operation. On each side of the casing *a* were planed two 45° dovetailed grooves, which were provided with gibs and screws for adjustment. In these grooves moved the blocks *h*, carrying the roller-holders *i*. These roller-holders were free to swivel on the screws, and each carried two small rollers, one of which rolled on the inside of cam *c*, and the

PUNCHES, DIES, AND TOOLS. 87

other bore against the piece x being bent. A flat spring on one side held the roller-holder continuously against the cam c. The rollers and holder were case-hardened, being made of machinery steel, as was also the cam c. The handles l were mounted on the sliding blocks h so as to move longitudinally, the holes for the screws being elongated for this purpose.

DIE WITH AUTOMATIC FEED FOR PUNCHING, SHEARING, AND DRAWING.

Figs. 155, 156, and 157 illustrate a die for punching, shearing, and drawing a piece of sheet metal into the form of a cup with irregular edges like Fig. 158. The principles of construction and operation differ very much from any others heretofore described. The finished piece being of an unusual shape, com-

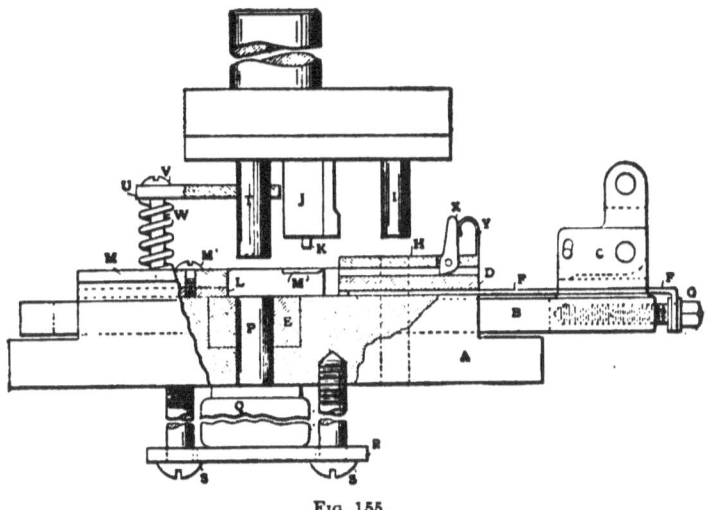

FIG. 155.

binations of mechanical movements are required to effect the desired result. The arranging of combinations of mechanical movements, guided by a knowledge of how a piece of sheet metal will act under different conditions, constitutes up-to-date diemaking in certain lines of work.

In the engravings, Fig. 155 is a side elevation, and Fig. 156 an end elevation of the die-bed and slide, the punch being left out of this view, as it would not add materially to the description without overcrowding the drawing. Fig. 157 is a plan, Fig. 158 (three views) shows the finished article. Fig. 159 shows the strip of metal, .011 inch thick and ⅜-inch wide, with two notches cut by the shearing-punches *I*. The metal to the left of the dotted line shows the outline of the blank that forms the piece, Fig. 158.

A, Figs. 155, 156, and 157, is designed to show a cast-iron die-bed carrying the slide *B*. The grip *C*, its connection to the press, and method of operation are identical with that described

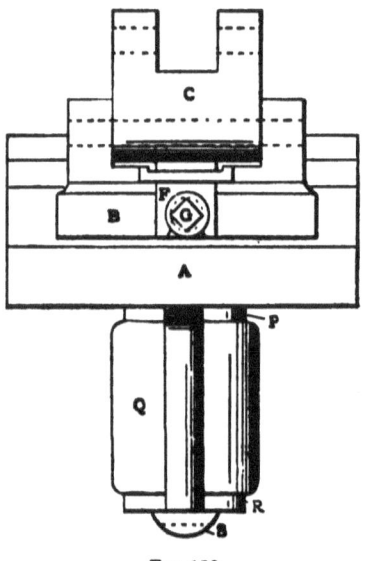

FIG. 156

in the die following this one in this chapter. In this instance it is necessary that the piece of metal that forms the finished article be severed from the strip before any action of forming takes place. The shearing die *D* is therefore raised above the level of the drawing die *E*, and slotted to accommodate an auxiliary *F* made adjustable by means of the shoulder screw *G*.

The strip of metal passing between the shearing die *D* and stripper gauge *H* has the notches cut by the shearing-punches *I*. Feeding the length of one blank, the final cut is made by the shearing-punch *J*. This punch is provided to insure the fact that the blank after being severed from the strip will not draw back, but will lie in front of the slide *F*.

Upward movement of the ram carries the strip of metal forward for another blank, and the blank already cut is carried over the drawing die *E*. The projections *M'*, which are part

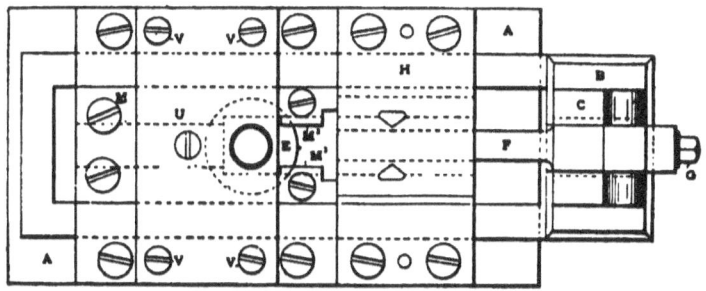

Fig. 157.

of the gauge *M*, are bevelled upwardly in front of the cutting die *D*, and are to insure the blanks taking their proper position.

The extent of the lateral feed is regulated by means of the screw *G* in connection with the slide *F* and the stop *I*, which is slotted into the gauge *M* and made adjustable by means of the screw M^1. Side-gauging is taken care of entirely without ad-

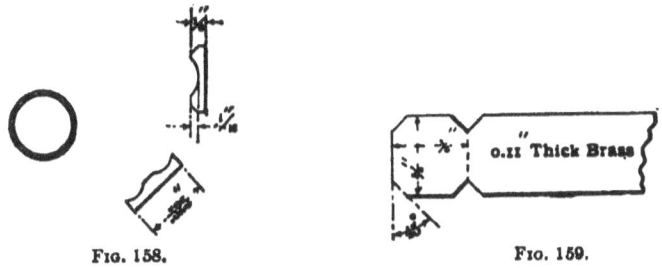

Fig. 158.   Fig. 159.

justment, as this depends on the width of material which is being worked, and to get good results from dies of this description it is always necessary that the width of material be uniform.

The eccentric *X* is to prevent withdrawal of the strip on the outward stroke of the slide. The arrangement of the pad *P*, actuated by the soft-rubber cushion Q, was made necessary by the fact that the shell 'had to be slightly concaved on the bottom; if the shell had been flat, an ordinary push-through drawing die would have answered the purpose. The action of the pad P insures the fact that all of the shells will stick to the drawing punch T. This necessitates the addition of the stripper plate U, which travels on the studs V and is held upward by means of the spiral springs W. The travel of the ram of the press being more than the travel of the stripper-plate, continued upward movement insures the stripping of the shell from the punch.

COMBINATION SHEARING, PIERCING, BENDING, AND FORMING DIE WITH AUTOMATIC FEED ATTACHMENT.

The combination shearing, piercing, bending, and forming die shown and described in the following, illustrates a principle that is applicable to a large variety of press work. The die, as shown in Figs. 160, 161, and 162, is provided with an automatic feed attachment that is simple, self-contained, and positive in

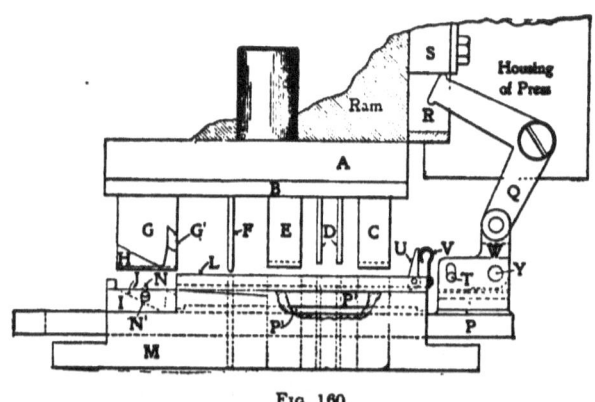

FIG. 160.

its action, and does not require any elaborate press connections. Where the production is not large enough to warrant the expense, this feed may be dispensed with, and the stock may be fed to the die by hand, without detracting anything from the

PUNCHES, DIES, AND TOOLS. 91

exceedingly good qualities possessed by the action of a die of this kind.

Referring to the drawing, Fig. 163 shows the piece to be made, a piece that the die-maker will recognize at once as a very an-

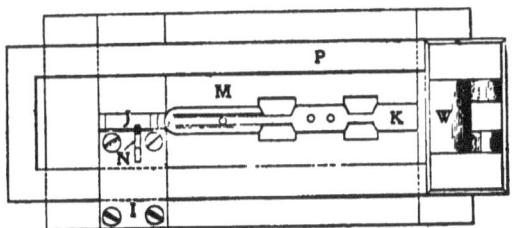

FIG. 161.

noying and difficult piece of bending; annoying, because of the difficulty of gauging the various operations when performed separately. The die illustrated will produce an article of this description complete in one operation, and the pieces will be

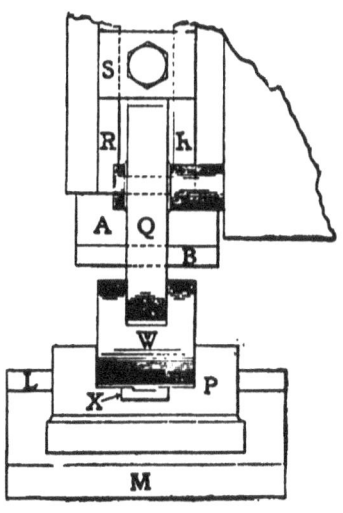

FIG. 162.

duplicates. Fig. 164 shows a section of the continuous strip of metal.

Fig. 160 is a drawing of the side elevation; Fig. 162 is an end elevation; Fig. 161 is a plan of the die and die-bed with

92   PUNCHES, DIES, AND TOOLS.

stripper gauge removed. M is a die-bed slotted out to receive slide P, the shearing, piercing, and bending die K, and the forming die J. The forming die must be the same width as the piece to be bent. The shearing, piercing, and bending die must be at least ¼ inch wider than the stock from which the piece is made. The shearing and piercing and bending of the small lugs shown in Fig. 163, are accomplished by the shearing-punch C, piercing-punches D, and the shearing- and bending-punch E.

The die ahead or to the left of the punch E is channelled on a bevel, as shown in Figs. 160 and 162, so as to offer no resist-

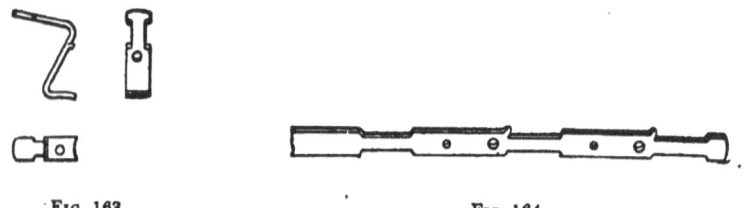

Fig. 163.   Fig. 164.

ance to the forward movement of the strip after the lugs are formed. The channel must not extend to the edge of the shearing die K that is next to the forming die J, as this cutting edge must be kept the full width of the strip.

The dotted lines near the bottom of the punches C and E, Fig. 160, show the cutting and bending faces, as in order to stand up to their work these punches should enter the die-bed and be backed up by it before the action of the shearing or bending takes place.

Between the bending-punch E and the shearing edge G^1 it will be necessary to leave the length of one full blank in order to utilize the pilot pin F, which is indispensable. The forming punch G has a backing-plate H bevelled at the bottom. The forming die J has a similar plate I on the opposite side. The punch G and the die J being the same width as the finished piece, it is brought to its place and confined, thus assuring uniformity in this direction. The piece G^1 which shears the article from the strip is made separate to allow of making repairs quickly.

The knock-out N is bevelled upwardly from the face of the

PUNCHES, DIES, AND TOOLS.

die J, and is held out by means of a spiral spring kept in place by means of the screw N'. When the stock is forced down into the forming die, the knock-out is forced back into the slot. When the pressure is removed, the finished piece is dislodged by action of the knock-out, and the press being set on an incline, the pieces fall off by gravity.

Perfect uniformity is assured in a die of this kind, from the fact that the strip is held by the action of the pilot pin F until the forming operation is almost complete, the shearing of the finished piece from the strip being almost the last thing that takes place.

The automatic feed shown is, we believe, rather novel. Several dies have been designed with similar feeds, all of which were very successful in operation, taking the strip from the reel. The production of these dies runs from 18,000 to 60,000 pieces each day of ten hours.

Fastened to the side of the ram of the press is a U-shaped piece R, which is adjustable up and down and is held in place by the clamp S, which is slotted so as to be also adjustable. This arrangement allows the feed to be adjusted to a nicety, which cannot be done with a roll and ratchet feed.

SECTION IV.

Cut, Carry, and Follow Dies, Together with Tool Combinations for Progressive Sheet-Metal Working.

AUTOMATIC SLIDE DIE FOR PIERCING, BLANKING, AND DRAWING IN ONE OPERATION.

THE first set of tools to be illustrated and described in this section is one of improved design for perforating, blanking, and shaping, or drawing, small lock cases used on cheap imitation leather satchel frames. Before the tools shown were made the lock cases required two operations; the first, piercing and blanking, and the second, drawing or shaping. These tools elimin-

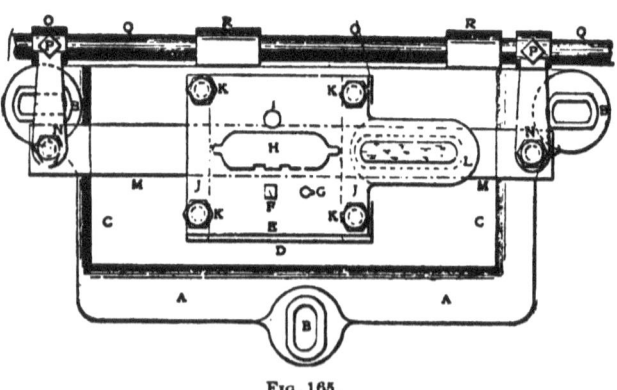

FIG. 165.

ated the second operation, the stock being fed and the finished article being produced at one and the same operation. Fig. 168 shows the flat blank, and also its appearance after being formed and drawn. The ends of the blank are semicircular except for the projecting lips $S\ S$, which are used to fasten the case to the satchel frame. The two notches at $T\ T$ are entrance ways for

PUNCHES, DIES, AND TOOLS. 95

the lock portions, U is the hole for the finger release, and V the keyhole.

Fig. 165 is a plan of the die complete, except for the cam motion or movement which works the blank-carrying slide. Fig. 166 is an end elevation, and Fig. 167 a side or front elevation, both partly in section. $A A$ is the bolster on which are located all of the working parts of the die. $C C$ is a raised portion of the bolster which is machined for the die blank D to locate on

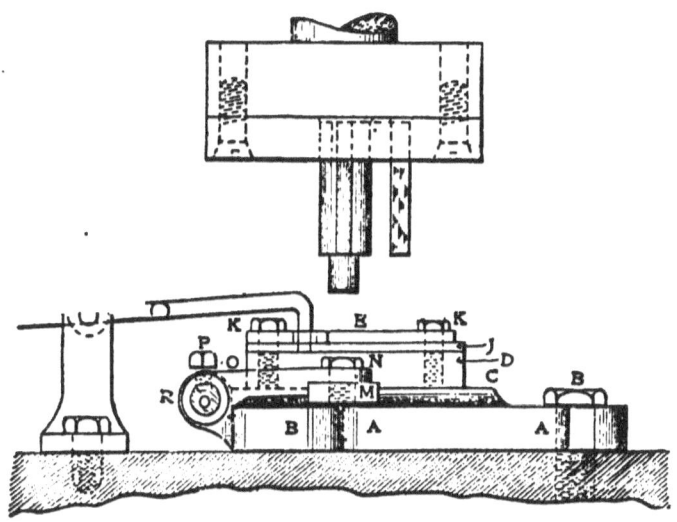

FIG. 166.

and finished with a channel cut longitudinally down its face for the carrying slide $M M$. In the die proper F and G are the piercing dies, H the blanking die, I a hole for the entrance of the automatic finger gauge shown in Fig. 166, $J J$ two gauge-plates between which the stock is fed, E the stripper, and $K K K K$ four cap-screws which locate and fasten the die to the bolster face. L is an extension of the stripper with a hole in it for the drawing and forming punch shown in Figs. 166 and 167. It will be noticed in the end view of the tools (Fig. 166) that the bottom of the die blank has a channel in it, a duplicate of that in the bolster face. This allows the slide $M M$ to fit so as to locate the die in perfect alignment with it.

The carrying slide *M M* is of tool-steel hardened and ground with a seat *X X*, Fig. 167, for the blank to drop into and a drawing-die portion, or hole, let into it at *U*. It is given a longitudinal reciprocating movement in the bolster and die by means of the connection *N N* at each end, and the circular rod *Q Q*, moving in bearings *R R*, and a cam movement on the end of the press crank shaft. The slide is adjustable in longitudinal reference to the circular rod *Q Q* by means of set-screws *P P* in the connections *O O*. The upper or punch section of the die, as shown in Figs. 166 and 167, requires no description.

In operating the tools it will be understood that the pressman

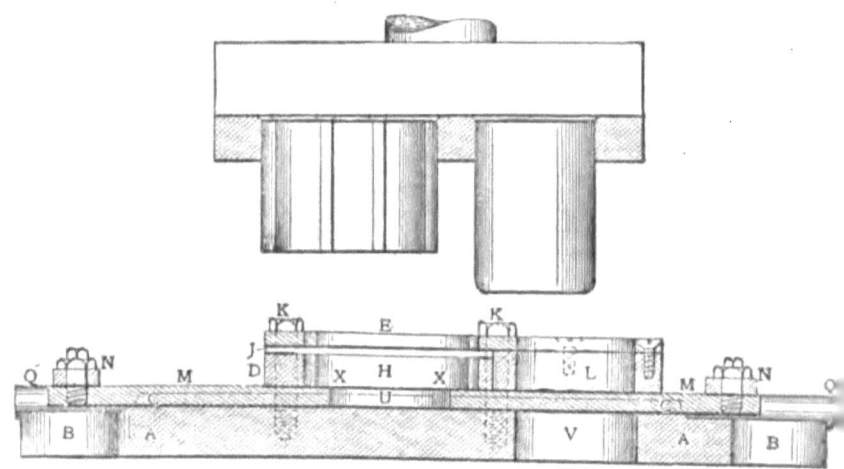

Fig. 167.

feeds the stock into the die between gauge-plates *J J* against the automatic finger gauge shown in Fig. 166, the press running continuously. First the piercing of the holes *U* and *V* is done, then at the next stroke the blank is cut and carried down into the die *H*, the punch descending far enough to allow the blank to drop on the face of slide *M M*, where it locates itself in seat *X X*, Fig. 167. At the next stroke the blank is moved forward automatically by the cam movement of the press until it is directly under the drawing and forming punch. The slide then remains stationary while the punch continues to descend and forms the blank in and draws it through the drawing die *U* in

PUNCHES, DIES, AND TOOLS. 97

the slide, from which it drops out into and through the clearance hole V, Fig. 167, to a box on the floor. As the punch rises the slide returns to its former position and the blank that has been meanwhile cut locates in the seat XX.

The simplicity of these tools and their rapid productive qualities, as well as the infrequency of bad work through their use, should cause the type to be extensively adopted for other work. As the construction of the die and slide is such as to make the carrying of two blanks at once by the slide into the drawing portion impossible, accidents cannot occur except through gross negligence.

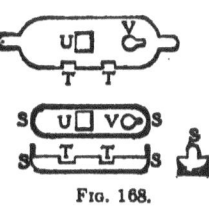

FIG. 168.

Almost any make of power press can be equipped with a cam motion for dies of this type. When run properly they will produce articles at the rate of from 60 to 75 per minute, or about 40,000 per day. When an automatic feed is used in conjunction with the tools, and stock cut to the necessary width and coming in rolls is used, such work as staples, tags, belt-hooks, small irregular-formed shells, metal buttons, eyelets, lock parts, metal novelties, and small artistic metal goods may be manufactured more cheaply than by any other means.

The automatic finger gauge shown in Fig. 166 and used with the die may be used to advantage in connection with all piercing and blanking dies, or where the operations required to produce the article succeed each other in the one side. It consists, as will be seen, of an adjustable bent rod or stop, the bent end of which rests on the face of the die while the punch is up. It is automatically raised from this position by the ascent of the punch after each stroke, allowing the stock to be fed forward for the next operation. Thus, by dropping back through the hole in the stripper and into the hole last punched in the stock, it acts as an accurate stop gauge, which does not impede the progress of feeding the metal, nor necessitate the raising of it after each stroke, as is usually necessary when stops or gauges of the common type are used.

CUT AND CARRY DIES FOR BICYCLE RIM-WASHERS.

The die shown in the accompanying engravings, Figs. 170, 171, 173, and 174 (not all to the same scale) is of the type known

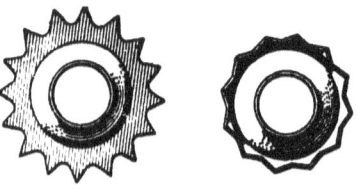

FIG. 169.

as "cut and carry dies." This die was to make commercial washers for bicycle rims, one of which, in two different stages of

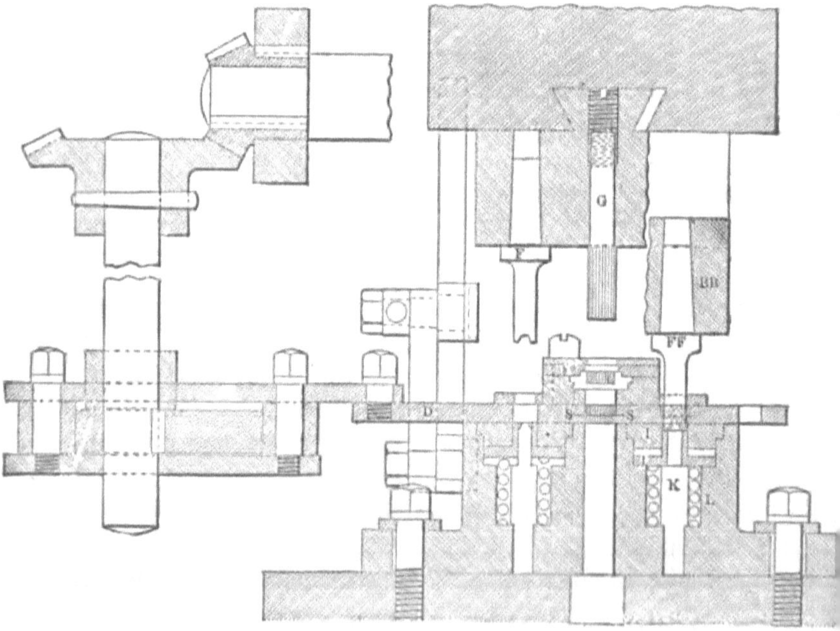

FIG. 170.

manufacture, is shown enlarged in Fig. 169. The washers were required to be of uniform outside diameter, and the diameter of the hole was allowed a possible variation of .003 or .004 inch. The teeth were to be uniform in length, the countersink in the centre, and of proper depth to let in the head of the nipple, and as there were two sets of formers there were at least eight places

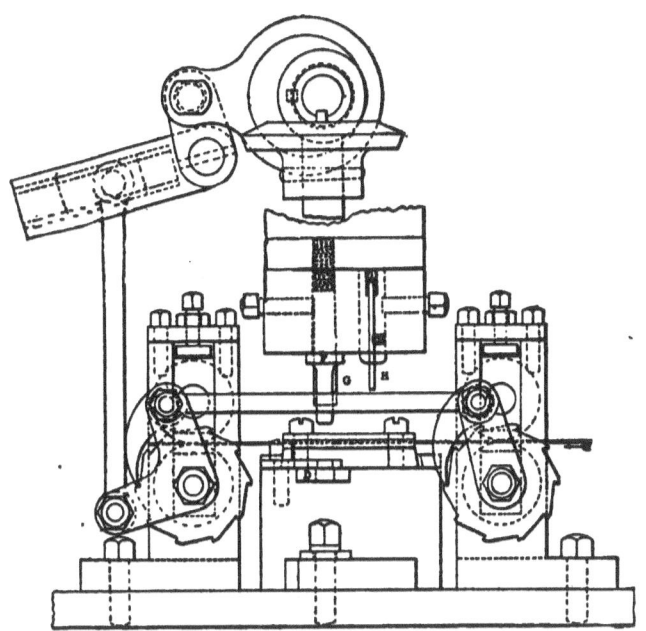

Fig. 171.

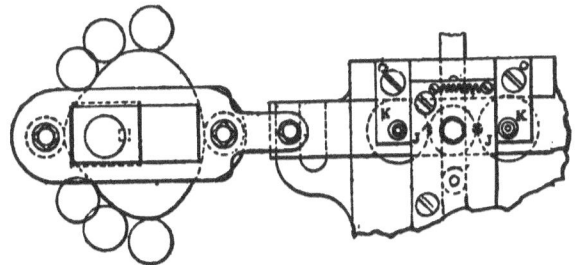

Fig. 172.

Fig. 173.   Fig. 174.

for errors to creep in. No attempt was made to run "old tin cans" through this die. The stock furnished was metal tape, in lengths of 100 feet or more, ⅝ inch in width and 1/16 inch thick.

The frame of the press is not shown, as it will not be difficult to surround the dies as shown with a suitable framework or of placing them on any suitable press with an open back.

The operation of this die is as follows: The steel tape is fed through the die by the rolls of the double roll feed and is pierced by the small punch H, Fig. 171. The slide moves up, during which movement the feeding mechanism has fed the strip along one space, and as the small and large punches are two spaces apart, nothing happens on the next stroke, except that another hole is pierced by the small punch. The next movement carries the tape under the large punch, where the washer is blanked out, and deposited on the shoulder in carrier at $S S$, Fig. 170, not by gravity but by the punch itself, which makes thus a sure job of it. The slide moves up about one-half before the carrier d starts to move, and while D is moving from one position to another the slide will finish the upper half of its stroke and come about halfway down again before D comes to a standstill. Fig. 172 shows the dwell of the cam to be 105°. At the fourth downward stroke of the press, supposing the carrier D to be in the position shown in Fig. 170, the large punch G will blank out and deposit a blank in the left of the two large holes in D. The small punch will pierce and the former $F F$ will force the blank out of D down into the ring I, which will turn up the points, and as it reaches the bottom of the stroke it comes against the inside ring J and the inside forming pin K which forms the countersink, the top block being cut away and let down to show the operation. In this stroke the former F will not have anything to do, but will have on the next down stroke.

As the slide moves up, the spring L forces J up, thus forcing the newly formed washer back into the carrier, and as it moves into the left-hand position the washer will drop through the centre hole into the box below. As the die is now full, a complete washer will be made at every stroke of the press, and as the press can easily run at 100 rotations per minute, 60,000 washers will be the output of ten hours' continuous work. At one time

PUNCHES, DIES, AND TOOLS. 101

one of these dies made a maximum record run of 6½ days, making about 390,000 complete washers without being ground. Details of this are shown in Figs. 173 and 174.

The engravings illustrating the construction of this die are sufficiently complete and clear to enable any mechanic of ability to adopt the principle in dies for the production of other similar parts.

A FOLLOW DIE FOR FIVE OPERATIONS.

The die illustrated in Figs. 176, 177, and 178 performed five distinct operations before the finished article shown in Fig. 175 dropped completed from the press. In many sheet-metal shops

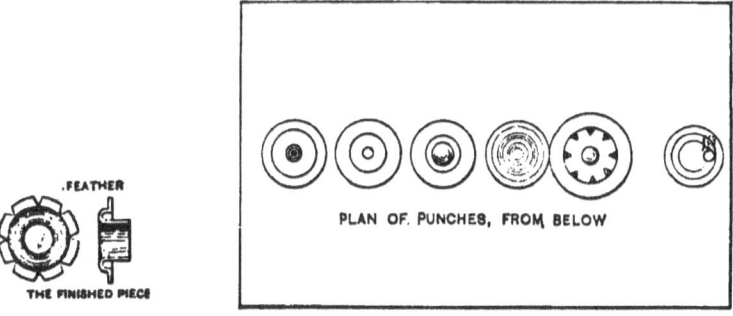

FIG. 175.   FIG. 176.

three presses and three sets of tools would be used to produce exactly the same results which are produced in this one die.

In constructing the die it was not deemed practicable to make it of one solid piece since one small flaw would, in this case, spoil the entire die. A die block of machine steel was therefore used, having recesses counterbored for the insertion of tool steel bushings. These recesses were accurately spaced in the following manner: One side and one end of the die block were machined perfectly square and a centre line drawn lengthwise on the face of the block. The location of the recesses was approximately laid out with lead pencil and the recess for the first die bored in the lathe, by strapping the block to the face-plate. Before loosening the straps by which the die block was held, the

102   PUNCHES, DIES, AND TOOLS.

parallels for set edges were strapped to the face-plate bearing against the finished edges of the block. The straps holding the block were then loosened and the block moved along the horizontal parallel sufficiently to allow of the insertion of a spacing-block, which had been previously made of the required thickness. The die block was then fastened and the second hole recessed. By repeating this operation, and adding a block each time until all of the recesses were bored, it was possible to space the die far more accurately than would have been possible by the time-honored method of laying it out with dividers. The punch-holder and the stripper were then bored in the same manner, using the same spacing method and blocks.

The bushings F, G, H, I, J, K were next made, and after being hardened they were lapped to size. The outside of the bushing was ground concentric with the hole by wringing the bushing on a piece of soft steel held in the chuck and turned to fit the hole in the bushing. The bushings were then forced into their seats in the die block and the die was completed. The punches were ground all over, to insure straightness, and they, in turn, were forced into the punch-holder. The drawing and

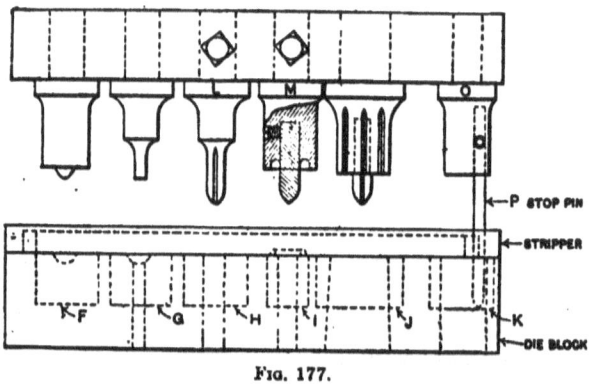

FIG. 177.

forming punches L and M were held with set-screws to prevent them from pulling out. In using a die containing two or more punches, considerable trouble is sometimes experienced on account of the variation in width of the stock to be punched. Should the stripper be planed to fit one of the strips of stock

PUNCHES, DIES, AND TOOLS.

very nicely, the chances are that the next strip will not enter the stripper at all. The part N, shown in the plan view, Fig. 178, is a novel and practical way in which this trouble is overcome. The stripper is planed out $\frac{1}{16}$ inch wider than the stock and is recessed to allow the spring guide N to slide freely when

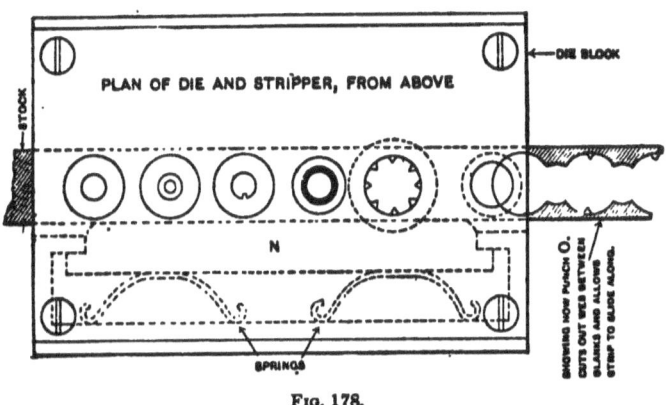

FIG. 178.

the stripper is in position in the die. By glancing at the sketch the reader can readily see how the springs keep the work or stock pressing against the gauge-plate or side of the stripper. The punch O does not perform any work pertaining to the finishing of the blank, but is used for cutting the web in the stock in order to allow the strip to move along until the next web touches the stop pin. As the stop pin P does not come out of the stock it is therefore impossible to "jump" the stock and make a miscut, which would mean disaster to the drawing and forming punches.

After setting up this die in the press, the punches, of course, descend five times before a single finished piece appears, but thereafter a finished piece drops out at each stroke of the press. The first punch, beginning at the left, indents the stock, and the face of the punch is so adjusted that it levels the stock also. The second punch pierces the bottom of the indentation. The next punch draws the stock and at the same time forms the feather shown in the finished piece. The fourth punch is the forming operation, and the last is the blanking.

104   PUNCHES, DIES, AND TOOLS.

DIES FOR CHAIN-PURSE BODIES.

Dies used for the manufacture of chain purses are of very interesting construction, so much so that a description of a set here will enable the reader to understand the accuracy and ingenuity utilized in the designing and constructing of tools and dies for the production of such work.

The purse bodies for which this set of dies was used consisted of a series of rings held together by four-pronged settings. A section of these purse bodies is shown in Figs. 179 and 180; the two lower rows show the rings as they came in strips from the punch, while the two upper rows are shown fastened together by the settings. One of these settings as it comes from the press is shown in Fig. 180. Both rings and settings were formed upon

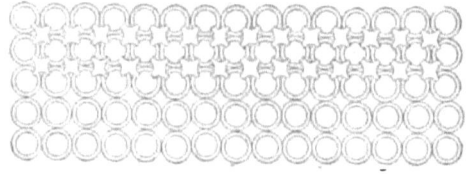

FIG. 179.   FIG. 180.

the power press, and we will first consider the tools used for making the settings or links that hold the rings together. Referring to Fig. 181, D shows the cutting die, which was made from a piece of tool steel recessed into the block C and flush with its upper surface. Over this was fastened the stripping-plate B containing the slot A for guiding the stock. Recessed very carefully into the body of the plate was the circular disk Z, which revolved around a central stud. Upon the edge of this disk were cut the pawl and stop-pin notches so as to index the dial eight times to a complete revolution. To the left of the dial was a slide P, which carried the indexing pawl M, and was provided with a return spring S and an adjustable stop W. The stop lever Q, which worked in a recess in the face of the plate, was actuated by the spring R, of sufficient strength to bring the dial into exact position in case the pawl should vary in its action. The eight holes, $X X$, were so located in the disk Z as

PUNCHES, DIES, AND TOOLS. 105

to come exactly central with the die as they were brought around by the indexing mechanism. These holes were counterbored at *Y*, the counterbore being a trifle larger than the diameter of the blank, so that it could be easily forced to the bottom, where it remained until indexed around under the drawing punch, which operated at *H*. This punch drew it into the shape shown in Fig. 180 and carried it through the hole *H* out of the machine.

The punch, shown in Fig. 182, consisted of three parts—the blanking punch *D*, which is shown in section in Fig. 185; the drawing punch *F*, and the cam *G*. Through the centre of the blanking punch was a small piece of wire, having a collar at its

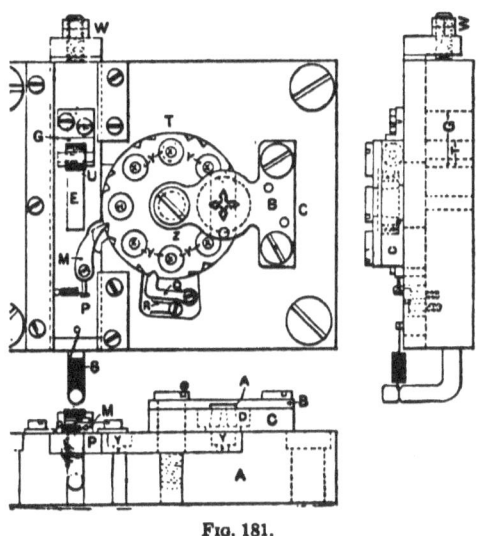

FIG. 181.

lower end against which bore a spiral spring. This spring pin was held into the shank of the punch by means of a headless set-screw. The object of this spring pin was to push the blank firmly into the recess in the dial and also to prevent the blank from sticking to the punch, as it would be apt to do if some such means of prevention were not provided. The drawing punch *F* was held in the holder *E* by means of a set-screw *K*, but the hole in the shank was considerably deeper than the end of the punch, so that if the dial should not index properly, and the punch should strike against it, F would slide up into E instead of being broken. The cam G was set so that it would engage the cam roll U, Fig. 181, and thereby index the dial. The hole at T was to provide an escapement for the chips which, before it was put in, caused some trouble in the working of the dial.

The operation was as follows: The stock, which came in strips, was fed into the slot A by means of an ordinary roller feed. As the blanks were cut out they were forced by the spring pin, in the punch, into the bottom of the recess Y, where they

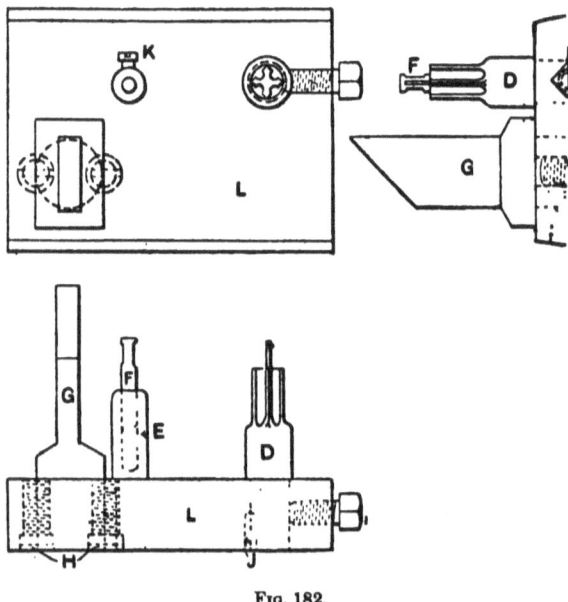

FIG. 182.

remained until indexed to H, when they were drawn by the punch F, Fig. 182, and fell completed out of the machine. With each descent of the punch X the cam G, engaged with the roll U, caused the plate P to slide outward sufficiently for the pawl to index the dial one notch. After the cam had ascended, the plate was returned to its original position by the spring S.

The punch and die for forming the rings are shown in Figs. 183 and 184 and form a comparatively simple tool. The stock in strips was fed, by hand, into the die at A in the direction of

PUNCHES, DIES, AND TOOLS. 107

the arrow and held firmly against the side K by means of the spring-gauge B. After each cut the stock was moved forward until the edge came to the points E, E, which were very accurately filed on the set edges. The stock first encountered the punches G G, which cut out the outside of the rings. Each side of the hole made by these punches was about .003 inch less than a quarter of a circle, so that about .006 inch of metal was

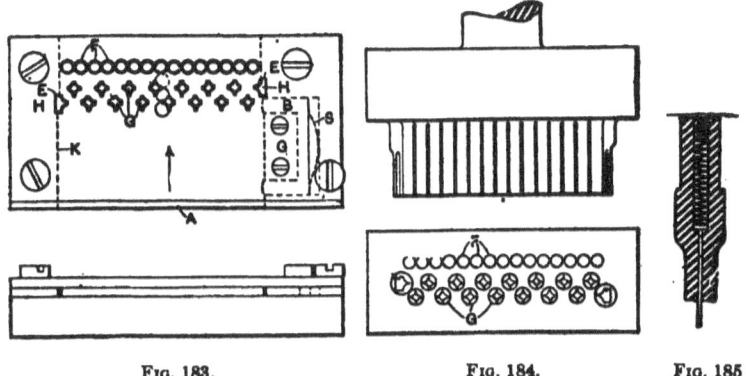

Fig. 183.   Fig. 184.   Fig. 185.

left to tie the rings together. After this punching they came to the round punches F F, which removed the metal from the centre of the rings and left remaining the plate of rings shown in Fig. 179. The settings were then put in place by girls and "set" in a foot press, after which the piece was rolled a couple of times in the hand and the stock between the rings was easily broken, leaving a flexible body.

BLANKING AND FORMING A SHEET-METAL ROLLER.

The combination press tools shown herewith were used for blanking and forming from 14-gauge stock the piece shown in Fig. 186; the piece is afterward bent to the form of a roller. It is customary to do the blanking in one operation and the forming in another, the blank being put into the forming die by hand or small pincers. Either way is slow and dangerous, and more than one operator has lost fingers handling this class of work.

108   PUNCHES, DIES, AND TOOLS.

With these tools one boy has made over 250,000 pieces, in less time than three boys did the same number in the old way.

The blanking punch is shown at *A*, Fig. 188, and the form-

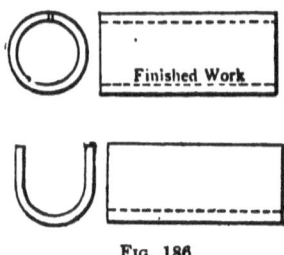

Fig. 186.

ing punch at *B*, Figs. 187 and 188. *C* is the cast-iron holder with a shank *D* to fit the press. Screws *E* hold the punches in place. The dies are held in block *F*, Fig. 187. Here *G* is the

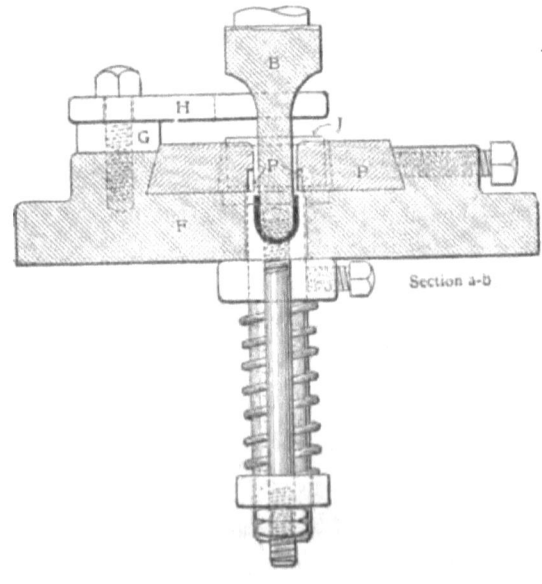

Fig. 187.

stripper support and *H* the stripper itself. *I* is a stop collar placed on ejector *J*, which part is guided in plate *K* resting on adjusting nuts *L*. The blanking die is located at *M* and held by

PUNCHES, DIES, AND TOOLS.   109

screws N; the gauge-pin is at O. The forming die (shown clearly in Fig. 190) is carried at P. The blanking die is of one piece of tool steel and is flat; but the punch is made concave to

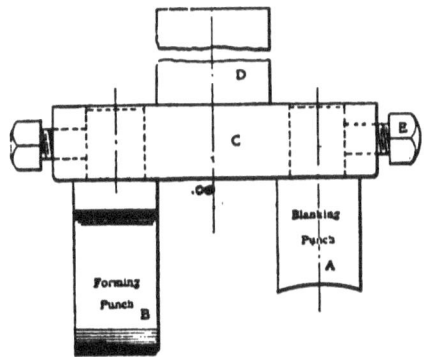

Fig. 188.

give a shearing cut. The blank, indicated in Fig. 191, is cut through but not punched free of the stock, being held at the sides, and in this way it is carried under the forming punch,

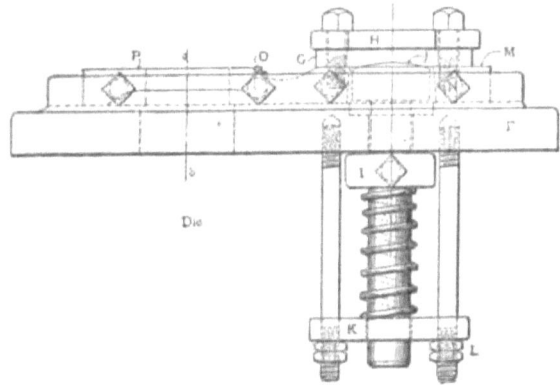

Fig. 189.

enabling the blanking of one piece and the forming of another to be done at the same time.

Owing to the shape of the punch the blank is convex, and the edge Q is about $\frac{1}{16}$ inch lower than edge R, so as to strike

110   PUNCHES, DIES, AND TOOLS.

the gauge-pin O, Fig. 189, the die being concaved at this point to allow the edge of the blank to drop against the pin.

Sometimes the blank when cut through does not stick at the sides and would drop through were it not for the ejector J,

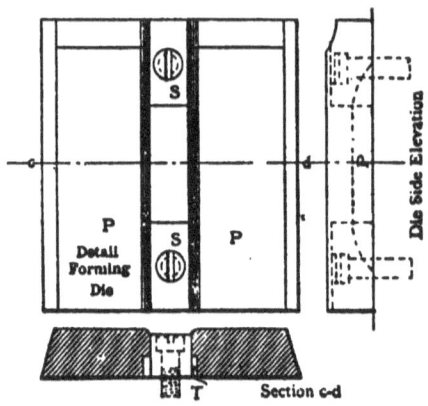

Fig. 190.

which, passing up through the die, presses the blank into the scrap again. A stiff spring resting on plate K and against collar I, which is made fast to J by the set-screw, keeps the top of J against the blank.

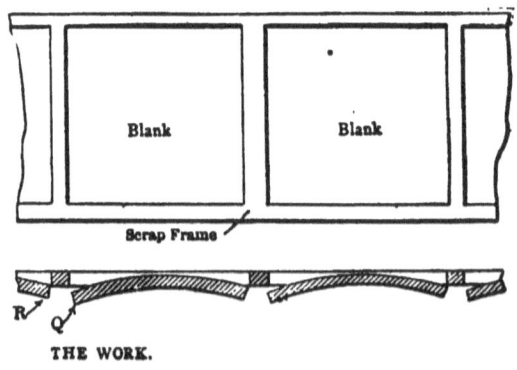

Fig. 191.

The forming die in Fig. 190 is made in four pieces, $P\ P$ being of tool steel with machine steel pacing-block $S\ S$ between. Clearance is indicated at T—this also doing the stripping. Dies

PUNCHES, DIES, AND TOOLS. 111

P P are made as hard as possible, after which they are passed over the fire to draw the temper a very little. When the die is worn it can be ground at the centre, edged, and made to size very quickly and cheaply.

A THREE-OPERATION FOLLOW DIE FOR THIN STOCK.

The die shown in the four views, Figs. 192 to 195 is used to bend up and cut off the piece shown in Fig. 196. This is of semi-hard copper, .007 inch thickness, $\frac{1}{4}$ inch in width, and with a diameter for the inside of the cylindrical part of $\frac{1}{17}$ inch. The tape to make these pieces from is procured in a roll. This

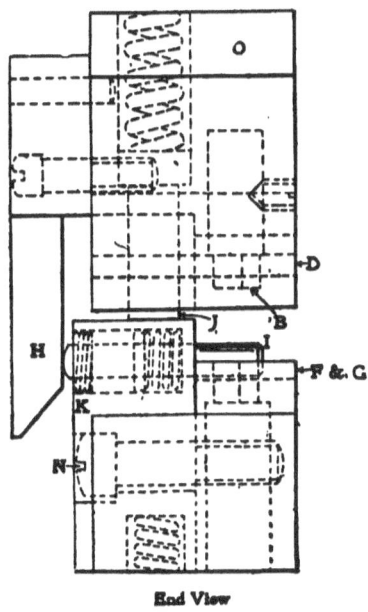

Fig. 192.

roll is supported on some sort of an axis so as to unroll readily, and is fed in by hand to the die between the guide pins *A A;* it is pushed along sufficiently so that the knife or punch *B* will trim the end. The spring pad *C* first catches it and holds it while the end is trimmed, and when the clearance shown at *E* is far enough down to leave the trimmed end free. The form D catches the tape, and, with form F, makes the lower half of the

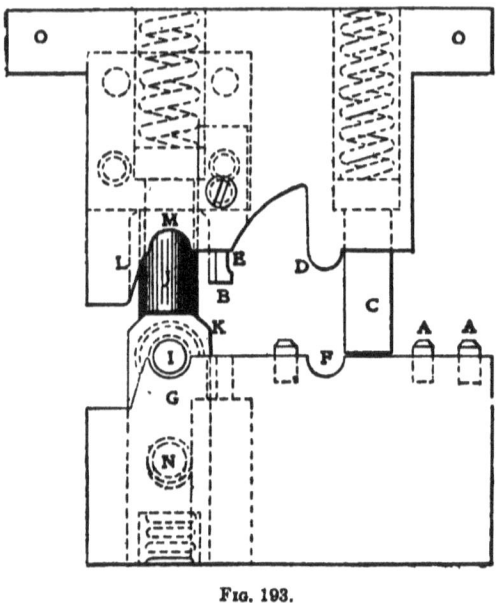

Fig. 193.

bend, with end enough left with which to make the upper half.

The tape is then helped along and the half-bend already

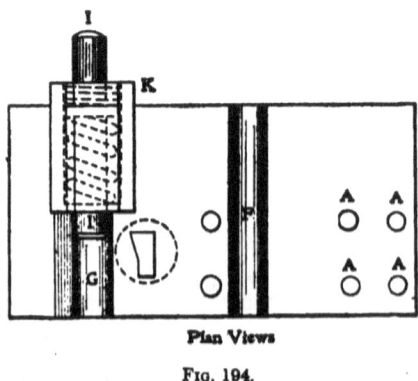

Plan Views
Fig. 194.

made is laid into the form G. Then on the next down stroke of the press the machine on H pushes in the pin I over it, and the pad J, having a stronger spring, pushes down the pad K, pin I, and all, thus securing and holding the tape, while the upper half of the bend is made and the end cut off the proper shape, completing the piece, and at the same time striking a half bend for a new piece on the remaining end of the roll. The office of the extending part L is to catch the end of the tape

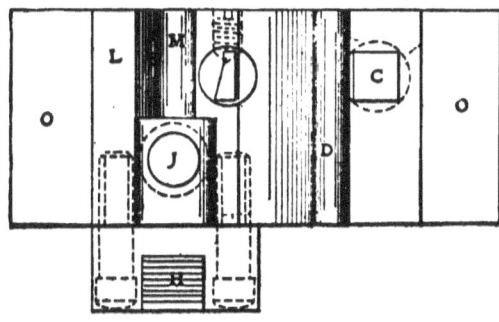

FIG. 195.

shown in Fig. 197 and start it over into the form M as the press comes down after the piece has been secured by I, J, and K.

In addition to the foregoing description of the tools it is also necessary to know that the spring actuating the pad C must not be stiff enough to mar the copper tape. The lug H should fall low enough to push the pin I in before the pad J takes hold. The pad J should be long enough to secure the tape before L

FIG. 196.   FIG. 197.

engages with its free end. The punch B is formed on a cylindrical shank, and held up in place and kept from turning by a set-screw, as shown, and the clearance at E must extend low enough to free the end of the tape before the form D touches it. The cutting die corresponding to B is worked through from the top to a hole bored in the bottom, within $\frac{1}{4}$ inch of the top. The screw N serves not only to hold pad K in place, but also

114 PUNCHES, DIES, AND TOOLS.

to limit its upward motion. The lugs on the upper die O O are used to screw and dowel through into a punch-holder, and the lower die can be held in any one of a multitude of ways. The projection of the pad C and its spring is omitted in the end view for the sake of clearness, but would appear in dotted lines behind the punch B.

GANG OR MULTIPLE BLANKING TOOLS.

Some of the most interesting methods of gang die-making are to be found in metal-button manufacturing concerns, and in the following is presented a method for making a gang die for producing a part for a cheap metal button. The blanks cut in the dies were of sheet steel and were called "fillers," or, in other

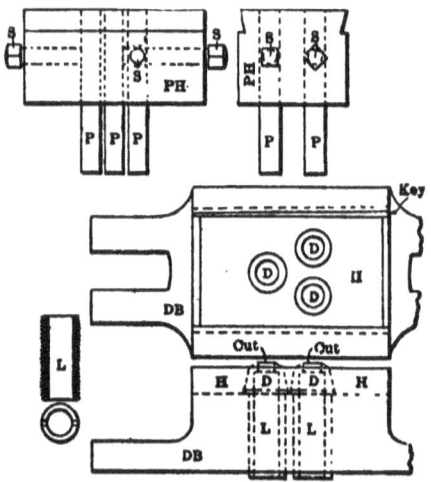

MULTIPLE DIE FOR BUTTON BLANKS

FIG. 198.

words, they were used to fill up the spaces between the fronts and backs of the buttons as they were assembled in a foot press. Fig. 198 contains diagrams of the tools to cut the blanks.

The cast-iron holder PH carries the three blanking punches P held by the set-screws S. The views of the cast-iron die-bed DB illustrates the method of holding the inserted blanking dies

PUNCHES, DIES, AND TOOLS. 115

D by means of the die-holder plate H (machinery steel) which is keyed into the die-bed D by means of a tapered key. $L\ L\ L^1$ represent the locating screws which hold the dies $D\ D\ D$ in place from the base of the die-bed DB. The blanks, after being cut, pass through the dies D and the locating screws L into a box. A sectional view of one of the locating screws is given. Both dies and punches are made from first quality tool steel (Styrian in this case). The punches are hardened and tempered; the dies are hardened, tempered, and ground.

In making the dies they were "crowned" on the top, which made it much easier to grind them when they got dull. The three dies in H were located so as to obtain the most number of blanks from the given strip of metal which only left a little waste of scrap. The bottom or bearing surface of the inserted dies had to be ground true after hardening. The open-back, single-acting press in which the tools were used was equipped with automatic multiple pawl rachet feeds, by which devices a small fractional part could be added or subtracted from the length of the feed. The several pawls were disengaged simultaneously by a lever connected with this useful attachment.

BLANKING DIE WITH GUIDE-PINS FOR THE PUNCH.

The remedy for overcoming the tendency to produce work that should be smooth with a burr around the edges through the punch shearing in the die is given in the following:

FIG. 199.

The blank, Fig. 199, was of cold-rolled stock, No. 4 B. and S. drill gauge (0.111 inch) with five holes, No. 50 B. and S. drill gauge (0.07 inch), these to be flat and very smooth, and above all to be made very cheap. The blank is shown in Fig. 199. Several attempts in the presses at command failed to give the required result, so a tool was made as per Figs. 200 and 201, which is perhaps rather patterned after a sub press, though better than that for this work. The base and the punch-holder are of cast iron,

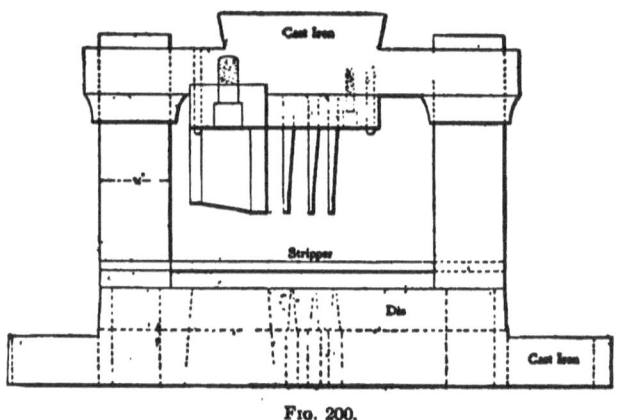

Fig. 200.

the four pins of tool steel hardened and ground and a good working fit in the holder, the holes of which had been bored when all were clamped together.

The die was made and keyed in the press. The punch-

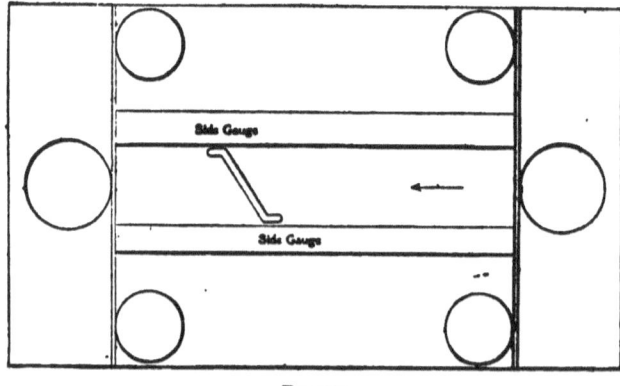

Fig. 201.

holder was adjusted to the pins. The block for the punch was screwed and dowelled on. The main punch was fitted with the holder on the pins, was hardened, and the small round punches

PUNCHES, DIES, AND TOOLS. 117

lapped to fit. It was keyed in the press, one with a narrow gate so that the pins could clear when at the lowest point. One boy with a finger feed and the press running 250 per minute cut out 450,000 without injury to the tools, and they would have gone much longer without sharpening if it had been necessary. The additional expense, in proportion to the better results secured, is nothing, and the tool-setter can never shave the punch even the slightest bit.

PIERCING, BLANKING, AND BENDING DIES.

In Fig. 202 is shown the bending die for producing the bent piece seen in Fig. 203. The material used for these pieces was

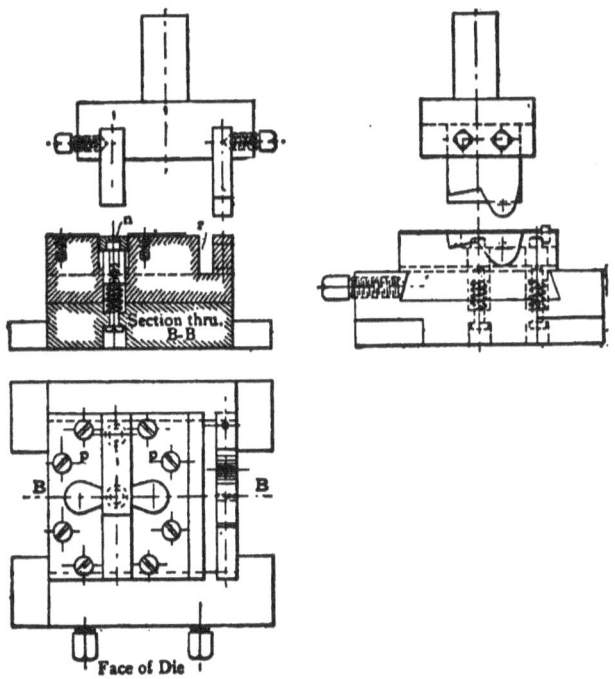

FIG. 202.

No. 16 B. and S. gauge (0.058 inch) half-hard sheet brass. The blank is shown in Fig. 203.

The die block of the blanking die, Figs. 202a and 202b, is

118   PUNCHES, DIES, AND TOOLS.

of tool steel worked out to the shape, then hardened and ground. The stripper plate is of mild steel fastened by screws to the die block and dowelled. The stop finger is similar to those in common use in most press-rooms that are up-to-date, and have been adopted by the majority of manufacturing concerns.

The punch-holder is a mild steel forging, machined as required; the piercing punches and the blanking punch are fixed into the mild steel plate in the ordinary manner, the plate being

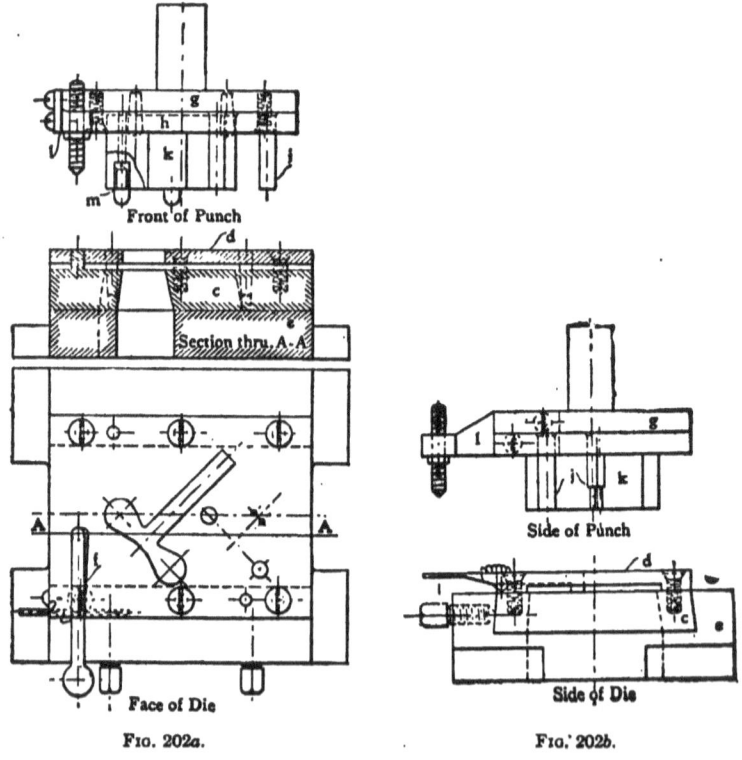

FIG. 202a.   FIG. 202b.

fastened to the punch-holder in the same manner as the stripper plate.

Particular attention is called here to the locating of the piercing holes and the blank slot in the die block.

These are placed at .45°, the reason for this being as follows: In the first and second bending operations it will be noticed there is a double bend, as indicated at $x\ x$ Fig. 203.

PUNCHES, DIES, AND TOOLS.

Now, if the blank were cut either with the grain or across the grain, there would surely be trouble in making one or the other of these bends. The material was cut in long strips, with the grain, as indicated in Fig. 204, and punched in the manner shown.

At *m* is shown the method of holding the guide-pins in the blanking punch *k*. The ordinary method of fitting these pins directly into the punch is unsatisfactory. In this case the punch is drilled and counterbored and then hardened, after which a

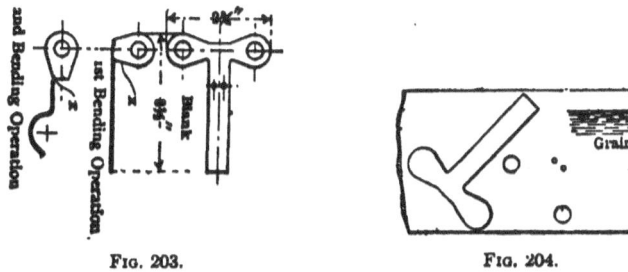

FIG. 203.   FIG. 204.

mild steel plug is driven into the counterbored hole, and this plug is then properly laid out and drilled to receive the guide-pin, thus insuring accuracy.

The construction of the bending dies is clearly shown in Fig. 202, and thus no lengthy explanation is necessary. *N* is the tension pad, whose action will be readily understood; *p p* are gauge-plates for locating the blank for the first bending operation; *r* is a clearance groove only. These dies are simple in construction, yet have given perfect satisfaction.

COMBINATION CUTTING-OFF, BENDING, AND FORMING DIE.

Fig. 205 shows a combination die for shaping the piece shown at Fig. 208. It is of composite construction, and was designed with a view to produce the article as cheaply as possible. The lower half of the die includes the pieces *I*, *J*, *K*, and *G*, which are of tool steel, while the piece *S* is machinery steel, and *X* is cast iron. *O*, which is a piece of angle iron, is fastened to the piece K by means of two screws, P. The upper part will be described further on.

The stock is cut to the desired width on the square shears, and after the stock has been cut, it is placed in the guide O, which keeps the piece in alignment while in the continuous progress of forming, as it is fed into the forming die K. Die K is machined to the desired shape, and is cut a little deeper than the piece J, to allow of the cushion-plate N being inserted

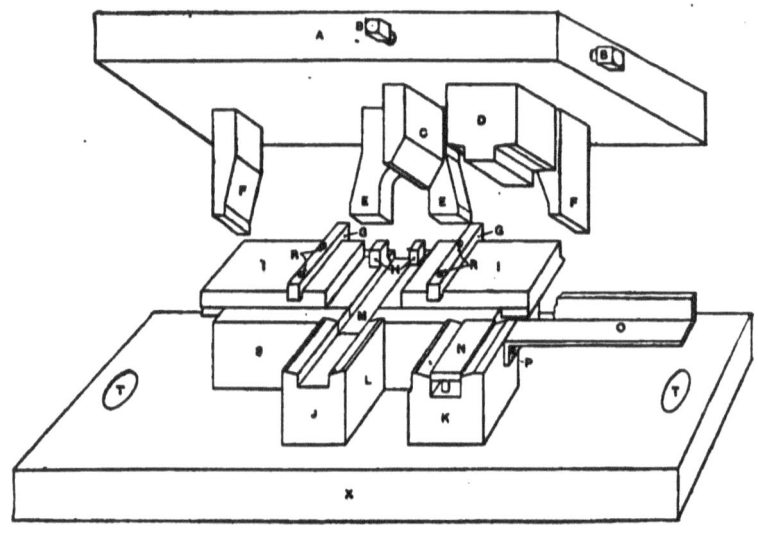

Fig. 205.

and still give the required depth of the piece to be formed. After the piece to be formed has been placed on the guide O, and is shoved further to the side of the forming block K, with about $\frac{1}{4}$ inch projecting over (that being about the amount of which the metal will draw in), the press is tripped, and we get the first form. The piece is then shoved along until it strikes the block J, on the side marked L, which measures off the exact length, $1\frac{1}{2}$ inches, for the next form. The press is then tripped again, and we get the second form as shown in Fig. 206 "showing continuous forming." It is then fed into the block J, which is the same shape as that of the piece formed, and the press is tripped again, which gives us another form, and at the same

PUNCHES, DIES, AND TOOLS.   121

time cuts the piece off, as shown in the sketch, Fig. 207. This die is worked in an inclined press, and after the piece has been cut off, it is fed by gravity into the closing-in die *M*. *L* is a cutting face, and *C* is the cutting-off punch. The block is slightly cut away, as shown, to allow of the metal being drawn as equally as possible from both sides. The cutting-off punch *C* is left about ₁⁄₁₆ inch longer than the form block or punch *D*, and it cuts the metal off before the metal commences to form; it is tapered as shown so as not to interfere with the other portion of the metal being formed.

After the piece has been cut off, and on the ascent of the ram of the press, the piece slides into the portion of the die marked *M*, and against the pins or fingers *H*, which prevent the piece from falling through the die until it has been formed. The block *S* is made with a dovetailed groove, and the pieces *I* are made a nice working fit therein; the fingers *F*, which are given the right degree of taper to produce the necessary travel of the sliding pieces *I*, are of tool steel and hardened, and have ¼ inch taper on the inclined parts. The pieces *G* are tool steel hardened, and are held in place by the screws *R*. The fingers *E* are also made of tool steel and are given the same degree of taper as the fingers *F*. The piece is fed into the die as described,

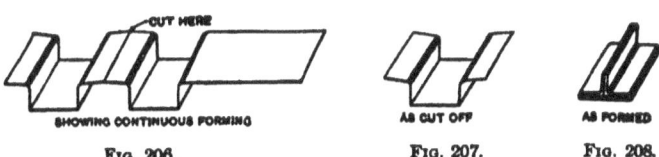

FIG. 206.   FIG. 207.   FIG. 208.

and on the descent of the ram the fingers *F* force the slides *I* inward and form the piece to the desired shape as shown. The fingers *E*, working against the pieces *G*, on the upward stroke force the slides back to their normal position, and the die being worked in an inclined press, the formed piece slides out of the back and a new one slips in. Thus the continuous forming continues until the strip is worked up. It will be noticed that there is about ¼ inch left straight on the fingers *E* and *F*, so that they are always in contact with their respective working parts. A portion of the fingers *E* is cut away as shown, so as to allow of the formed pieces sliding through.

This die is fitted to a press having 1¾ inches throw, and is fastened to the bed by means of the holes *T*. The parts *F*, *E*, and *C* are held in position by set-screws *B*, while the other parts are held by means of fillister-head screws and dowel-pins, as one cannot depend on a screw to keep a thing perfectly in place. The parts *F* and *E* must work in harmony with each other and both have the same degree of taper. *A* is of cast iron, and *J*, *K*, *G*, and *I* are all hardened and drawn to a dark straw color. There is also a shank fitted to this die which is not shown, which holds the upper half of the ram of the press while in operation. It will be noticed in the engraving that the piece is formed a little less than right angles, so that in closing-in there is no chance for the metal to spring apart and leave the piece open at the top. There must also be allowed twice the thickness of metal either in the slides *I* or in the bottom of the die *M* for the metal while in operation of closing-in. After the third stroke of the press a finished article is delivered at every stroke, and it can be made for about three cents per hundred. This die should prove suggestive for adoption of modifications of its construction for the production of a large variety of sheet-metal articles which have to be produced at a low cost. Fig. 208 shows the finished article.

FORMING AND EMBOSSING DIE FOR A SPRING CLIP.

To make the spring clip shown in Fig. 209 the dies shown in Figs. 210, 211, and 212 were used.

Figs. 210 and 211 show front and end views of the punch and die, with the detail of construction in Fig. 212. The die shoe is cast iron. *A* is a pad in which the blank is embossed, formed to a required arc, and the end marked *K*, Fig. 209, is bent to a right angle. The pad is of hardened tool steel.

The tool-steel piece *B* forms a wall around the pad and is left soft; a pin in each end retains it central in the shoe.

The holes *D D* are for the pressure pins that actuate the pad

PUNCHES, DIES, AND TOOLS. 123

from the buffer; the large tapped hole in the centre of the shoe is to receive the buffer stud.

C is a stationary knee of hardened tool steel. Its function is to give the motion to the "kicker" E, seen in the punch. F F

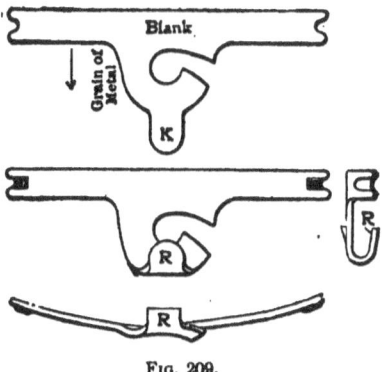

Fig. 209.

are gibs. The kicker E, as the punch travels down, strikes the knee C on the angle L, and it slowly begins to close in toward the punch G. The punch, while E is travelling down L, is

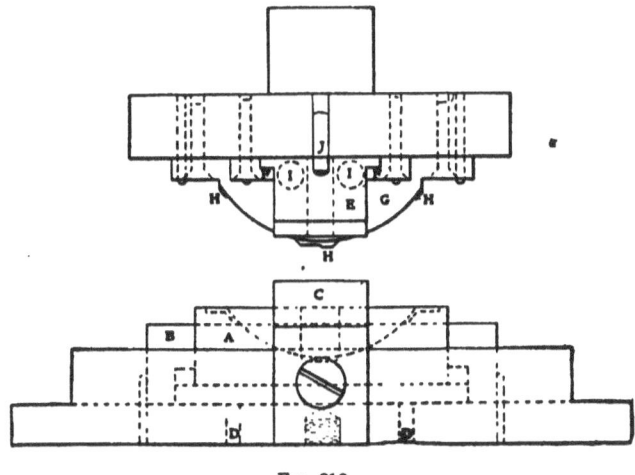

Fig. 210.

forming the blank in the pad and bending the hook R, Fig. 209, at a right angle. When E arrives at M the punch G has completed forming the blank with the exception of the embossing at the ends and middle and curling the hook. The pad now rests on the bottom of the shoe. Now, at the dotted line P in

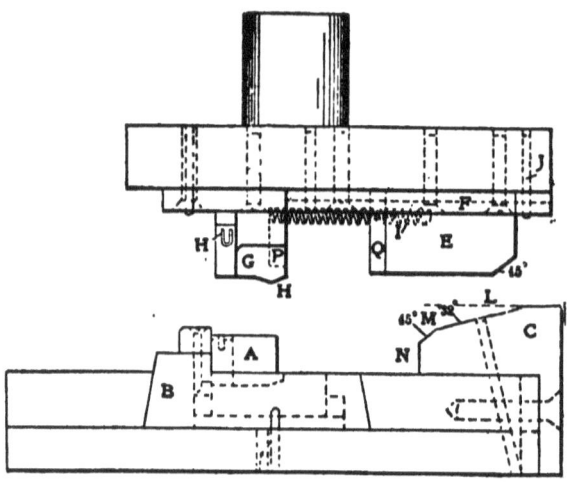

FIG. 211.

punch G, Fig. 211, it will be noticed where the hook part of the clip is curled around. This opening is just large enough to

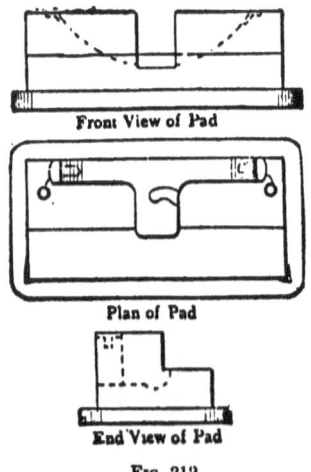

FIG. 212.

admit the rib Q plus the thickness of one metal—.041 hard brass. At the short angle M, Fig. 211, the kicker E really does

PUNCHES, DIES, AND TOOLS. 125

its work; it has been loafing while sliding down L, so as to keep from touching the blank until the right time. After E passes the 45° angle M, the kicker passes on down at N; this movement allows the punch G to finish the job of putting in the embossing. As the punch ascends the blank clings by the hook to the opening at P, Fig. 211, and is easily taken care of by the operator. The radius of the circular part of the punch G is 1¾ inches. The radius of the finished blank is 3⁺⁺ inches.

In this set of dies no allowance had to be made for the spring of the metal for the embossing. It was figured that it would be necessary to allow some, and so the pad was made accordingly, but that was not necessary, and consequently a new pad had to be made, as the first did not fit.

COMPOUND PUNCH AND DIE FOR LEATHER WASHERS.

Fig. 213 shows an engraving of a compound punch and die for blanking leather washers for the filler valve on a gasoline stove tank.

The engraving shows the punch and die one-half size. A, B, C, F, G, I, and K are of machinery steel, while D, H, and J are of tool steel, and L of cast iron. A, which represents the shank of the punch, is bored out large enough to accommodate the spring shown, and is held in position by the tension plug I, which is bored out so that the shank of the knock-out B passes through it and keeps in alignment with the portion which travels in P. It will be noticed by the engraving that the lower portion of A is bored a little larger than P where C fits; C acts as a backing for the punch D. The piece C is made a nice fit into A, while F acts as the punch-holder and is made a nice driving fit into A and H, and the punch D is made a nice fit into F. G, which is called the knock-out or kicker, is operated by four pins, $e\ e$, etc., which pass through holes drilled in F and C, and work against B. As the punch ascends, the spring R forces the blank or washer from the punch, and leaves it on the upper face of the die, and as soon as the punch and die are separated the blank is ready to slide off, if worked in an inclined press; otherwise it

126   PUNCHES, DIES, AND TOOLS.

must be picked off by hand. *H*, which is the outside cutting die, is held by four fillister-head screws *M*, and as *H*, *F*, and *A* are all snug fits, there is no danger of its shifting. The lower half of the die is made as follows:

L, which is of cast iron, is made to fit a 4-inch shoe, and is bored to receive the piece *J*, which is held in place by the four fillister-head screws *N*; *J* is a snug fit in *L*. The stripper *K* is

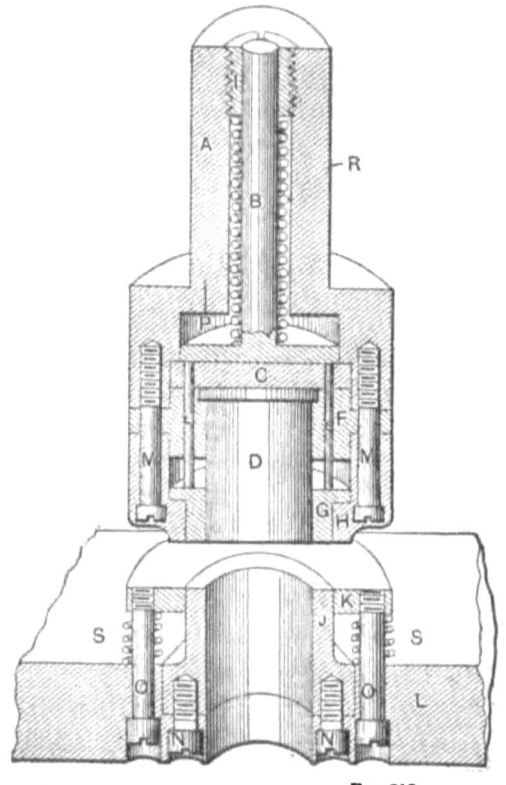

Fig. 213.

held in place by the four fillister screws *O*, and operated by four springs *S*.

It will not be necessary to give any further instructions in regard to the punch and die, as the engravings show the construction more plainly than could be described. Fig. 213 shows the washer as produced by the tools. It may be further stated, however, that no clearance at all was given, as it was necessary that the washers should be of a definite size. It was thought when making the tools that by giving no clearance whatever, there would be no danger of changing the size after grinding, and as the punch did not enter the die over $\frac{1}{16}$-inch, it gave excellent satisfaction.

When making compound dies of this construction, it will be found that by giving no clearance good results will be attained when blanking thin stock, as it does away with the burr, and there is no chance for the changing of the diameter of the blank after grinding the die.

COMBINATION DIE FOR PUNCHING, PIERCING, AND SPLITTING LABELS.

The set of tools illustrated in Figs. 214 to 216 were used for the manufacture of labels shown at Figs. 217, 218, and 219, and were used in a single-acting press.

As will be seen in Fig. 218, it was necessary to cut the blank, pierce three holes, and split the stock between two of the smaller holes on a $\frac{1}{16}$-inch radius to allow the enclosed material to be deflected for the insertion of an endless hoop. The flap was then bent back into place and secured by a small gummed label pasted over the juncture.

Fig. 214 shows the die, the first part to finish being the cutting ring A. The corrugated edge was made by dividing the circumference into an equal number of parts and drilling every other one, the remaining stock being filled up, with hardly any clearance, to an arc of a circle having the same diameter as the drilled holes, thus producing the outline as shown in Fig. 219. The cutting ring was then hardened and ground on the base (which was slightly recessed) and up a short distance on the outer circumference where it fits the cast-iron plate D. It was also ground on the outer edge which was left about 15°.

The main punch, E, Fig. 215, was now worked up and finished by shearing it through the cutting ring A.

Pressure plate B, Fig. 214, was turned up with a $\frac{1}{8}$-inch annular groove, $\frac{1}{4}$ inch from the lower edge and deep enough to clear the inside points of the cutting ring A. It was then in-

128 PUNCHES, DIES, AND TOOLS.

serted from below into *A*, and the cutting edge scribed off and worked up.

Punch-holder *C*, a ⅛-inch steel disk, was next clamped to the plate *B*. The two largest holes were drilled and reamed through both and then it was used for a drilling jig on the

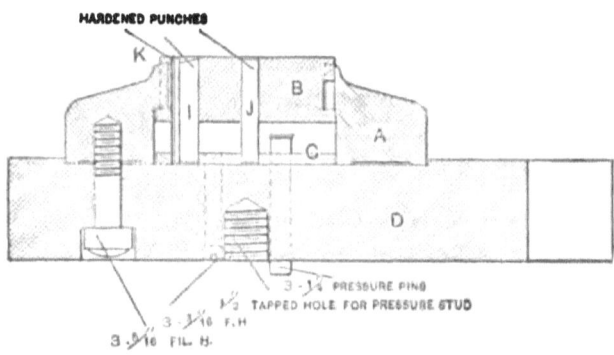

Fig. 214.

punch *E*, after which the largest hole was plugged with the same quality of steel to prevent the drill from running out, and holes were drilled for the two smallest punches *K*. The pressure-plate *B* was now used to drill the small holes in the punch *E*, which had also been plugged the same as the plate *B*. The

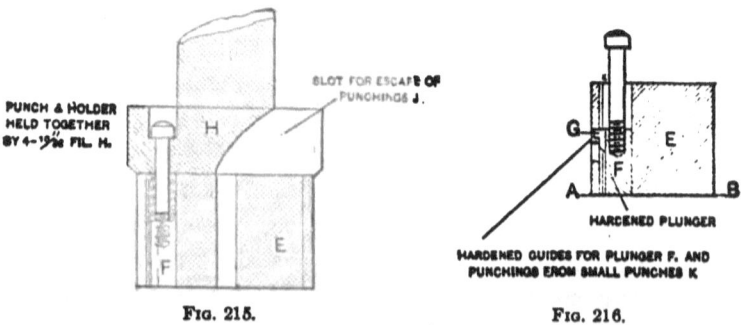

Fig. 215.   Fig. 216.

short plug used in the pressure-plate *B* was used as the plunger *F*, Figs. 215 and 216, and the plug used in the punch *E*, Fig. 215, was the long punch, *I*, in Fig. 214, with the top faced off on an angle to enter the main punch *E*, which acts as a die, splitting the stock to the small holes previously punched by the

PUNCHES, DIES, AND TOOLS. 129

punches K. Plunger F is faced off to the same angle as punch I, but not quite up to the edge of the small holes which form the piercing dies. The punchings are expelled through the side of the punch E, by the agency of the bevelled guides G, Fig. 216.

Plunger F, Fig. 216, was kept from turning by two right-angle guides G, set into small holes above the discharge holes in

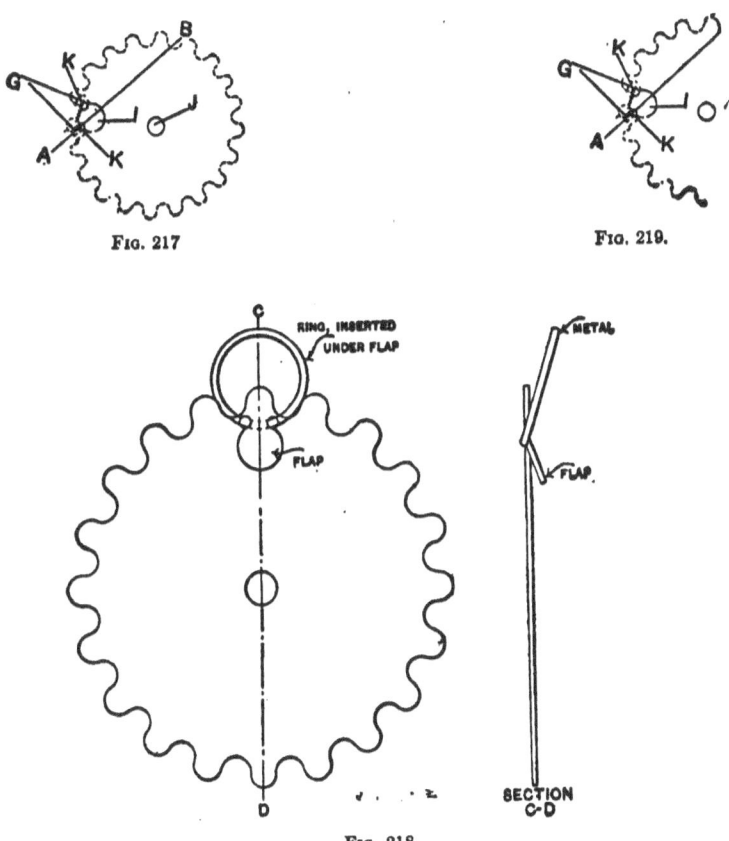

FIG. 217.

FIG. 219.

FIG. 218.

the punch body E, from the inside, the plunger when in place keeping them from working out. The guides G were bevelled to facilitate the expulsion of the small punchings from K, Fig. 214.

The punches K, K, I, and J were riveted over the punch-

9

plate *C*, Fig. 214, the tapped holes of which were spotted by using die plate *D* for a template.

Allowing the punchings from *J* and *J* to work up through the punch is a decided advantage, since there is no need of cleaning out the scrap from the finished punching, as happens when using strippers, which force the stock back into the piece. This saves considerable time in assembling on this class of work.

What may be new to some mechanics is the use of compressed air for carrying away the work after punching, in lieu of gravity, as on the inclined presses. It was successfully made use of in producing the articles worked out in this set of tools on the upright press. The air was compressed on the down stroke and allowed to strike the face of the die on the up stroke, thus carrying the article away, which of course was light.

SECTION V.

Notching, Perforating, and Piercing Punches, Dies, and Tools.

PUNCHING SMALL HOLES IN TOOL-STEEL BLANKS.

FOR the notching, perforating, and piercing of accurate small sheet-metal parts, tools of careful and positive construction are required. Frequently the construction of these punches and dies involves the combining of many ingenious arrange-

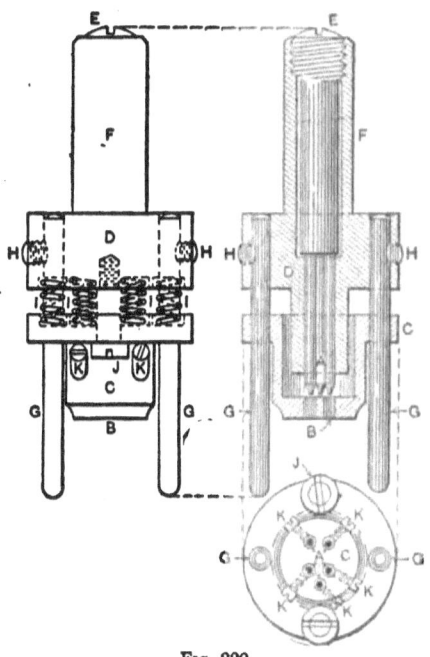

FIG. 220.

ments for assuring the accuracy of the operation and the perfect duplication of the product. Where the notching or piercing is very small, great care is required in so fastening and guiding the punches as to obviate the ever-present tendency to shear in

132   *PUNCHES, DIES, AND TOOLS.*

the dies, to bend under pressure, or to snap off when passing through the work. In this section examples of various types of notching, piercing, and perforating dies are described and plentifully illustrated, so that the mechanic may adopt their construction for similar work or adopt the principles in modification of the designs.

The tools described in the following and illustrated in Figs. 220, 221, and 222 were used for punching five holes .050 inch

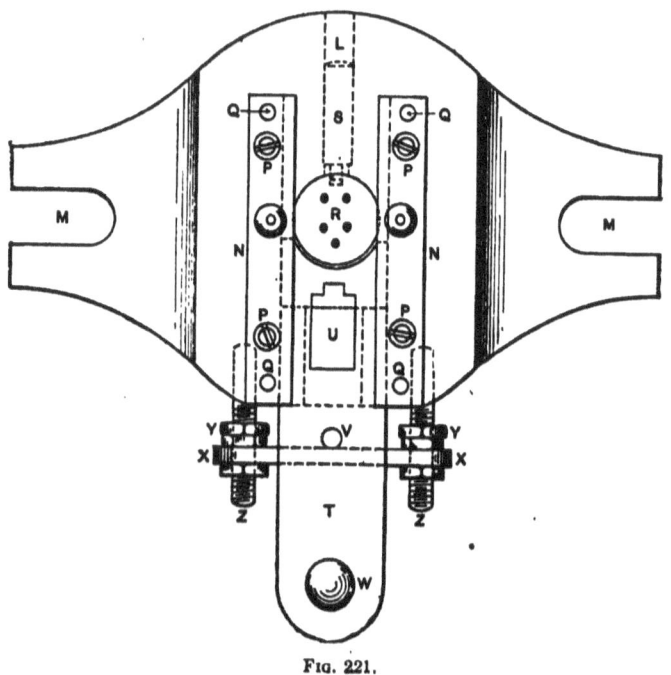

Fig. 221.

in diameter through tool steel $\frac{1}{16}$ inch thick. Now it will be conceded that the accomplishment of this work presented unusual difficulties, the exceedingly small diameter of the holes in comparison with the nature and thickness of the material worked upon precluding the use of tools of standard design and construction.

Figs. 223 and 224 illustrate the two parts for the piercing of which the tools were built and used. As will be seen from the black dots five holes were punched in each. The part shown in

PUNCHES, DIES, AND TOOLS. 133

Fig. 224 is a small dovetail piece of tool steel which is afterward riveted to the part shown in Fig. 223, thereby forming a slide which is used in a machine of world-wide use in printing and publishing concerns. After assembling by riveting, the article is hardened, tempered, and ground, and the dovetail surfaces are shaved to accurate measurements, the allowable limit of error being only 0.0005 inch at any point.

Fig. 220 shows three separate views of the upper or punch portion of the tools. The construction is such as to overcome as much as possible the tendency of such frail punches to buckle, snap off, or shear the dies, and is as follows:

D is the holder of machine steel in which are located and fastened the five punches A; E, the adjustable backstop, for the

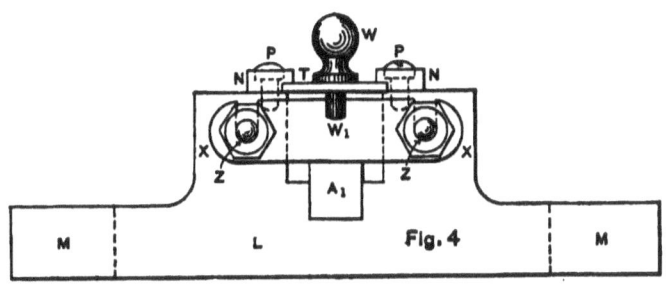

FIG. 222.

punches to bank against; and G, the two die-aligning studs. C, the punch-supporting pad and stock-stripper combined, has five holes let through the face B, into which the punches fit snugly. Five fillister-head screws, K, fasten the punches in the seats in the holder. Two studs, G, are fastened and located in accurately-reamed holes by the two headless screws H. Eight small coil springs—four of which only are shown—located in countersunk seats in the holder and punch-supporting pad, keep the pad C tightly against the blank while the holes are being pierced; they also cause it to strip the blanks after each stroke. It will be seen that the construction of the punch throughout is such as to insure a perfect alignment with the dies, and to give a stiff and positive support to the five punches up to the point where they enter the stock. The two shoulder-screws J prevent the combined supporting-pad and stripper C from descending too far.

The lower section—the die—is shown in Figs. 221 and 222. These are plan and side views respectively. The construction is interesting.

L is a bolster of machine steel, R the die in which the five piercing dies are contained and which are indicated by the black dots in the plan view. The manner in which this die is fastened and located within its seat in the bolster so as to remain positively fixed is shown clearly in Fig. 221. A keyway is let into

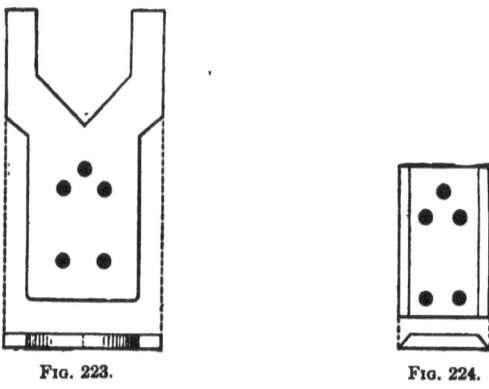

FIG. 223. FIG. 224.

one side of the die R, which is engaged by the key-ended stud S. The hole C' in the bolster is tapped from the under side for the greater part of its length for the hollow headless screw B', upon which the die R rests. Thus, when it is necessary to grind the die face of R, the headless set-screw B is utilized to force the die to a height even with the surface of the bolster. Returning to the engravings, Figs. 221 and 222, N are two toolsteel gibs in which the slide T, for locating the work on the die face, moves. These gibs have reamed holes at O for the aligning studs G to enter. The gibs are fastened and located on the bolster by means of four screws and four dowels; Q, X, Y, and Z indicate part of the adjustment stop which serves to adjust and fix permanent the movement of the slide T. W is a small knob which the operator utilizes when working the slide back and forth during the punching of the holes. V is the stop-pin which prevents the slide from coming out too far from the gibs;

PUNCHES, DIES, AND TOOLS. 135

and W^1 is the pin which prevents it from moving in beyond the proper locating point for the blanks.

When the tools are in operation the punch is set so that the two aligning studs G are in the holes O, in the die gibs. The die is clamped in position, and the stroke of the press is regulated so that the aligning studs G do not leave the guide holes during the up-stroke of the ram. The operator then slips one of the small blanks shown in Fig. 224 into the locating seat U, in the slide T, pushes the slide forward until the pin W^1 rests against X, and then he stops the press treadle. On the up-stroke of the press the work is stripped from the punches by C

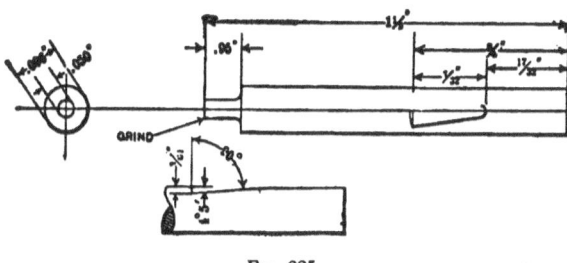

FIG. 225.

and remains in U. The operator pulls the slide back, and the blank drops into the exit chute indicated at A^1, Fig. 222.

Fig. 225 shows detail drawings of the punch. These punches are made in lots of five hundred at a time, from drill rod of the exact diameter of the punch shanks; a small accurate box tool in the screw machine turret reduces the punch ends to size.

Over 100,000 blanks of both shapes, shown in Figs. 223 and 224, are produced per year in the works where these tools are used. To use the tools for punching the five holes in the larger blank shown in Fig. 224, a different slide is substituted for the one used for the small blank.

SPECIAL ATTACHMENT FOR NOTCHING ARMATURE DISKS.

The drawings 226 to 231 show an attachment which is in actual use and is suitable for any punching-press with sufficient space between the slide and the bolster to receive it. It

136   PUNCHES, DIES, AND TOOLS.

can be easily and quickly adjusted for any diameter of disk between 3 and 14 inches, and can be built for approximately $75. This attachment was made and used in a shop making a variety of electrical apparatus and building twelve different sizes of armatures, none of which were made in lots of over fifty, and it gave good satisfaction from the very start. The dies were made of Novo steel 2 inches wide and ½ inch thick. The appearance of the attachment with a disk in place when on the press can be readily imagined from the drawings, as can the

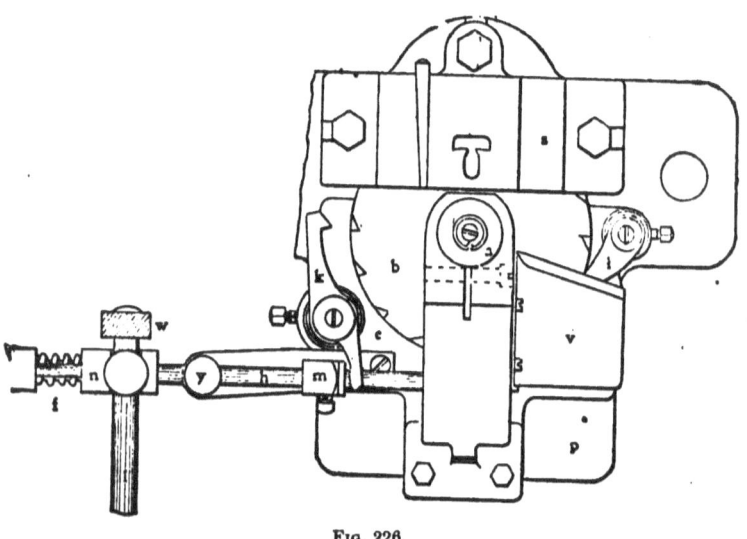

FIG. 226.

means of transmitting the motion from the slide of the press to the index spindle of the attachment be understood without much description. On the back side and the upper end of the long lever, W, Figs. 226 and 229, is a stud and a roller which is held against the inclined edge of the templet by a coil spring (not shown). This templet is about ⅜ inch thick, is held under the clamping nuts on the slide, and is slotted for adjustment. As the slide moves up and down a vibrating motion is thus given to the lever.

Referring now to the drawings, Fig. 226 is a plan, Fig. 229 a front elevation, Fig. 227 a plan with section on line A B, Fig. 229; Fig. 230 is an end elevation with the portion back of the

spindle *a* journalled at its lower end in the plate *e* and at its upper end in bracket *d*. It carries two disk wheels, *b* and *c*, dowel-pinned to rotate together, Fig. 230, the upper one keyed to the spindle and each having ratchet teeth, these pointing in opposite directions, as shown in Fig. 227. The upper wheel, *b*, may be regarded as the index wheel and the lower, *c*, as a ratchet wheel for imparting the movement to the index wheel. The movement of the ratchet is transmitted through the pawl-carrying rocker *r* and rod *h* from the lever shown partly at *W*, Figs. 226 to 229. The rocker *r* does not turn directly on the spindle *a*, but on a concentric boss on plate *e* surrounding the journal. This plate *e* carries the pawl *i* to prevent back-movement of the

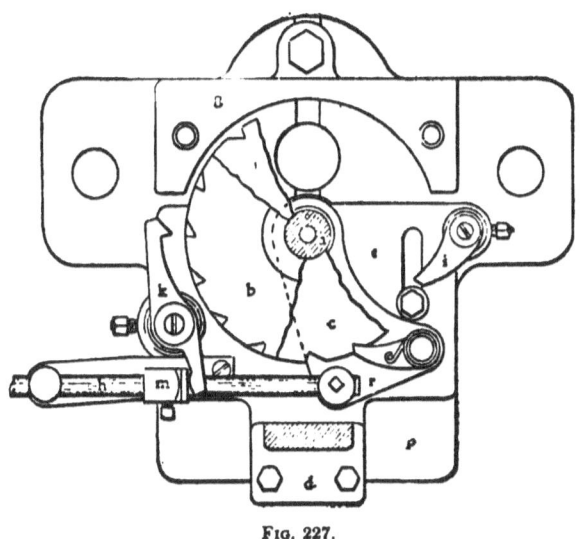

FIG. 227.

ratchet, and also a stop pawl *k* so placed that it engages the teeth of the index wheel *b*, being normally held into its notches by a spring and withdrawn therefrom by the adjustable tapper *m* coming in contact with its projecting finger on the return stroke of rod *h*, which withdraws it sufficiently to clear the teeth of the wheel and enough more to allow the next forward stroke to rotate the wheel at a little to the right before it is let back into the position to again engage, thus leaving the spring to force it into place in the next succeeding notch when it arrives.

138 PUNCHES, DIES, AND TOOLS.

The radial face of the notch is brought up against the stop k with some pressure, and to insure this being always constant the index is given a tendency to rotate a little too far by adjusting the templet above the slide to give the rod h a movement somewhat greater than required to move the index one division, this surplus movement being taken up by the coil spring f. The block n is a sliding mounted on a rod h for this purpose, and is re-

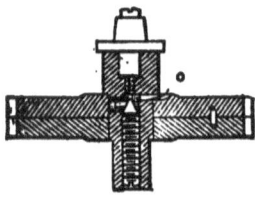

Fig. 228.

tained in its vertical position by the headed pin x attached securely to n and sliding freely through the stationary block y. Block n has a universal joint connection to the lever w through the long pin l.

The die is keyed in a machinery steel shoe s adjustably clamped to its support g which is permanently secured to the base plate p. It may be noted that the entire indexing mechan-

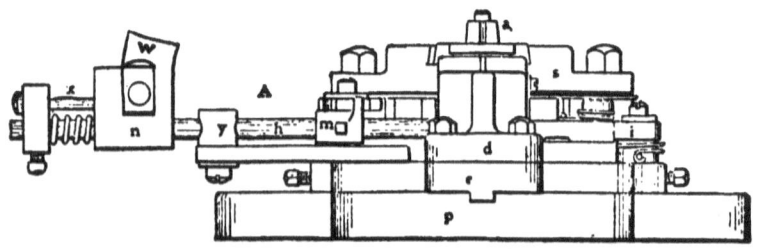

Fig. 229.

ism is contained on the plate c, and this is provided with a tongue fitting a groove in the base plate p, and is also slotted for the clamping screw, thereby allowing for an adjustment to and from the die for the different diameters of disks to be notched.

Fig. 228 is a sectional view of the spindle a. The pin o fits a radial key-seat in the upper wheel. The pin t fits freely in a

PUNCHES, DIES, AND TOOLS. 139

radial hole and can be forced outwardly by turning the vertical headless screw shown, thereby clamping the wheel securely to the spindle. The upper end is bored to receive the shanks of hardened steel heads kept in stock for the different sizes of shaft holes in the disks to be notched. These are held in place by means of a screw through the centre and dowel-pin, as shown.

To lead oil to the lower journal a small hole should be drilled down from the top, coming out at the lower shoulder in

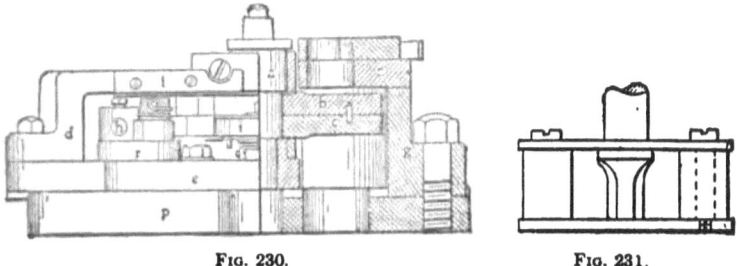

FIG. 230.   FIG. 231.

some place not to interfere with the key and clamping-pin described.

The punch, Fig. 231, is provided with a stripper-plate and two rubber cylinders for actuating springs.

When the ring is set for the larger sizes, above 6 inches in diameter, the slugs from the die will fall directly through, but when set as shown for a small size they fall on the index wheel, and are carried around and collected by the knife-edged plate v and can be brushed off as they accumulate.

It will be noticed that the opening in the die is T-shaped; thus square or irregular shaped disks with the shaft hole punched in them can be notched and trimmed at the same time. The trimmings may be brushed off with a pine stick without stopping the press.

ARMATURE BLANKING AND PIERCING PUNCH AND DIE.

The disk, Fig. 232, is made of $\frac{1}{32}$-inch sheet steel and is $7\frac{3}{4}$ inches diameter; at the centre is a $3\frac{3}{8}$-inch hole. The die, Fig. 234, is fastened to the ram of the press, and the punch, Fig. 233, to the bed. The die is composed of a steel ring *a* formed on the inside to the shape of the blank, while in the centre of the ring

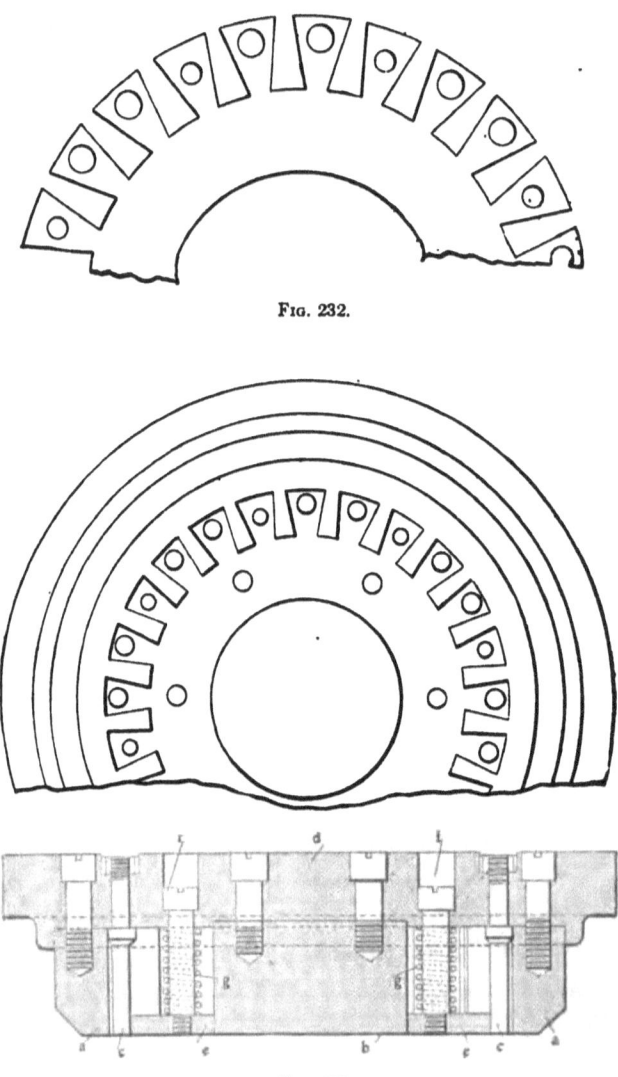

Fig. 232.

Fig. 233.

is a round steel punch *b*, and near the edge a number of small round punches *c*; all these parts being fitted to the cast-iron plate *d* and fastened, as shown, with screws and nuts. A steel

PUNCHES, DIES, AND TOOLS. 141

stripper *e* fitting loosely inside of the die is held in place with screws *f* and springs *g*.

The punch, Fig. 233, is composed of a steel ring notched at the edge, as shown at *h*, to suit the die, and bored out at the centre and provided near the periphery with a number of small holes for the punches in die ring *a*. The punch proper is fastened to the cast-iron plate *i* by means of the screws shown, and is provided with a steel stripper *j* held in place by screws *k* and

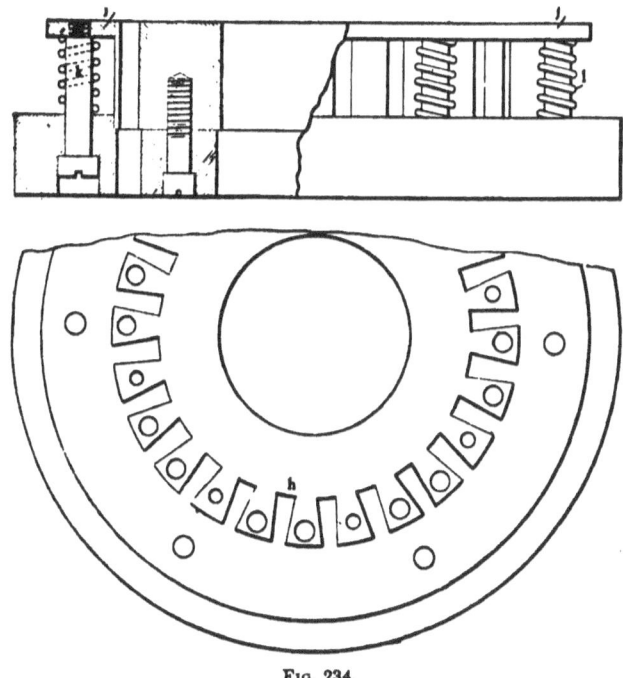

FIG. 234.

springs *l*. When the ram descends the die is forced over the outer cutting edge of the punch, fastened on the bed of the press, cutting out the outside of the blank; at the same time the round punches inside the die ring form the centre hole and the small holes near the edge of the blank. When the ram goes up, the strippers actuated by the springs eject the blank from the die and lift the waste material at the outside of the punch, leaving the finished blank and the waste loose at the top of the punch to be removed by hand.

SLOT INDEX DIE FOR ARMATURE PUNCHINGS.

The drawings in Figs. 235 to 239 will serve to show a die known as a slot index die, used for punching the slots in armature disks. The die is for punching the slots only, two operations being performed on the disk before coming to this die, namely, the punching of the hole and keyway, and the blanking of the outside to size, a large circular die being used for this.

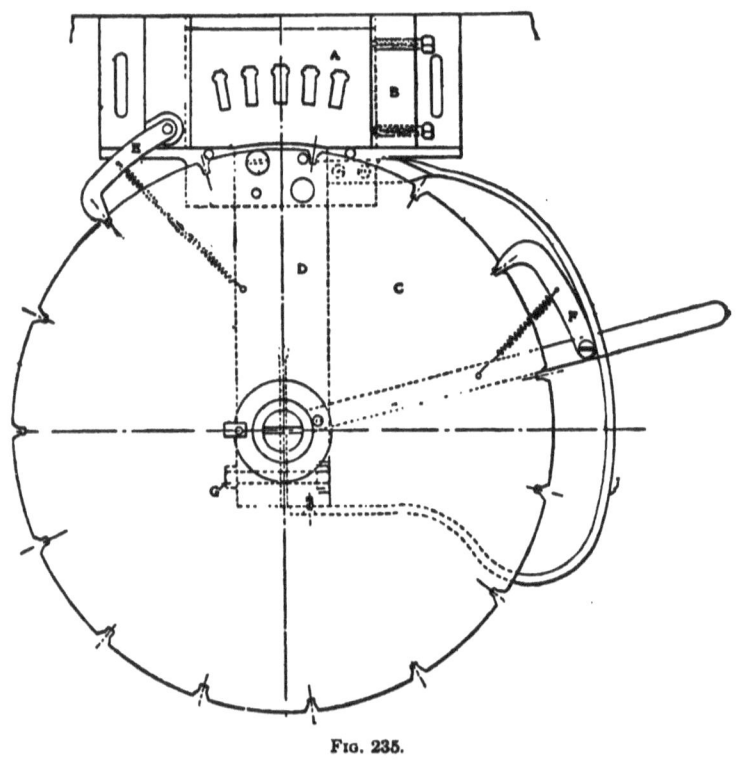

Fig. 235.

These operations are simple, and no description is considered necessary.

In Fig. 235, A is the die block made of tool steel. The five slots are drilled and filed to size, a one-half degree taper being given to let the scrap through freely. The die is hardened and the top and bottom are ground true. B is the gray-iron die-shoe, planed and fitted as per drawing. C, the index plate, is

PUNCHES, DIES, AND TOOLS. 143

of mild steel turned true and the hole threaded to take in the phosphor bronze bushing *H* shown in Fig. 237. This plate is also notched as shown. The disk in this case is to have 75 slots; the die having five slots, therefore the plate should have one notch for every 5 slots; or, in other words, the plate has 15 notches. The centre of the plate is turned to fit the hole in the

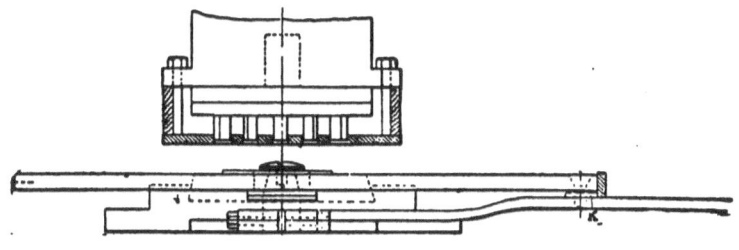

FIG. 236.

disk, a key being fitted into it to fit the keyway in the disk. *D*, the index arm, is of mild steel and is fastened by screws and dowelled to the gray-iron shoe *B*. This arm is also bored and threaded to receive the centre stud *I*, as shown in Fig. 237. It also has a saw slit about one-third of the distance from the end and a clamp bolt *G* through the end for clamping stud *I*. The

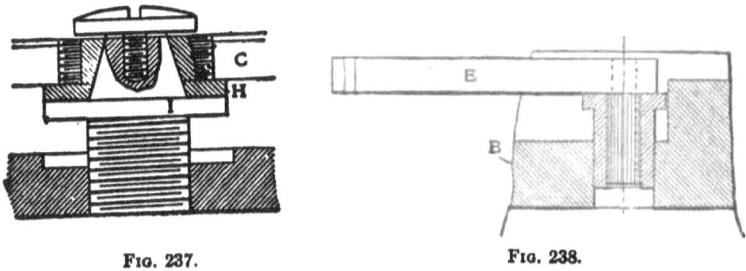

FIG. 237.   FIG. 238.

lever *E*, Fig. 235, acts as a stop, the spring tension being quite sufficient to hold the lever in the notch as shown, in this manner holding the index plate in position. *P* is the operating lever and *J* the supporting rail for it.

The operation is performed thus: The disk or blank, Fig. 239, is placed in position on index plate *C*, Fig. 235, and the first five slots are punched. Then a steady pull on the lever

144 PUNCHES, DIES, AND TOOLS.

F will release the stop lever E and the plate will turn until the lever E drops into the next notch, when the next five slots are

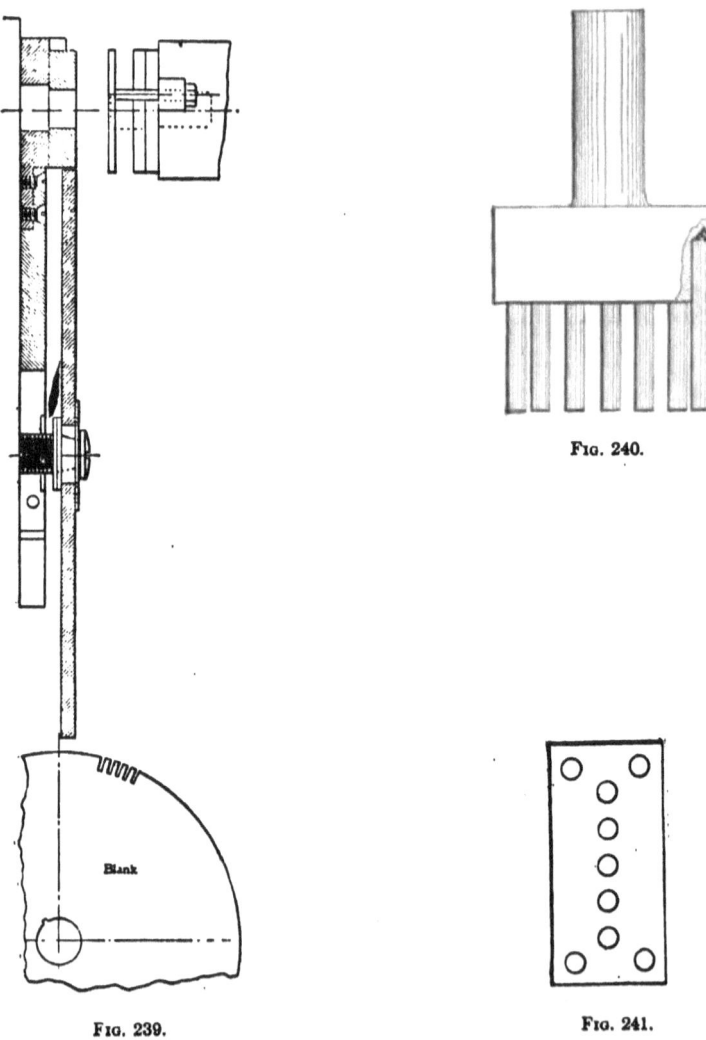

Fig. 239.   Fig. 240.   Fig. 241.

punched. This operation is repeated until the circle of slots is completed.

It will be noticed in this construction that provision has been made for adjusting the different parts. The cutting edges of the

PUNCHES, DIES, AND TOOLS.

die will wear, thereby necessitating grinding off the face of the die. This consequently will bring the die lower than the index plate C, also lower than the levers E and F. To adjust the index plate C, release the clamp bolt G, Fig. 235, and turn stud I, Fig. 237, until the plate C is in the same plane as the face of the die, then again fasten clamp bolt G. The lever E is lowered by driving the bushing L into the gray-iron casting or shoe B, shown in Fig. 239, until in the proper position in relation to the index plate. The lever F is taken off of the piece K to bring F in proper position.

These dies are made in various sizes up to 20 inches radius and give good satisfaction. They are here shown as made at a large modern electrical concern in England.

CONSTRUCTION OF MULTIPLE PUNCH FOR THIN METAL.

The following method is valuable for making multiple punches cheaply for piercing thin metal.

When the occasion arises for making a punch for cutting a series of small holes in this metal, say of a about $\frac{1}{16}$ inch diameter, the most economical way to do the job is to use pieces of Stubs steel wire fitted to a cast-iron punch-plate, as shown in Figs. 240 and 241.

The method of setting these punches is somewhat odd, but it is all right and works perfectly satisfactory on punches for thin metal. Holes were drilled in the body of the punch-plate relative to the die and about $\frac{1}{64}$ inch larger than the wire used for the punches. These holes were then filled with soft solder and the plate levelled in a vise. The punches were well tinned on the setting ends and placed in the die, which fitted them tightly. After heating the plate so that the solder would run, the die containing the punches was placed over the plate and the punches tapped to the bottom of holes. After cooling the tool was ready for use. Experience has failed to demonstrate that any of the punches ever come loose when fastened in this manner.

A COMPOUND PIERCING DIE.

Figs. 242 and 243 show a compound piercing die for piercing a series of holes laterally in the shell of the punching *A* and also a single hole in the centre. The drawing and forming die for producing this punching was of a construction similar to various ones illustrated in Section XI. As will be seen, in Fig. 244 the

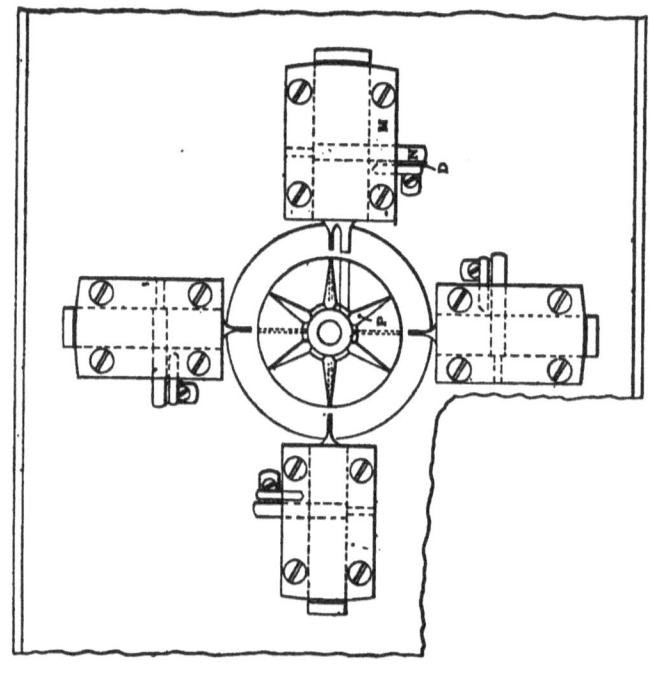

FIG. 242.

punching is an aluminum shell ¼ inch deep and 1⅞ inches in diameter, and has a six-pointed star raised from the bottom.

In designing this die it was necessary to combine accuracy and long life, as it was an expensive and delicate thing to repair. The reader will find, therefore, that all cutting edges have a provision for wear. There are five holes to be punched on the sides, and this is accomplished by a cam arrangement. *D*, Fig. 243, is a cross-section of a bevelled ring set into the punch-plate. It was made in ring form to insure more accuracy in setting the die, and it acts as a guide for the central punch. It is bored to a running fit for the lug blocks B, and is timed so that when the slides C are driven in, the punch K is just entering the metal.

E is one of the punches which are arranged so that they can be taken out and ground and put back, adjusting them for

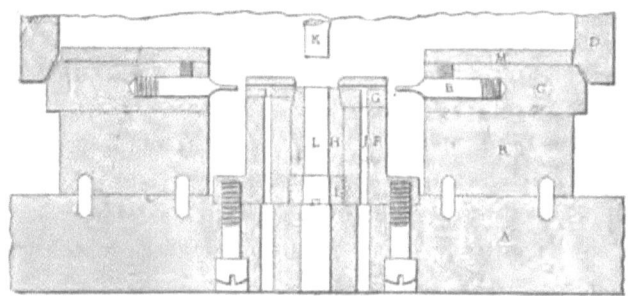

Fig. 243.

length by a thread cut on them, and holding by a blind set-screw. G is one of the dovetailed bushings let into the die. This one is larger than the rest, as it has two holes let into it. The other three are all of the same size. The holes are reamed from the back, and the punchings fall down through the clearance holes

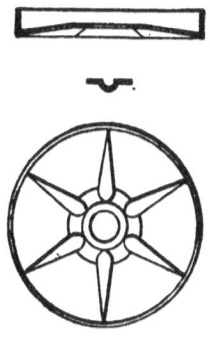

Fig. 244.

J. Although no trouble was experienced with any sticking of the punchings, pins were driven in with one side bevelled right over the clearance holes, thus preventing any chance of it. H is a central bushing or die for the punch K, and is supported by a screw I with a clearance hole.

148   PUNCHES, DIES, AND TOOLS.

Now, when H gets worn on the cutting edge, it can be taken out and ground, and adjusted by the screw I. In doing this it throws out the dovetailed blocks G by means of the taper on the upper end of H. These can in turn be ground very easily, and when done the die is again in condition to work perfectly.

In the plan it will be noticed that in the slide supports there is driven a small steel pin, and also in B there is a pin N. When the punch is let into the die, the pin N advances with the slide, there being a slot cut in the support to allow the necessary play. This bends the spring O against the pin in the slide support, thus delivering enough stress, when free, to return the punch to its initial position. The spring is simply a piece of sheet steel, tempered when bent, and screwed to the cast-iron base plate in proper position.

PIERCING, BLANKING, AND BENDING DIES.

The blanking and piercing operations in the production of the part shown in Fig. 245 were accomplished in a "follow" die of the construction shown in Fig. 246. The three small holes at the left-hand or piercing end of the die are located so that at

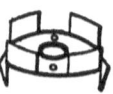
Fig. 245.

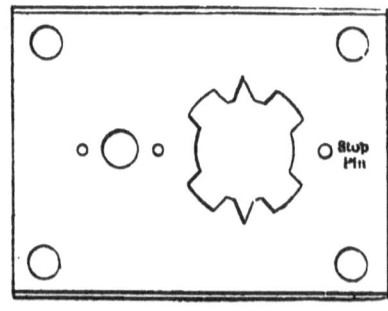
Fig. 246.

the next operation of the stock they will be exactly central over the blanking die, permitting a pilot pin in the centre of the blanking punch (not shown) to enter the centre hole of the punching.

The combination bending, drawing, and stamping punch and

PUNCHES, DIES, AND TOOLS.

die to finish the work at one operation were made as in Figs. 247 and 248. The blank was inserted in a slight recess A, fitted with two gauge-plates B, to insure the correct position of the piece.

The extending lugs C on the punch first bent the sharp points of the blank down over a rather stiff spring ejector in the centre of the die. Continuing in the downward stroke the circular body of the punch forces the blank into the round portion D of the die, which is twice the thickness of the blank, larger than the punch, and has slightly rounded edges, thus bending the four

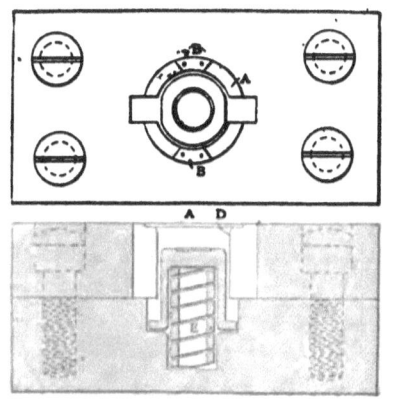

FIG. 247.

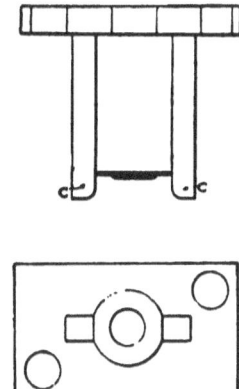

FIG. 248.

lugs to the opposite side of the blank from the points, and drawing them into cylindrical shape, and, finally, when the spring ejector has bottomed, embossing a depression in the centre of the blank, corresponding to the countersink in the top of the ejector. On the withdrawal of the punch the blank is thrown out of the die by a stiff spring E, which was preferable to rubber, as its life is much longer.

As the space of the spring was limited, a small piece of drill rod was caught in the lathe chuck, its size being proper to fit the ejector. A square thread of the proper pitch was cut to give $\frac{1}{8}$-inch stock in the spring, removing the stock from the centre of the spring with a small drill and then a counterbore, finally cutting it off to the proper length and hardening in oil.

PUNCHING FOUR HOLES AT RIGHT ANGLES.

The drawings, Figs. 249, 250, 251, and 252, show a punch and die for piercing four holes at right angles to each other in the piece shown enlarged in Fig. 253.

J is the bed, which is milled away on the outside to make it

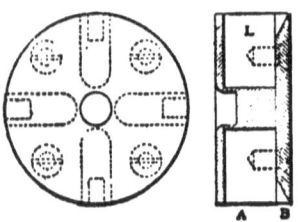

FIG. 249.

handy to get at and lay out and drill holes for the pins *D* which fasten one end of each of the links *C*. The inside is bored to

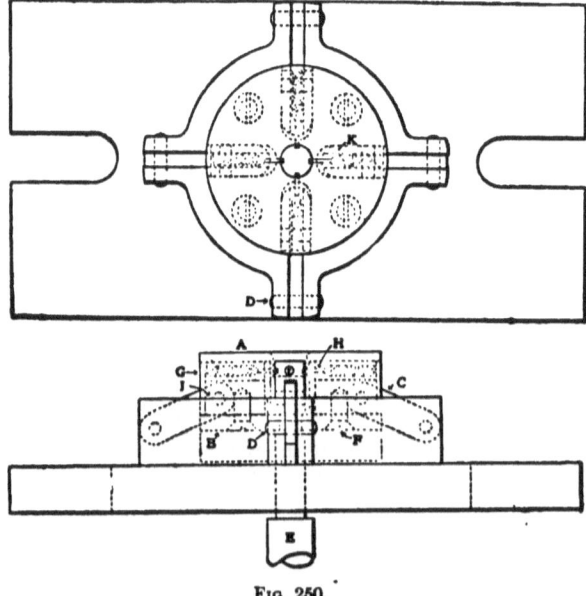

FIG. 250.

receive the pieces *A* and *B*, Fig. 250, and deep enough to give motion required to carry the punches through the blank. The

PUNCHES, DIES, AND TOOLS.

piece A is then milled out to receive the slides or punch-holders, which are fitted nicely to it, and held in place by plate B. The slides K are milled on the bottom to receive the links C, which are pinned to them by the pins I. They are also drilled as shown in the section of the punches, Fig. 252, H, held in place by the screws G. The punches H also support in the small holes A, and never travel back far enough to come out. Fig. 251 shows a piece fitted to the plunger of the press, the end of

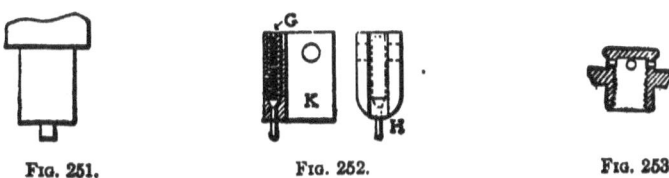

Fig. 251.   Fig. 252.   Fig. 253.

which is turned down to an easy fit for the taper hole in the blank to be pierced, and just long enough to clear the punches as they come through.

The blank to be pierced is placed in the hole in the top of A, which is just deep enough to allow Fig. 251 to bear hardest on the plunger A. The plate B is fastened to A by the four screws F and is milled out to clear the links as the plunger comes up. When the blank is in place and ready to be pierced the press is brought down, which forces the plunger A down, also causing the slides K to travel forward until they have pierced the blank. The plunger is then brought back by the pin E, which is operated by a lever, not shown.

SHEET-METAL PERFORATING.

Perhaps the most important application of the ratchet feed on the power-press is for perforating sheet metal. Fig. 254 is a sketch of a complete punch-and-die equipment. The top bolster, made of machine steel, is fastened to the plunger of the press and serves as a backing to which the punch-plate or holder is fastened by the screws; it is also a stop for the punches. The punch-holder, also of machinery steel, is drilled to receive the punches, and upon the top the punch holes are countersunk to

152   PUNCHES, DIES, AND TOOLS.

receive the heads. The punches being placed in these receptacles, the plate is, as stated above, securely fastened to the top bolster. Then we have the punch member complete. The punches, projecting as can be seen from the drawing, are placed in a stripper, which is drilled to match the punch-plate. This stripper-plate is to be fastened rigidly by whatever means the press builder has provided, so as to stand a certain distance

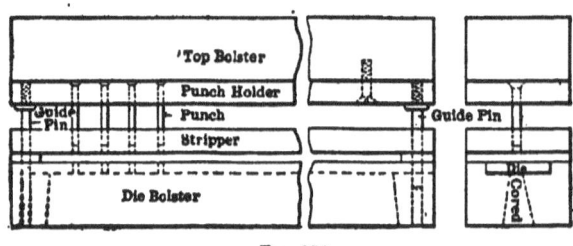

FIG. 254.

above the die, leaving just sufficient space for the stock to be perforated.

These little punches are cut to length from rods of tool steel, the heads being upset in a split block or heading tool, Fig. 255, this operation being performed either in a press or with a hand hammer, preferably in the press.

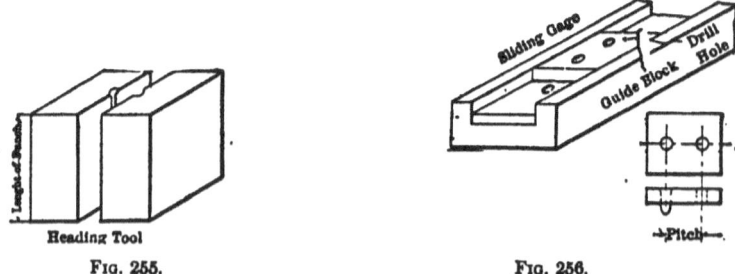

FIG. 255.   FIG. 256.

The punches when headed are to be hardened. Some prefer to harden these little tools in brine, and again others use an oil bath, the color being generally a straw; this is determined by experience and a thorough knowledge of the class of work to be done. When hard, the cutting face of the punch is ground; in very large hole-work the punches are advantageously ground

PUNCHES, DIES, AND TOOLS. 153

so that the edges cut first, and again their lengths are often made different so that every other punch is of the same height, allowing half the number of punches to enter the stock first and greatly reducing the initial strain. The punches should fit quite tight in the punch-holder, but are guided by the stripper.

The die, the length of which is governed by the press capacity but may be here considered as about 30 inches, is made of tool

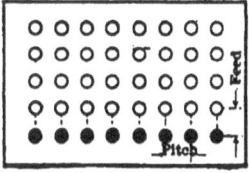

FIG. 257.

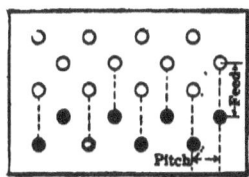

FIG. 258.

steel. In some instances, and according to some manufacturers, it is made of unannealed stock, and is used in this state; other die-makers, nevertheless, harden the dies. These dies are sharpened by grinding the face, as in other flat dies. The shearing may be helped by rounding the face of the die so that the cross-section shows a curved top, highest at the line of holes; this is only practical in large work and is advised with caution, as in some cases anything but a flat die will buckle the stock.

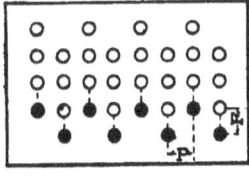

FIG. 259.

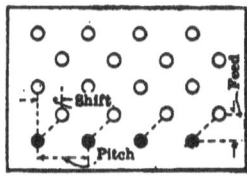

FIG. 260.

The die, which is about $\frac{1}{8}$ inch thick and 2 inches wide, is set flush in the gray-iron bolster and held in place by the screws set at about 4-inch pitch. It is rather impractical to give any rule for the size of screws and parts used, as the conditions upon different presses are various. This die member is next securely fastened to the bed of the press by screws tapped into the bolster. To save drilling the bolster, a slot is cast through its centre and reinforced by webs.

In the drilling for this work the stripper is drilled first and a rig like Fig. 256 is often used. The two holes in the sliding gauge are spaced equal to the desired pitch, and when one hole has been drilled in the stripper, the plug is dropped into it, which gauges the next hole, and so on. The stripper, when drilled, serves as a gauge for spotting the die and also the punch-holder.

The dies are sometimes laid out with a single row and often with a double row of holes. In the "regular" layout, with a single row of holes and with a feed equal to the pitch, a rectangular pattern is produced, as in Fig. 257. When the holes are staggered in the die, as in Fig. 258, and with the feed shown, the stagger effect is perforated sheet. With the same staggered die but half the feed, the holes are in regular lines and closer together, as in Fig. 259. With a single row of holes in the die, the stagger effect is produced in the sheet, as in Fig. 260, by alternate shifting of the sheet in connection with the forward feed. Some presses are provided with both of these movements.

SECTION VI.

Composite, Sectional, Compound, and Armature Disk and Segment Punches and Dies.

AN ACCURATE SECTIONAL BLANKING PUNCH AND DIE.

For the production of extremely accurate sheet-metal parts which require to be produced in so exact duplication that an extremely narrow margin of error only is permissible—often .0005 inch being the limit—dies of sectional and compound con-

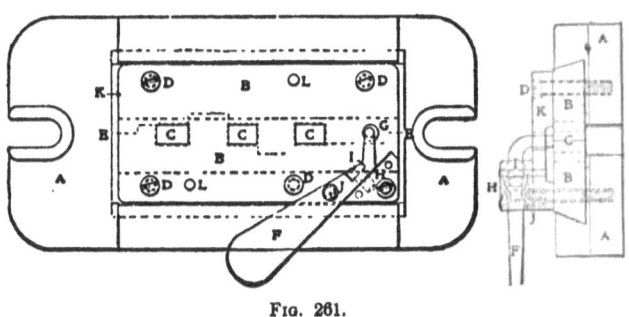

Fig. 261.

struction are necessary. The proper methods to pursue and the most suitable design and construction to adopt should be known to all who may have occasion to make tools for this accurate and extremely interesting class of work. In this section, numerous designs, constructions, and methods are explained and illustrated, their number being sufficient to allow the practical man to adopt the same designs, or modifications of them, for the producing of almost any shaped piece of sheet metal to the most accurate dimensions.

The press tools illustrated in Figs. 261, 262, and 263 are for punching small tool-steel blanks $\frac{1}{16}$ inch thick, which afterward form part of an extremely accurate precision article. The tools are accurately constructed and produce three blanks at a time.

In Fig. 261 are shown a plan and an end view, respectively, of the die assembled. *A A* is the gray-iron bolster in which the die-sections *B B* are dovetailed and fastened by fillister-head

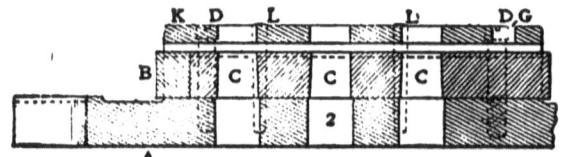

Fig. 262.

screws *D* and dowels *L L*. The dotted lines from *E* to *E* show where the sections interlock. *C C C* are the blanking dies proper. The manner in which the die sections *B B* are constructed to interlock at *E E*, insures a rigid and strong composite

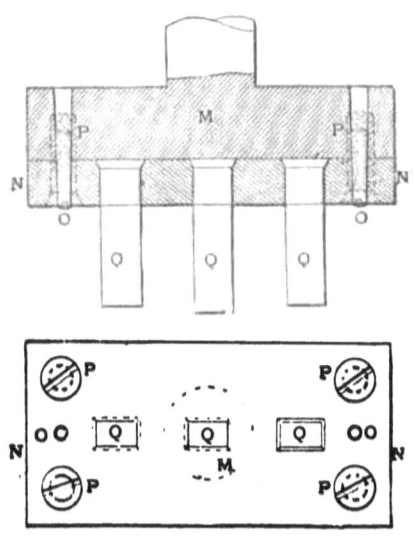

Fig. 263.

die which will maintain a perfect alignment with the punches and obviate all tendency to shear during its life. The die sections *B B* are exact duplicates of each other, making it easy and convenient to work out, to grind, or to repair when necessary.

PUNCHES, DIES, AND TOOLS. 157

In the plan and end views can be seen also the stop for regulating the feed of the stock after each stroke. *H* is a block fastened to the top of the stripper *K*, *I* a screw which adjusts the stop *F*, and *J* a small spring which keeps the end of the stock at *G* against the die face. With this stop it is not necessary to raise the stock when feeding it forward after each stroke; all that is required being for the operator to press down on the extended end of *F*, and after giving the stock a push forward to let go when the stop locates the stock in position for punching again.

In Fig. 263 are a vertical section and a plan of the punch. This comprises the usual gray-iron stem or holder *M*, a machine-steel pad *N N*, located and fastened by two dowels *O* and four

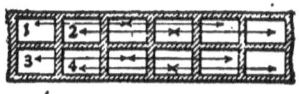

FIG. 264.

flat-head screws *P*, and lastly three tool-steel punches *Q*. The central punch is made a little more than one thickness of metal longer than the other two, so that the stock will be worked upon successively, also so that the central punch will have punched its blank and entered the die before the other two engage the stock, thus locating it positively and preventing warping and uneven spacing.

Fig. 264 shows a portion of the stock as punched. It comes to the press in long strips and of sufficient width to allow of punching two rows of blanks. It passes over the die four times. At the first passage the three blanks shown on the line starting at 1 are punched, at the second blanks on line from 2, at the third blanks on line from 3, and at the fourth stroke those on the line starting at 4. Fig. 262 shows a section of the die construction.

A DOUBLE-SECTIONAL BLANKING DIE AND PUNCH.

It is frequently said by some first-class die-makers that sectional dies do not pay. Nevertheless, quite often it is found more advantageous to make them in sections than otherwise, on

158　PUNCHES, DIES, AND TOOLS.

account of the difficult contour or angles to be machined and filed. Then, again, if one spoils a section in machining, filing, or even in hardening, it can be replaced more easily. Sectional dies are used in a great many shops for double-acting cupping-press tools and also in compound dies connected with sub-press, both of which tools lend themselves readily to the operation of such tools.

In the manufacture of what are known to the trade as "school aids," which include geometrical surfaces and solids, a

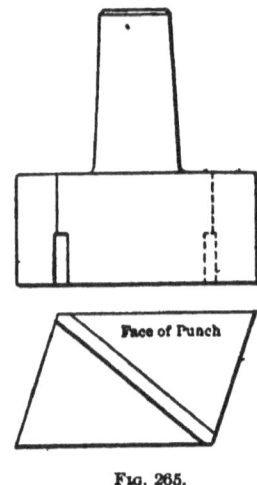

FIG. 265.

set of sectional dies is used which will be of interest to the reader. The solids are made of well-seasoned hardwood, and the surfaces from what is commercially called "leather board," .028 inch thick. These tools, herein described, perhaps differ somewhat materially in detail from sectional dies used for other work, and thus a short description of them will help.

Fig. 265 shows the blanking punch, which has 1° taper from its face back to the shank, thus allowing a small amount of stock to be shaved off in the screw press in accurately fitting the punch to the die after it (the die) has been hardened. The punch is *always* left soft in doing this class of work—"leather-board" blanking.

Fig. 266 shows the die with its three sections, D^1, D^2, and

PUNCHES, DIES, AND TOOLS. 159

D^2. The four independent locating keys that have their respective seats milled midway between the top and bottom of the sections D^1 and D^2 are shown in the drawing by the dotted line and the same in the centre section D^3, also represented by dotted lines.

Above the die face are shown the gauges with the stripper upon them. The gauges are made of $\frac{1}{16}$-inch stock, and the stripper from $\frac{1}{32}$-inch cold-rolled stock. Both are held in place by $\frac{1}{4}$-inch hexagon screws. The blanking part has an internal clearance of one-half of a degree.

Perhaps the strangest part of the whole construction of this die was that the three sections were each made of a different make of tool steel. This was done to find out the real value of

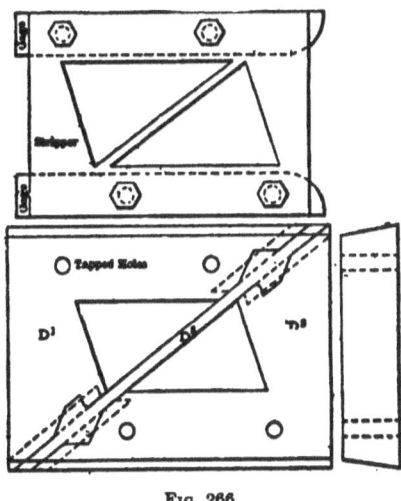

FIG. 266.

each. The steel used in D^3 was a brand that could be depended on, as there had been secured quite a varied experience with it. D^1 and D^2 were made from two new blanks, each of whose merits were "painted in glowing colors" to the users.

The open-back, single-acting blanking presses in which dies of this construction were used, are equipped with an automatic multiple-pawl, ratchet feed, by which device a small fractional part can be added to or subtracted from its length of feed. The several pawls are disengaged simultaneously by a lever.

A SECTIONAL TRIMMING DIE FOR TOOL-STEEL PARTS.

The trimming of tool-steel blanks as here shown in Figs. 270 and 271, $\frac{1}{32}$ inch thick and $\frac{1}{8}$ x $\frac{5}{32}$ inches all over, to very accurate dimensions, requires press tools of accurate construction and also tools that may be easily and cheaply repaired in case of any part becoming defective. A set of tools of this class is shown in Figs. 267, 268, and 269, the tool-steel blanks trimmed by them being illustrated in the "before and after" shapes in Figs. 270 and 271.

The blank to be trimmed is left square on the ends, is milled all over, has a channel milled in both sides at A and B, respectively, and has a section punched out at C. These trimming

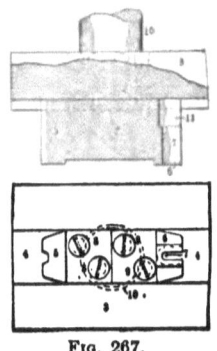

Fig. 267.

tools notch the piece at D and E and trim the ends to the symmetrical shapes indicated.

In view of the nature and thickness of the stock to be punched, the accurate dimensions demanded in the finished product, the fact that thousands of the pieces were required weekly, and the low piece-work price to be paid for the operation—two cents per hundred—it was thought economical to design a set of tools cheaply and easily repaired, which would insure the interchangeability of the products, and lastly be rapid producers.

F is a machine-steel bed or bolster in which all the die sections and working parts are fastened. H, I, J, and K are the four hardened, tempered, and ground tool-steel die sections, which act to trim or shear and notch the ends E and D. These sections are let into accurately machined channels or seats in F,

PUNCHES, DIES, AND TOOLS. 161

are fastened by fillister-head screws—as shown in the plan view—and are side and end banked against set-screws, *P, N, Q,* and *O*. *L L L L* indicate narrow slots milled into the base or holder *F* to the depth shown, and with their inner sides an exact distance apart, thus serving as set edges and locating or banking surfaces for the ends of the die sections *H* and *J*. The centre of *F* at *Y* was worked out to a clearance size for the trimmed blank to drop through after punching.

The other parts of the die which require mention are the reciprocating slide or pad *S*, upon which the work rests previous to the descent of the punch, the locators *V* and *R* for locat-

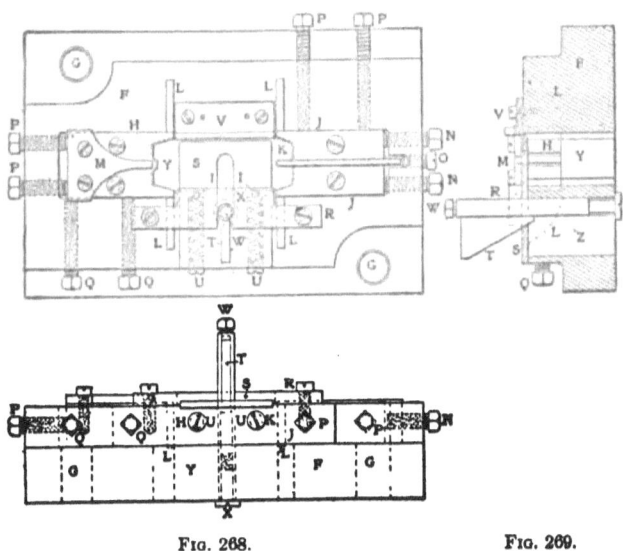

FIG. 268. FIG. 269.

ing the work edgeways, and the arrangement for working the slide back and forth by the action of the press ram.

The vertical cross-section of the die shows the construction of the various parts, and also how the slide works, while the plan shows the springs *I I*, the inclined-faced actuating surface *T*, the adjustable set-screw *W*, and the headless spring-adjusting screws *U U*.

The upper or punch portion of the trimming tools is illustrated in Fig. 267. *E* is a machine-steel stem or holder, of which 10 is the stem proper; 4, 4, a square-bottomed channel milled

11

162   PUNCHES, DIES, AND TOOLS.

down its entire surface, in which the two punch sections 5 and 6 are fastened by fillister-head screws.

The tools being set up in the press, and the set-screw W, in T, being adjusted so that the face of the punch slide will come in contact with it and cause the slide S to move back out of the way before the punch enters the die, a blank is placed on the pad S by the operator and located between the plates V and R, and endways against stop M. The press being then stepped, the

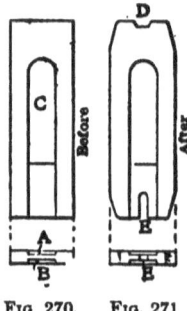

FIG. 270.   FIG. 271.

punch descends, the face of the ram strikes the set-screw W and forces cam-stud T downward, thereby causing pad S to draw back from under the work and leave a clear working space for the punch to trim the ends and force the part into the die. Upon the up-stroke of the press ram the strings $I\,I$ cause pad S and the cam-stud T to resume their former positions, and the die is ready for another blank.

COMPOUND ARMATURE SEGMENT DIE.

Illustrated in Figs. 272, 273, 274, 275, and 276 is a die that may be described as follows:

The gray-iron die-plate F, Fig. 273, is taken first and top and bottom are planed, also the slots for side pieces D. Then it is placed on the radial planer and the centre is dug out for the mild-steel plate used for holding the blades, as shown at M, Fig. 280, also in section through $C\,C$.

The mild-steel plate M is fitted into the gray-iron die-plate F and screwed and dowelled into position. Then the plate F is

PUNCHES, DIES, AND TOOLS. 163

again placed on the radial planer, and the slots are planed for the curved tool-steel die pieces *E* and *G*.

These tool-steel die pieces are first heated and bent to the approximate radius, then the tops and bottoms are planed, after which the circular edges are planed. The dovetails in *E* are slotted and the pieces split through the centre of dovetails, an allowance being made for grinding to fit after hardening. The screw holes are next drilled and tapped in *E* and *G*, and they are then hardened and drawn (or tempered). The top and bottom

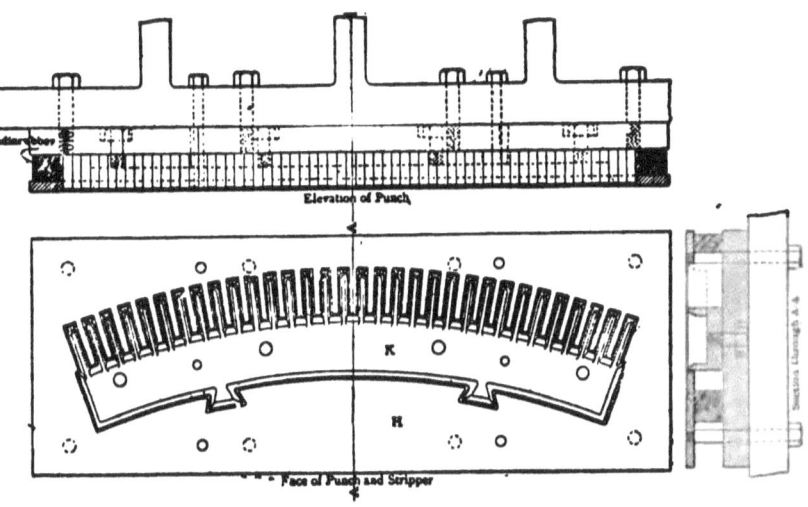

Fig. 272.

are then ground on a surface grinder. The pieces are now ready for the radial grinding operation, which is done by first placing them top side down and grinding at the points indicated by arrows in section through *C C*, Fig. 273. The dovetails in *E* are fitted to a templet, after which both *E* and *G* are placed in die-plate *F* and fastened in position. The plate is then put on the radial planer and the die is ground true to fit the templet. The side pieces *D* are finished on the surface grinder. The method of holding the tool-steel blades in the mild-steel plate *M* is shown in Fig. 276. The half-blades at each end are dovetailed into the side pieces *D*. Rubber is used under the inside stripper for bringing up the punching.

164   PUNCHES, DIES, AND TOOLS.

The punch K, Fig. 272, is a tool-steel forging in one piece. It is planed top and bottom and on the outer arc also. The templet is then placed on the face of the punch, and the slots, the inside arc, and sides are accurately scribed off. Then the punch

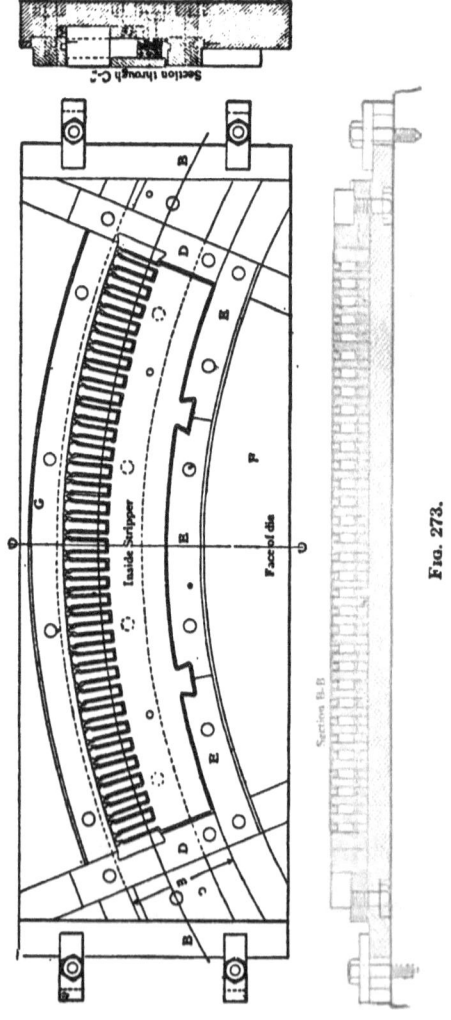

Fig. 273.

is slotted to lines, very little filing being required if the job is properly done. The inside is then dug out, as shown in Fig. 272, and more clearly in Fig. 275, leaving a wall about ½ inch

PUNCHES, DIES, AND TOOLS. 165

thick around the edge of the punch; this wall is milled at an angle almost to an edge. The solid portion between the slots also is milled at an angle, about $\frac{1}{32}$-inch flat surface being given all around the cutting edge. This punch is not hardened; therefore, after being in use and showing signs of wear, it is taken out and the cutting edges are hammered (or flattened). Then it is again placed in the press in the proper relation to the die

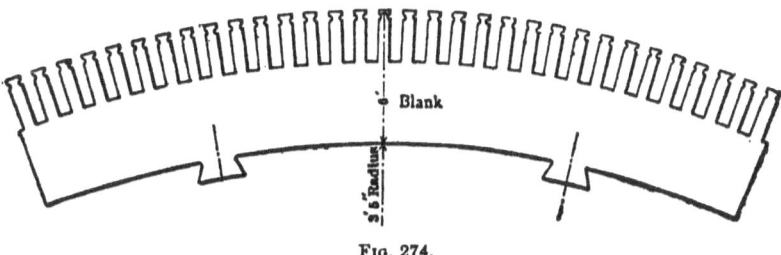

FIG. 274.

and brought down into the die, the edges being thus accurately sheared to size. An outside stripper is provided for this punch, rubber being used under the stripper for clearing the punch scrap material.

Fig. 274 shows the blank punched complete in this die at one operation. Eight of these segments are in this case required to complete the circle.

One point more should be noticed. It will be seen that the

FIG. 275.   FIG. 276.

heads of the cap screws for holding the die pieces are flush with the bottom of the die-plate shown in the section through $C\ C$, Fig. 273; also the cap screws holding the punch to the plate, Fig. 272, are flush with the bottom of the plate. This is done to avoid all possible chance of the pieces working loose while in

PUNCHES, DIES, AND TOOLS.

the press. Dies of this type are made in sizes from 20 inches, outside radius upward.

DOUBLE SECTIONAL BLANKING DIE FOR TOOL-STEEL BLANKS.

The double blanking tools here illustrated were designed for the punching of the slide portion of an accurate precision article. The material of which the slides are composed is tool steel, and as the parts are required to be machined accurately tools of accurate and durable construction are necessary.

From the engraving, Fig. 277, showing a portion of the stock from which blanks have been punched and the scheme of blanking—that is, the position of the blanks in the strip and

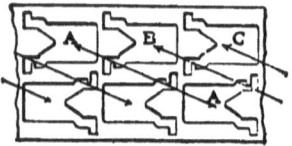

Portion of Punched Stock

FIG. 277.

their relative positions to each other—can be easily understood. Thus A A are punched at one stroke of the press, B B at the next stroke, then C C, and so on until the entire strip has been worked up. As the blanks are required to be punched to unusually accurate dimensions, and as the stock is tool steel, a sectional die is used. The manner in which the sections are made to locate and interlock, the ease with which they may be worked out, and the comparatively inexpensive manner in which any of the parts may be duplicated and replaced in case of breakage, shearing, or chipping, will appeal to the practical man and will at the same time, no doubt, prove suggestive for adoption in the production of similar accurate press work.

In the die, Figs. 278 and 279, G is the bolster of gray iron, H H are bolt channels, and I I is an accurately machined seat for the die parts to locate in. The die proper is composed of six parts: D D, E E, and F F, which are located and fastened within the seat *I* in the bolster by means of the six fillister-head screws *J* and the two set-screws *Q*. *P* is a steel plate in which are combined the stripper and gauge-plates, which are shown by the longitudinal dotted lines. An oblong hole *O* is let through the stripper for the stop pawl *N*. *K*, *I*, and *M* are the bracket, the dowel, and fastening screw for stop. With a stop of this construction the stock to be worked is pushed forward after each stroke and then pulled back sharply, the front end of the last hole punched setting against the straight face of pawl *N*. The

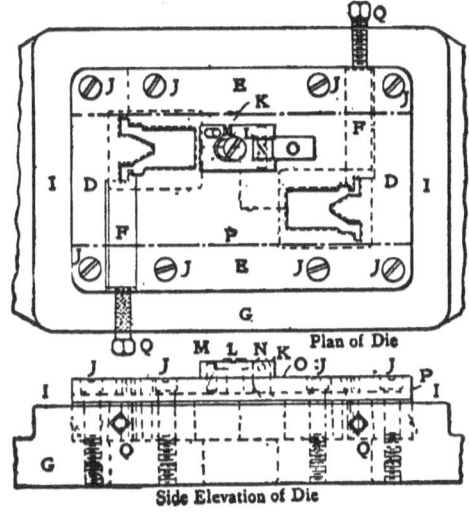

Fig. 278.

sections *F F* of the die are finished with one inclined or tapered face so that when the die parts are assembled and fastened within the seat *I* in the bolster by the screws *J* there will be no tendency for parts *F F* to rise under the working pressure of the punches' up-stroke.

The punch, Fig. 280, consists of the gray-iron holder *R*, machine-steel pad *S*, six flat-head screws *U*, two locating dowels *V*, and the two punches proper *T* of tool steel hardened and drawn to a straw temper.

Much accurate work and careful sizing was required in the making of the die parts *D*, *E*, and *F*; for instance, the giving of 1° clearance all around the cutting edges, the grinding and lapping of the locking and locating surfaces after hardening, etc. Lastly, it will be noticed that *D* and *D* are exact duplicates, like-

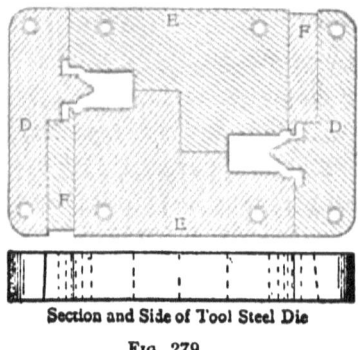

Section and Side of Tool Steel Die
Fig. 279.

wise *E* and *E*, and also parts *F F*. By arranging the composite construction in this manner the machining, grinding, lapping

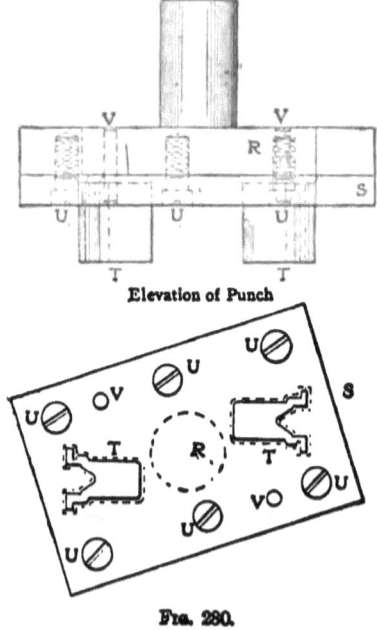

Elevation of Punch

Fig. 280.

and locating of the sections are considerably simplified and expedited; furthermore, the replacing of a sheared, broken, or

PUNCHES, DIES, AND TOOLS. 169

chipped section then presents no difficulties of unusual magnitude.

MAKING A COMPOUND ARMATURE-DISK DIE FOR SMALL DISKS.

It will interest many to know the following method of making compound dies for very small armature disks.

The die is illustrated in Figs. 281 to 285 and is used for cutting a disk $1\frac{1}{2}$ inches diameter. Accurate spacing, accuracy in

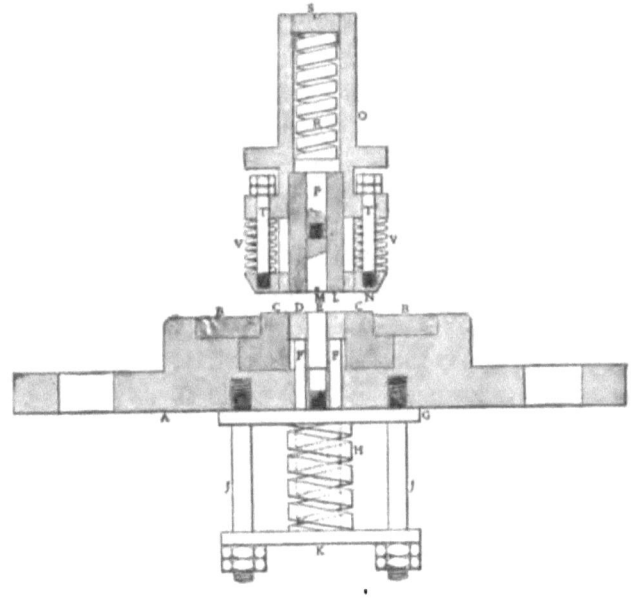

FIG. 281.

the size of slots, flatness, and freedom from burrs are the requisites. It has been found that these are more easily secured by making up the die of segments, as shown in Fig. 282. Figs. 281 and 282 show sections of punch and die, and plan of die, respectively. This part is sufficiently clear in the engravings.

In making these dies, first mill the punch and "push-out" D, Figs. 281 and 282, to shape. If you have not unlimited confidence in your miller and dividing head, it will be safest to mill

170 PUNCHES, DIES, AND TOOLS.

opposite grooves at the same setting—that is, cut top and bottom without shifting, holding the piece in the chuck.

For the segments, make a templet and make fly-cutters A A, Figs. 283 and 284. Plane the steel for segments to the shape of Fig. 283. Bolt the cast-iron piece B, Fig. 284, to the miller table and true it with a good sharp side cutter.

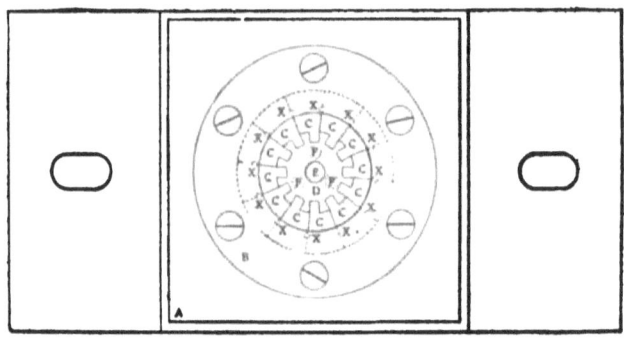

FIG. 282.

Secure B, Fig. 283, in position shown, clamping with straps fitting the holes D. After milling the first shape proceed as illustrated in Fig. 281. Make the cast-iron ring D, Fig. 285, ½ inch thick, and tap for screws shown. Plane a piece for segments to shape shown in Fig. 285; cut into lengths for segments,

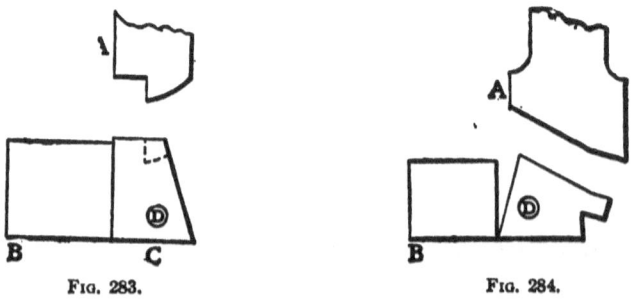

FIG. 283.   FIG. 284.

put the punch L on a mandrel, and secure the segments in position, with the ring D and screws E, and turn one end to size. Put the die block A, Fig. 281, on the face plate and bore out to fit the ends of segments already turned, and turn out to fit the ring B. Drive mandrel out of punch L, and fit a shouldered brass or copper plug into the end of L. Drive segments and punch into A, and finish turning and drive segments to place. Drill holes and put in the twelve pins x, Fig. 282. Fit ring B and put in screws.

Now you can drive out the punch, provided you did not forget to put in the brass plug before you drove it in. The push-out M, Fig. 281, is screwed into the spring-actuated plunger P, so that it may be taken out for convenience in setting. The push-out D is removed, and is simply dropped into place on the pins F, F, F, after setting. By pulling up the bolts T, T, Fig.

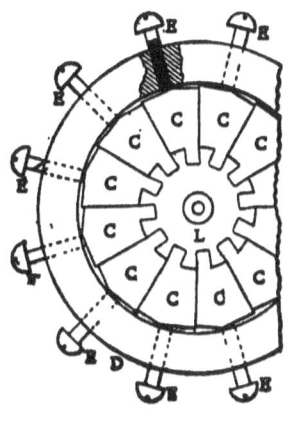

FIG. 285.

281, with the nuts, the stripper is raised to allow the punch to enter the die in the setting.

H and R are springs of flat steel tempered. There are eight of the springs V. S is a screw plug for adjusting the tension on spring R. The die should be hardened and drawn just enough to start the color. The punch may be drawn to a medium straw color. The push-out D should be case-hardened, if made of soft steel.

Tire steel is a good kind to use for strippers, especially when there is not too much work to be done on them. It is highly satisfactory in most respects, sufficiently hard without tempering, stiff and tough; but, like most good things, has its faults, the most serious one being that it is hard to work.

ANOTHER COMPOUND DIE FOR SMALL ARMATURE DISKS.

In the following a method of constructing compound dies for small dies rather different from that described in the foregoing is given.

Fig. 286 is a side elevation and plan of the die. *A* is the die, which should be bored in the lathe and then be turned on a mandrel, enough being left to grind. A jig with the eleven holes equally spaced, and with projections to fit into the hole in the die, can now be used to drill and ream the holes for the notches at the periphery. Now place on centres in the milling machine and cut through to the holes the desired width. File radius *B*

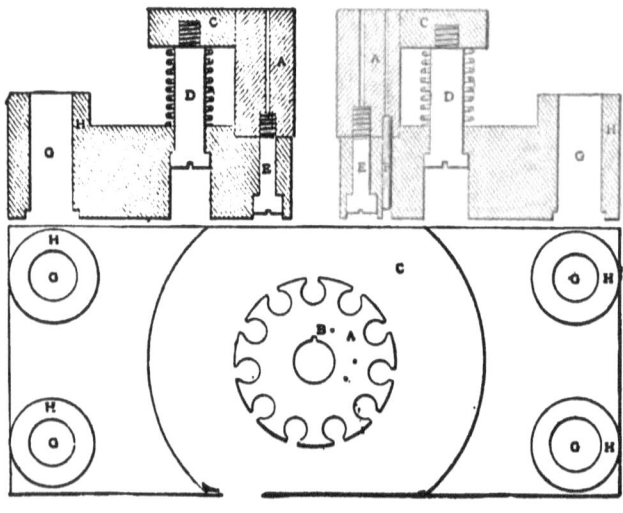

Fig. 286.

on side of the centre hole, harden and draw to light straw color, and grind to size.

C is the stripper, retained in its proper position by six screws *D* and actuated upward by springs. *E E* are four screws to hold the die in cast-iron holder. *F* is a dowel pin to prevent the die from turning. *G* are four 1-inch holes to sub-press die, and are bossed to give bearing and support to the pins.

Fig. 289 is a side view of the punch. When construct-

PUNCHES, DIES, AND TOOLS.

ing these tools place the outside cutting punch *I*, Fig. 289, in the lathe and turn and bore almost to size. Dovetail the slots at an angle of about 10° on a side; taper them 2°, making them large at the top or cutting edge of punch. Harden and grind to allow the die to enter, and also grind on outside to fit recess in holder.

Fig. 287.   Fig. 288.

The centre punch *J* is turned, hardened, and ground, being previously dovetailed to receive piece for cutting notch in blank. *K* is a stripper, supported by outside punch and operated by springs. *L* is a dowel pin to keep the punch from turning in the holder when in operation. *M M* are holes for sub-press pins.

Fig. 287, *n* are parts of punch that are to be fitted after the

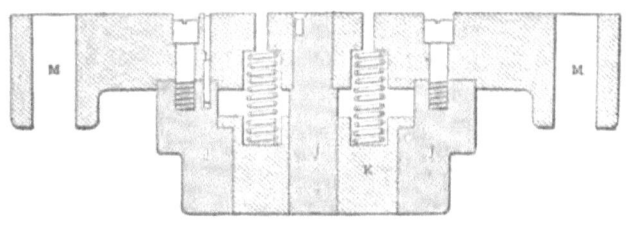

Fig. 289.

ring is ground, being driven into the dovetail slots and $\frac{1}{64}$ inch below surface. Enter the die and mark out the shape; number and remove work down to line; replace and prove work. After finishing harden and drive the pieces into their proper places. As the slots taper, a tight fit will of course be obtained every time.

174 PUNCHES, DIES, AND TOOLS.

The next move is to fasten the die and punches into their respective holders with strippers removed. Enter them; level and support with parallels; strap together. Now drill and ream the four 1-inch holes for the sub-press pins. The pins should be made of tool steel, hardened and ground to a perfect fit in the punch-holder, and left a little large for about 2 inches from the head. Drive up from the bottom into the die-holder. Assemble your punch and die, and if your work is properly done they will set once and for all.

When it is desired to re-sharpen your die, carefully drive out pins and replace when ready for use. The cutting punches and outside of the die should be ground straight; the inside should have a taper of .005 to 1 inch. Fig. 288 is exact size of sheet-iron blank produced by this die, though we have sizes much larger produced by dies on this plan.

COMPOUND DIE FOR ACCURATE PIERCING AND BLANKING.

The tools illustrated in Figs. 290 to 294 were made for punching a blank from bright-rolled Bessemer steel, $\frac{1}{8}$ inch thick, $\frac{1}{4}$ inch wide, and the length shown in Fig. 295. We think it will be clearly understood from the drawings, Fig. 290 being the plan and elevation. The cross-head or plunger A is recessed to receive the die B, Fig. 293, after it is hardened and ground. The die is tight-fitting, being held by the screws shown in Fig. 293, and is made in sections, the most of the work being done on the milling machine. The die is also provided with a shedder C for shedding the blank, and which is forced down by pins D, which are in contact with the springs E, the shedder having motion enough to allow for grinding and the thickness of a blank. FF are the adjusting screws or plugs for varying the tension on the springs E. G, shown in Fig. 291, is the holder for the piercing punches; it is a fit in the recess of the die, and is screwed and dowelled to the head A. The posts J after being ground and fitted nicely in the bed K are screwed down tightly by the nuts shown. The adjusting nuts I are reamed out so as to be a good fit on the posts J, also centring the head which is ready for the

PUNCHES, DIES, AND TOOLS. 175

babbitt which makes the sleeve *H;* the head being grooved so as to prevent the babbitt from turning while being pressed down by the adjusting nut *I.* After the babbitt has been poured, the nuts *I* are reamed just enough larger to give a little clearance. The

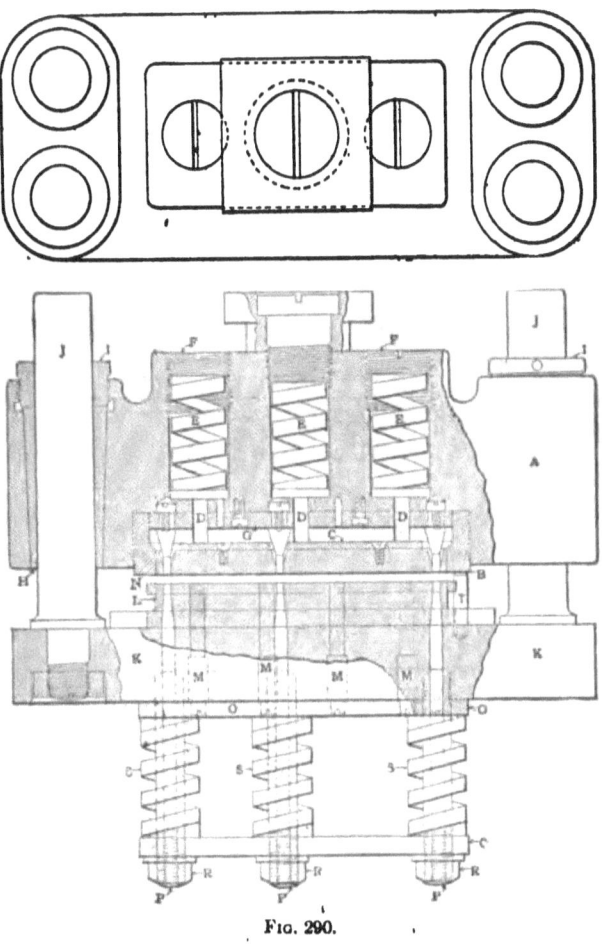

Fig. 290.

die-bed *K* is then planed out to receive the punch, Fig. 294, which is fastened to the bed by four screws. The other eight holes are $\frac{1}{4}$-inch clearance holes for the pins *M*, which are screwed to the shedder *N*, which sheds the stock from the punch, are turned to fit the counterbore in the bed which is just deep enough to allow the shedder to come even with the face of the

176 PUNCHES, DIES, AND TOOLS.

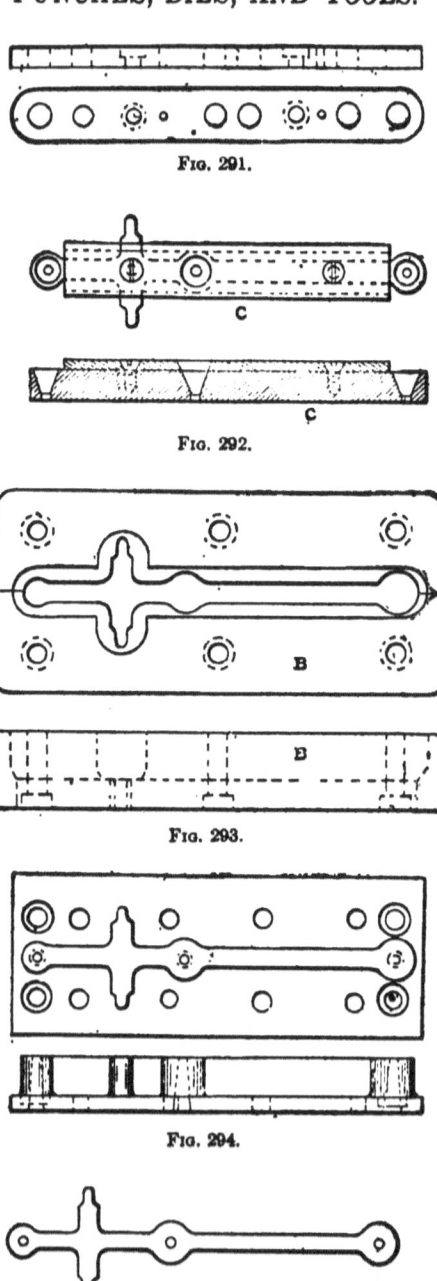

Fig. 291.

Fig. 292.

Fig. 293.

Fig. 294.

Fig. 295.

PUNCHES, DIES, AND TOOLS.   177

punch; while the other ends are also turned to ¼ inch and are slotted for the screw-driver, the bottom shoulder acting as a stop for the plate D, which has three ½-inch holes just free enough to let the plate slide freely on the studs P, which are drilled entirely through lengthwise, to allow the scrap from the piercing punches to drop, and are screwed into the bottom of the bed. The plate Q is fitted in the same manner, and is forced down upon the nuts R until the compression on the springs S is sufficient.

The operation and action of the die and the production of the work require no further description, as the foregoing and the drawings supply all that is necessary.

COMPOUND ARMATURE-DISK DIE WITH SPECIAL STRIPPER.

Armature disks, as shown in Fig. 296, were produced complete in large quantities with the compound die with special stripping arrangement attached, as shown in the other engravings in Figs. 297 to 300. Work of this description is usually

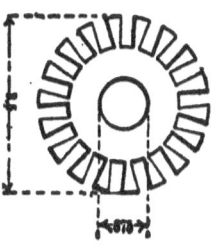

FIG. 296.

produced on presses having upper and lower die knock-out arrangements, thereby doing away with strippers in the dies; but the tool which is the subject of the last article in this chapter was made to be used on a press that has no knock-out attachment. Fig. 297 shows plan and section of the upper die; Fig. 298, a plan and section of lower die; Fig. 299, end elevation of complete tool; and Fig. 300, the special stripping attachment. It was designed principally to overcome the difficulties usually associated with dies of this description; that is to make them strip properly, which is rather difficult when the press is not fitted with knock-outs, on account of the large amount of cutting edge which necessitates a considerable amount of power behind the strippers, and also because there is not room wherein to locate sufficient springs or rubber; that is, without making the tool very complicated.

There is often a difference of opinion about the best way of stripping. Some will contend that a stationary stripper would be the best for a particular job, when others prefer movable

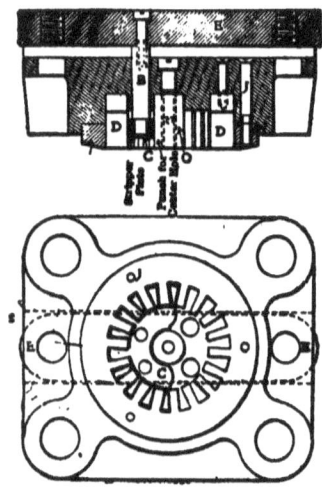

FIG. 297.

ones, and sometimes either would do equally well. Stationary strippers often distort the work, whereas movable ones straighten and hold down the stock before the tools come into operation; this is often of much importance when great accuracy is required. When springs or rubber, or perhaps both, are used, a considerable amount of power is often used up in compressing them before the actual work commences. This is a point that has to be considered, especially when the press is being taxed up to its limit of power.

As will be noticed, the body carrying the die for the central hole in the disk is at the bottom to allow of the scrap falling through the hole in the press body. The tool is sub-pressed by means of the four guide pins H, which are fixed in the lower die and are a good sliding fit in the upper die body. The blanking ring A for the outside diameter of the disk is located in a recess accurately turned in the body, and is held in position by the three screws J. The die for the slots is built up in segments D, so that if a part breaks it can be quickly replaced without destroying the entire ring. Each segment is held in position by

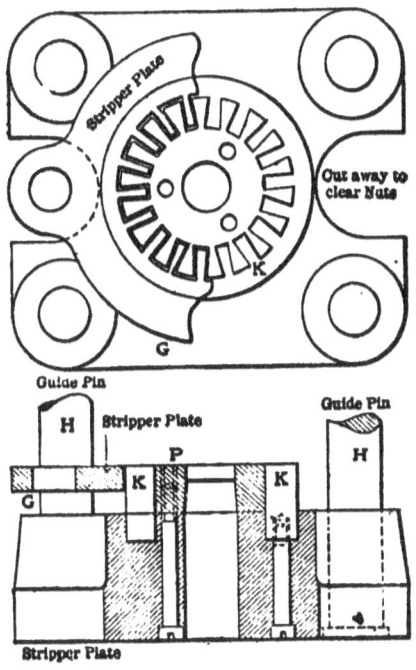

FIG. 298.

a cheese-head screw from underneath. The punch O is for the central hole, and P is the die for the same; see Fig. 298. The hole is parallel for about ¼ inch to allow of grinding without altering the size. Below this it tapers off to allow the scrap to fall clear. The punch for the slots is built up in segments K, which are located and held in position in the same way as the die. Now we approach the stripping arrangement, which is the principal feature of this die. It is shown complete in Fig. 300.

180   PUNCHES, DIES, AND TOOLS.

A bar E is located in a slot extending right across the back face of the top die. To this are attached two pins FF, carrying the stripper G for the bottom die. This stripper is accurately worked out, so that it will slide nicely between the segments

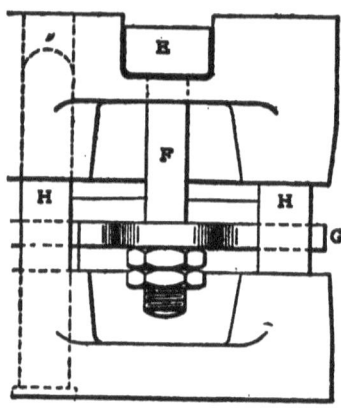

Fig. 299.

comprising the punch. It is kept in position and adjusted by the lock nuts. C is the stripper for the top die. It is also attached to the bar E by means of four pins B. The projections are rather weak at the root, so to prevent their breaking off, the

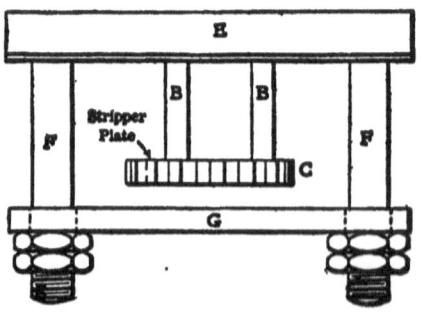

Fig. 300.

stripper must be made thicker than would otherwise be necessary.

The action of the arrangement is as follows: On the downward stroke, the strippers come down with the die until the stripper C rests on the stock. It then recedes into the upper die by an amount which is the difference between the thickness of the bar E and the depth of the slot it rests in, its upward movement being limited by its coming in contact with the bolster. At this point of the stroke the bottom stripper is of course below the face of the punch, the bottom die body being cut away to clear the lock nuts while in this position. On the upward stroke the bottom die stripper G, coming in contact with the scrap, strips it and at the same time, as the stroke is continued, brings down the bar E, carrying the top die stripper C to the bottom of the slot, thus ejecting the disk. The scrap and disk now lie loosely on the die, and so can easily be removed at the will of the operator. It is not necessary to describe the methods employed in making the various parts of the tool, as we have done this so thoroughly in other preceding articles comprised in this chapter that the reader can readily understand the methods and practice followed out in this case. The parts that need the greatest care in making are the punch and die segments. All parts, with the exception of the bodies, screws, and stripping arrangements, were hardened.

SECTION VII.

Processes and Tools for Making Rifle Cartridges, Cartridge Cases of Quick-Firing Guns, and Nickel Bullet Jackets.

PROGRESS IN DEVELOPMENT OF DRAWING PROCESSES.

THE development of the art of sheet-metal working, and the improvements made in the tools used, have been so rapid during the last few years, that only those who are directly engaged along these lines have been able to keep pace with the construction of the tools used in such work. Thus, we find that tools which yesterday were thought to represent the height of perfection, in regard to longevity and production, are to-day obsolete because of others having been designed and put into successful operation for accomplishing many times the work of the former and at a lower cost.

Of all the different processes of sheet-metal working, however, it is in the drawing and forming of small sheet-metal parts and articles that we find the greatest innovations being introduced in the tools for their production; and therefore it is the writer's purpose to illustrate in this section and that which follows it, and also to present descriptions so that each and every point will be clear, sets of tools used for the production of accurate modern drawn work. The matter should be of interest and prove of great value to most sheet-metal goods manufacturers and to all die and tool-makers.

In the small-arms arsenals of the principal governments of the world, much has been accomplished in the drawing of sheet metals to accurate dimensions and interchangeable shapes that is not generally known outside of those in government service; especially is this so in the manufacture of cartridge shell jackets, and bullets for rifles and rapid-firing guns. Therefore in

PUNCHES, DIES, AND TOOLS. 183

this section we devote the space at our command to describing tools and processes of the most up-to-date class that are used for the production of these articles.

SET OF PUNCHES AND DIES FOR DRAWING BRASS CARTRIDGE SHELLS.

The 30-30 and .303 cartridges belong to the most modern class of high-powered, smokeless powder ammunition, and owing to their high velocity and great penetrating power they have become favorites with experienced and successful game hunters. From the fact that there attends the firing of smokeless powder an unexplained deterioration that causes brittleness to the shell after the first discharge, very few of them are reloaded. From this cause alone the demand is so greatly increased that the requisite number to supply the market for one year for these two sizes alone runs up to almost incredible numbers. While the 30-30 and .303 differ very little in outward diameter and general outline, and the tools correspondingly, for the sake of clearness it has been thought best to choose the latter size to describe and illustrate in the following. All figures and references apply to the .303 only.

Although the accuracy of the diameter and shape of all rifle cartridges is imperative in order to allow the interchange of them, and to make it possible to use any make of cartridge in the same rifle, and the operations involved in their manufacture are numerous, it may be of interest to know that the cost of a .303 shell is only about three-quarters of a cent.

Figs. 301 to 312 inclusive illustrate the evolution of a cartridge from the first operation to the last.

Whenever the nature of the operation will allow, an automatic feeding and removing device is employed to carry the work to the dies and take it away, the friction and ratchet feed dials being the common medium of handling the work. Between each drawing operation annealing takes place, and this is accomplished by enclosing the cups in round iron drums with nearly air-tight covers. After heating they are allowed to cool gradually without submerging in water.

The work is next put into open tubs, set on an incline and revolved slowly, while a mixture of sulphuric acid and water is

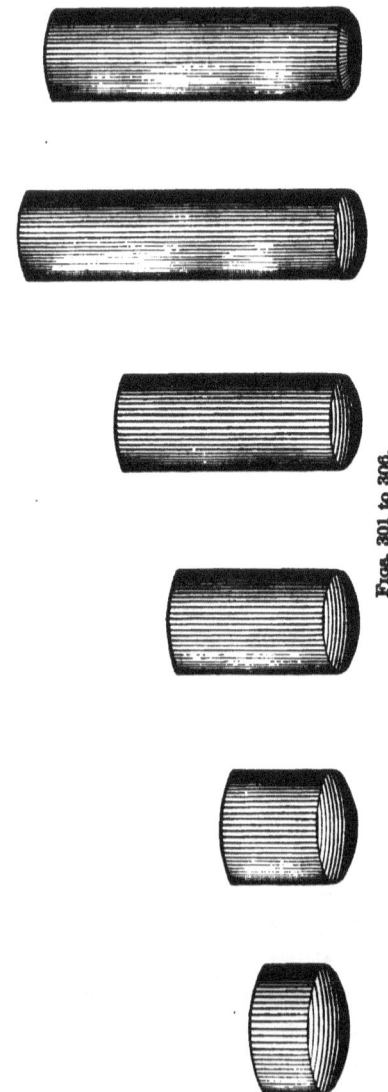

Figs. 301 to 306.

poured on them. After the scale has started and the solution is drawn off they are washed by showering with water.

PUNCHES, DIES, AND TOOLS. 185

The neck of the drawn shell, before reducing, is annealed in a machine with a friction dial feed that carries the work

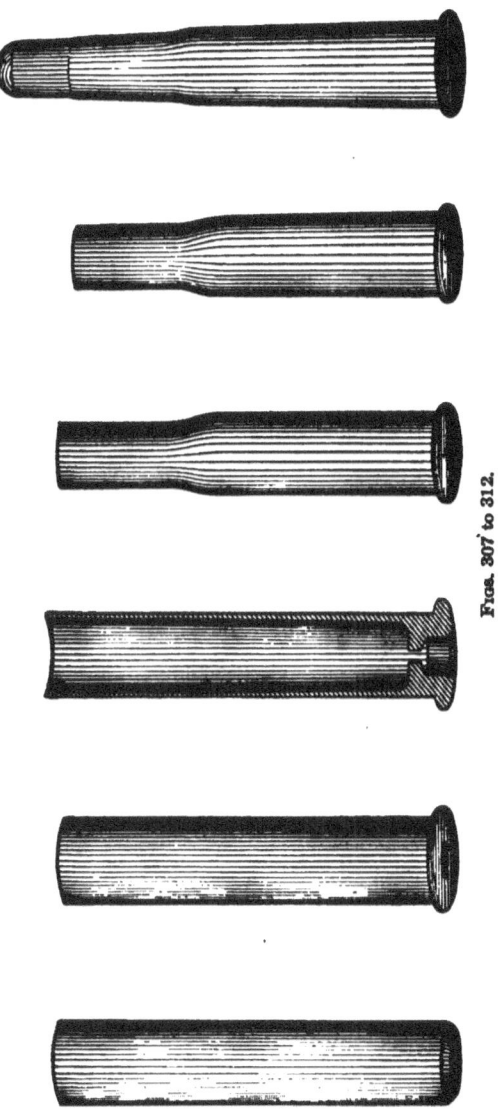

Figs. 307' to 312.

in succession past a gas flame and deposits it in a receptacle provided.

DRAWING AND RE-DRAWING THE CARTRIDGE SHELL.

The cup is first blanked and drawn in a double-action press of the conventional type from sheet brass .090 inch thick. Following this the re-drawing of the cup and consecutive operations are counted, so that, really, the first one referred to here is actually the second drawing operation. Figs. 301 to 312 illustrate the evolution of a bullet from the flat blank to the perfect article.

All the drawing dies are made from drop forgings, no finish whatever being given them on the outsides. They are held in a jig and drilled, and then reamed with a roughing and finishing reamer, leaving about $\frac{1}{16}$ of an inch land. The die is next turned bottom side up and counterbored sufficiently to have a finished surface of $\frac{1}{16}$ inch on the under side and the same amount of land in the hole for a drawing surface.

Following the work in the drill press the dies are hardened and lapped to size. When the small-size dies are worn out they are annealed and reamed out one size larger, and so on until worn out in the first draw completely.

For the first and second re-drawing operations two dies are run together, one above the other, making altogether a total of seven separate re-drawing dies and five punches. The dies are located on the die-bed in a shoe that holds them down securely and permits a limited sidewise movement to allow for self-centring. The punches are turned, their ends shaped in a box-tool with a separate forming tool for each size, and hardened and ground to the required diameter.

The first and fifth punches, first top and bottom and fifth re-drawing dies are shown in Figs. 313, 314, 315, 316, and 317. The second, third, and fourth re-drawing tools are omitted, as they are of the same general dimensions, a difference existing only in the sizes of holes in the dies and the diameters of the punches. The working parts of these intermediate tools correspond to Figs. 302, 303, and 304, illustrating the appearance of the work at various stages.

Following the fourth-redrawing operation the cups are length-

PUNCHES, DIES, AND TOOLS.

trimmed in an automatic machine by simply feeding the work down a hopper with the ends all one way. The object of this, the first trimming, is to remove the irregular edges and give a uniform length, which lessens the liability of pushing the bottom out in the fifth operation of re-drawing. From the trimming machine the work is run through another automatic machine and

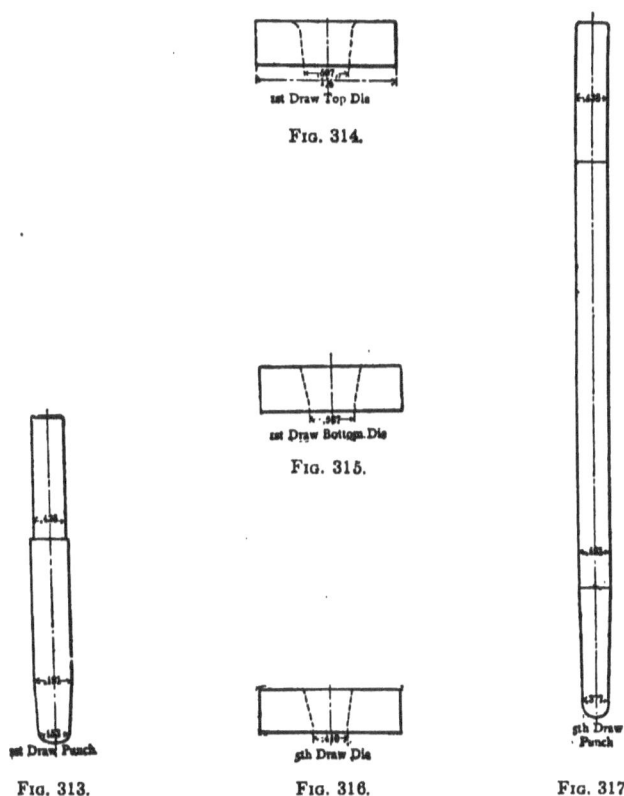

Fig. 314.

Fig. 315.

Fig. 313. Fig. 316. Fig. 317.

"operated," this operation being the rough-forming of the seat for the primer. The tools are operating die, punch, and bunter, Figs. 318, 319, and 320. The cups pass to the machine by feeding down a hopper as in the preceding operation; the punch by the action of the machine takes a cup and pushes it into the die from the bottom and holds it there until after the

188   PUNCHES, DIES, AND TOOLS.

bunter comes forward and forms the pocket, one cup following another in constant and rapid succession.

The cups are now fifth re-drawn, again length-trimmed, and headed. The set of tools for this latter operation is shown in Figs. 321, 322, and 323, and the operation is similar to that of "operating," the same machine being used. Heading refers to upsetting the head, lettering, and completing the pocket for the primer.

For piercing the hole through the seat of the primer pocket there was arranged on the bed of the press an accurately moving

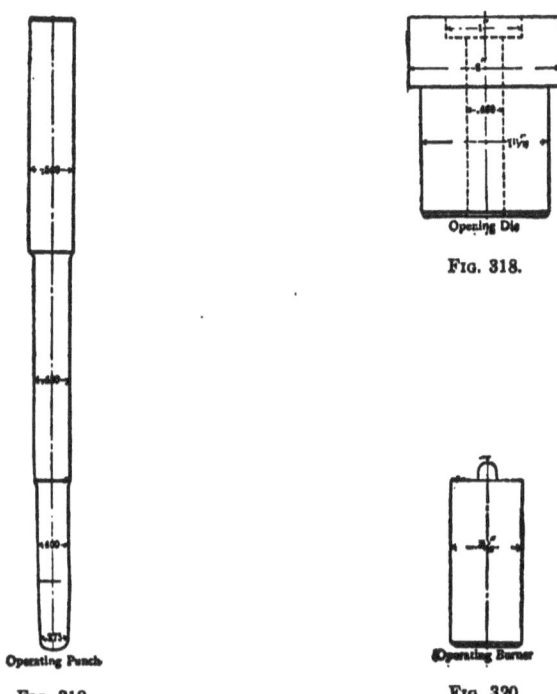

Operating Die
FIG. 318.

Operating Punch
FIG. 319.

Operating Burner
FIG. 320.

and positively controlled dial, driven by a pawl with a lock entering the dial notches. Mounted firmly in the dial were sixteen piercing pins, their outside diameter finished to fit the inside of the shell, and a No. 49 drill-size hole in each of them to receive the piercing punch; a large hole, as may be seen in Fig. 324, is drilled up from the bottom for clearance for the scrap punch-

PUNCHES, DIES, AND TOOLS. 189

ings. The punch consists of a straight piece of drill rod properly tempered. The auxiliary operation of removing the shell from the piercing pins will be referred to at the end of this article.

After annealing the neck, as previously described, the reducing takes place. This is completed in one operating by two

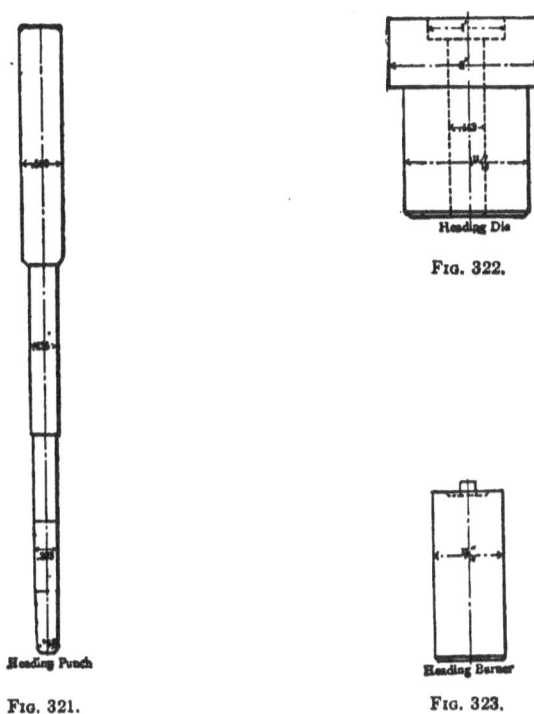

Fig. 322.

Fig. 321.

Fig. 323.

dies, working side by side, fastened in the press ram; their size and outline can be readily understood from Figs. 325 and 326. The dial feed is here again employed to advantage. The shells are hooked in the dial holding by their heads, and when brought under the first reducing die are reduced tapering, and then finished to shape and neck-sized by the finish-reducing die. A shell is completely reduced at each stroke of the press, and passing around still further, it drops through a hole in the press bed. The shells from the reducing of the neck pass on to an automatic machine resembling the one for length trimming, and

190 PUNCHES, DIES, AND TOOLS.

the head is formed to the specified diameter and thickness. A circular forming tool is employed for this purpose, which of course allows constant sharpening without changing the shape.

The shells are now fed for the sixth time through another automatic machine, similar in design to the one for rough length trimming, and finish length trimming or mouth trimming, as it is called. The same principle of feeding down a hopper is here employed to advantage as before.

It may be stated that the means applied to the machines for holding the shell are either a draw-in or a push-out collet of the conventional type, according to the fancy of the designer. Usually, before the heads are stamped, the cups enter the collet

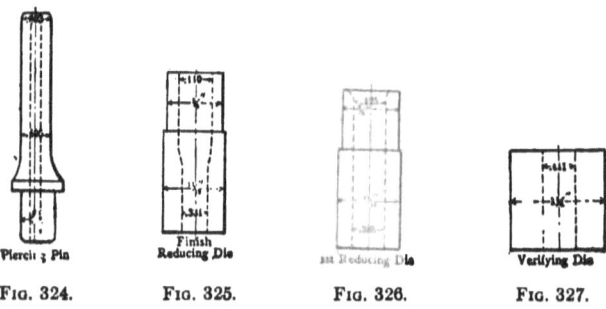

Pierch ; Pin　　Finish Reducing Die　　1st Reducing Die　　Verifying Die
Fig. 324.　　Fig. 325.　　Fig. 326.　　Fig. 327.

one after another automatically, and passing through the spindle, come out at the rear end. The machines are all equipped with feeders and appurtenances for carrying the work to the collets.

The work, after passing through the series of operations, is often slightly enlarged or bulged, and to correct any imperfections, if they exist, and bring them to a uniform size verifying dies are used. Fig. 327 illustrates such a die.

The twelve dies for this operation are seated in a dial of an inch or more thick, and as it rotates the shells with their small necks are easily and quickly placed in the dies by the operator. A flat-ended punch pushes the shells down in the dies to the head; a continued movement of the dial carries the work past the punch, and on the next stroke of the press ram the shell is lifted up and out of the die about $\frac{1}{4}$-inch and is immediately

PUNCHES, DIES, AND TOOLS. 191

gripped by the device shown in Fig. 328, and called a "pick up for shells."

The device is securely fastened to the press ram in a position to come directly over a die when the dial reaches a stopping point beyond the punch; the tube continuing up a satisfactory distance, makes a turn down, thus forming a "U"; the end, extending away from the press, permits the work to fall into a box.

The body A of Fig. 328 is a machine-steel forging, finished on the outside, with the centre hole bored a little larger than the greatest diameter of the shell. B B are two pieces of tool steel

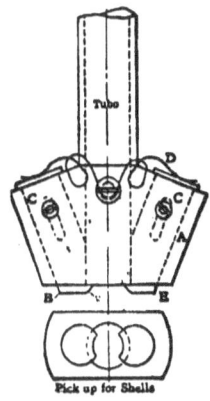

FIG. 328

fitted to A and slotted to allow the screws C C to pass through; thus preventing any rotatory motion, but permitting an easy up-and-down action. The wire spring D is provided to keep the pieces B B down and make a necessary grip of the work. On the down-stroke of the press the round shoulders E come in contact with the head of the shell and being forced up on an incline, open and permit the head to pass by the shoulder; the spring D, constantly pressing down, causes the shell to be gripped below the head and held vertically. When the ram comes down again, the lower and projecting end of the shell comes in contact with the head of the one in the die, and consequently is forced up in the tube, and the operation being constantly gone through, the shells pass up and then down in the tube and drop into the box.

PUNCHES, DIES, AND TOOLS.

The last operation on the shell, that of burring the inside of the mouth, is accomplished by holding the work by hand against a rapidly rotating slim taper countersink.

The successive operations shown in Figs. 301 to 312 are as follows: 301, cup made by blanking and drawing; 302, first re-drawing; 303, second re-drawing; 304, third re-drawing; 305, fourth re-drawing; 306, "length-trimmed" and "operated"; 307, fifth draw and "length-trimmed" again; 308, headed; 309, section of shell; 310, reduced; 311, mouth trimmed and head formed and burred; 312, the cartridge complete.

MAKING CUPRO-NICKEL BULLET JACKETS FOR 30-CALIBRE CARTRIDGES.

The bullet jacket for the 30-calibre cartridge manufactured by the United States Government is an alloy of copper and nickel known as cupro-nickel; it is blanked out and drawn up and shaped by a series of operations whose sequence is indicated in Figs. 329 to 344. The stock is 0.036 inch thick (19) gauge, and the disk punched out in the double-action presses is about

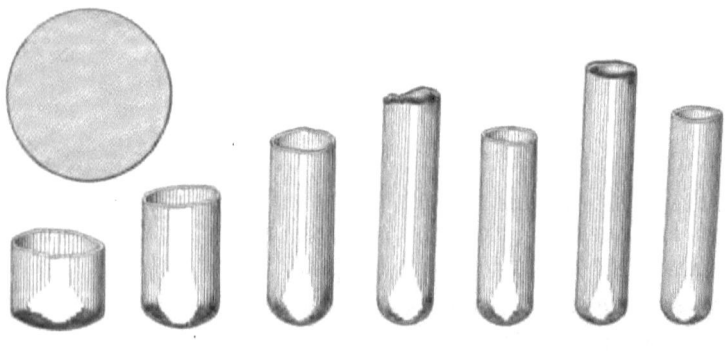

FIGS. 329 to 336.

1 inch diameter. After three successive draws, the surplus metal is trimmed off the mouth, and a final draw is then made, leaving the jacket still a little long, to allow for a final end-trimming to exact length. The row in Figs. 329 to 336 shows the work as it advances through the stages mentioned. The

PUNCHES, DIES, AND TOOLS. 193

draw presses used for the drawing operations are equipped with automatic feeding hoppers. The bullets or lead and tin composition cores are moulded in the lead shop at the arsenal at Frankford, Ill., which is equipped with seven oil furnaces with four lead pots each to a furnace. The moulds are long enough for sixteen slugs each, and one man will pour about 500 pounds of slugs per day. From the lead shop the slugs or cores are taken to the bullet-assembling machines, and there the jackets —after being cleaned by tumbling—are shaped and provided with their lead centres. The row of samples in Figs. 337 to 344 show the work accomplished by the assembling machines.

These assembling machines, or presses, require two operators each, and perform six operations on as many jackets (one on

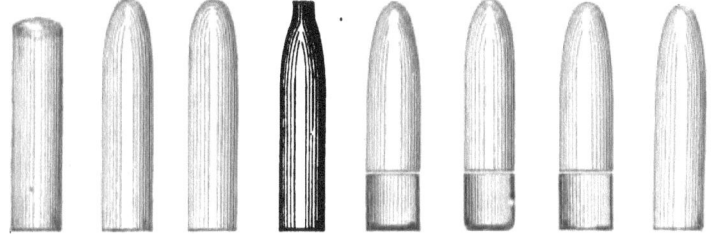

FIGS. 337 to 344.

each jacket) simultaneously. Two dials rotating in a longitudinal plane carry the work under the punches, the dials being provided with a series of chambers near the periphery, and the bullet dial at one side overlapping the jacket dial, so that at this point the composition core can be forced out of its chamber in the upper dial and down into the jacket beneath in the lower dial. The dials have ratchet-and-pawl rotation, and at each return movement of the pawl the punches carried by the ram of the press perform their work. As the jacket dial brings the cupro-nickel case under the first punch, the point of the jacket is partly formed, and the next punch at the following stroke of the press completes the point. At the next movement the slug is forced down into the jacket, the open end of the latter is crimped slightly, and the next punch completes the crimping operation; the last punch forces the now jacketed bullet down

13

through a sizing hole and out of the machine. In the re-sizing machine the rear end of the bullet is tapered slightly, and following this operation an "o" is stamped on the flat end.

The mechanism of the loading machines consists essentially of two horizontal dials—one for shells and one for bullets—a powder magazine and charger, and a series of punches operated by a vertically reciprocating ram. The dial at the right is notched to receive the shells, which are led into the notches by a continuously rotating disk upon which they are placed open end up by the operator's right hand; the dial at the left, which is geared to the shell dial, is drilled out to receive the bullets, these being dropped, one in each chamber, by the left hand. The rotative movement is by pawl and ratchet, the dwell preceding each advance being sufficiently long to enable the powder charger and the series of punches to perform their operations.

The powder charger is a reciprocating slide containing a chamber to hold just the required amount of powder; the slide is moved to and fro through a guide under the magazine, and at each dwell of the shell dial it advances until its chamber is over the empty shell or case and deposits the charge therein, then recedes in its seat under the magazine for a fresh supply or charge. A vibratory motion given the charger prevents the powder from lodging and insures the full charge being deposited each time. As the charged shell advances to the next position, a punch seats the powder and another punch connected with a bell gives warning if through any accident a case has passed the magazine without being properly charged. As the shell comes under the overlapping bullet dial, a punch forces the bullet down into place, and the loaded cartridge then passes out of the machine. Before loading the bullets are dipped in a lubricant of Japan wax and graphite, and the shells are thoroughly cleaned. Blank cartridges are loaded here as well as ball, and the capacity of each machine is about 16,500 per day of eight hours.

The sheet-steel, funnel-shaped magazine, while quite bulky in appearance, seldom contains more than 2 or 3 pounds of powder; it is made of the height sufficient to guard the operator from injury in case of explosion. The powder is actually held

PUNCHES, DIES, AND TOOLS. 195

by a straight pasteboard tube, say, 2 inches in diameter, with a flaring top, which extends to the wall of the metal funnel, and on the column of powder in this tube rests a metal ball attached to the end of a cord whose outer end, after passing through a sheave, is secured to a lighter ball, suspended at the side of the magazine. As the supply of powder lowers in the tube the ball descends with it and the ball at the outside rises; thus the operator can gauge the powder supply in the magazine and knows when to replenish it.

ASSEMBLING CARTRIDGES IN CLIPS.

After inspection the cartridges, whether blank or ball, are placed in their clips, as shown in Figs. 345 to 348, and this operation is also performed by machinery. The clips and springs are first assembled by special machinery and then placed in the magazine of an automatic cartridge-and-clip-assembling machine. Five cartridges are placed in each slit of the turret of the ma-

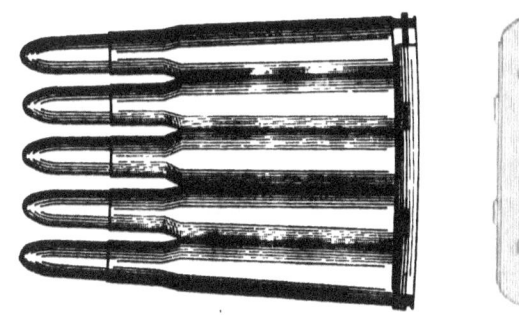
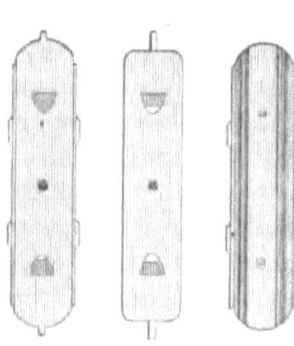

FIGS. 345 to 348.

chine as it is indexed around, and upon arriving opposite the magazine a clip is drawn forward and slid over the five heads. Upon the clipped ammunition arriving opposite the chute at the rear it is lifted from the turret and dropped out of the machine by a vertically reciprocating arm carrying a pair of spring fingers.

It has already been stated that the ball ammunition is .01 inch longer than the blank to prevent accidental assembling of a stray ball cartridge in a clip of blanks. Should such a cartridge get into one of these machines engaged on blank work its head would project above the other cartridges and the clip guide far enough to prevent the clip being drawn forward, and the machine would then stop. There are quite a number of these assembling machines in use, and each handles about 100 cartridges per minute. They were invented by Major Lissak, of the Ordnance Department, who also devised many other special machines used at this arsenal. One of these which should be of interest in connection with gauging operations noted above, although designed for facilitating the inspection of empty cases rather than loaded cartridges, must be seen to be appreciated.

THE DRAWING OF CARTRIDGE-CASES FOR QUICK-FIRING GUNS.

The development of the quick-firing gun has at once necessitated, and been rendered possible by, improvements in ammunition with a view to quick loading. Quick-firing guns differ from ordinary guns in having the propelling charge and the means of ignition contained in a metal case. The projectile may or may not be attached to the case, forming a complete cartridge, as this depends on the size of the gun. In ordinary guns the projectile, the propelling charge, and the primer or means of ignition, are all separate, the charge being usually contained in a combustible silk cloth or serge case. The advantages which metal cases present as compared with combustible cases are: (1) They are quicker in loading, since the primer forms an integral part. (2) The same reason reduces the probability of a missfire. (3) The sponging-out of the gun, to avoid the possibility of the burning remnant of a combustible case prematurely igniting the next charge, is avoided. (4) The expansion of a brass case under fire enables it to act as a gas-check, rendering the use of an obturator unnecessary.

Simultaneous loading with "fixed" ammunition, in which the projectile is attached to the cartridge-case, is practised with quick-firing guns up to about 3 inches diameter, above which size complete cartridges would be too unwieldy. Between 3

PUNCHES, DIES, AND TOOLS.   197

inches and 6 inches, therefore, separate loading is the rule, with the projectile separate from metallic cartridge case. Above 6 inches the gun ceases to be called quick-firing, and combustible cases with separate loading are used. Separate loading for larger quick-firing guns is desirable, not only because of the excessive weight of a complete cartridge, but also because of the danger of storing loaded and fused shell in the same magazine with loaded cases. With separate loading, the projectile may be placed near the gun at leisure, the cartridge-cases not being taken from store until the last moment. Such different conditions govern the storage, transport, and use of projectiles and of cartridge-cases that it is undesirable to attach them together.

Cartridge-cases for quick-firing guns are universally made of brass, this material having been found to possess the qualities best suited for this exacting service. Of all the numerous alloys, that of copper and zinc, commonly called brass, ranks as one of the most important. At one period the generic name of bronze was given to this alloy as well as to that of copper and tin, to which it is now applied. The two alloys, copper-tin and copper-zinc, are each characterized by well-defined properties, and each should retain its proper name of brass and bronze, respectively. No other metal or alloy, not even excepting iron, presents such widely varying qualities, or so great a field of application. Commercial brass consists of two parts of copper and one of zinc, and is used, with certain exceptions, in all countries for cartridge-cases, not only for rifles but for quick-firing guns. The exact composition is 67 per cent of copper and 33 per cent of zinc, with a margin of 5 per cent above or below for either metal. The French artillery department, which is noted for the care with which its specifications for cartridge-case metal are drawn up, not only specifies the above-named proportion and variation, but requires that the constituent metals shall be of accepted brands, and of known origin, the sources of supply of copper being limited to the following: "Calumet and Hecla," "Tamerack," "Ovscila," "Atlanta," "Franklin," "Quincy," "Wallaroo," and that manufactured by electrolytic deposition. The brands of zinc specified are: "Vieille Montagne," known as "Extra Pure Fonte d'Art," "Oeschger Mesdach," "O.M. Art

Zinc," and that of the Royal Austrian Company of Spain, known as the "R. O. A. Refinado." Subject to certain limitations, the use of scrap brass is also allowed.

The entire manufacture of metallic cartridge-cases involves a series of operations which, with the exception of two or three, consist in cold-drawing. The brass used is capable of extreme deformation when cold, but cannot be worked hot. After being formed into a cup-shaped disk, the metal is subjected to successive drawings, the object of which is to diminish the diameter and thickness and increase the length, the volume undergoing no sensible alteration. At each drawing the metal is deformed to a point short of the breaking point, every drawing operation

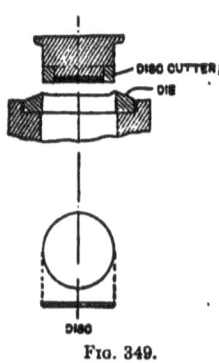

FIG. 349.

being followed by annealing until the desired form is obtained, namely, a long cylinder with thin walls, and closed at one end.

The earlier operations, while the cartridge-case is still short, are carried out in a vertical press, but when the length is such that the manipulation and the withdrawal of the punch become difficult, the operation is continued in horizontal presses. The two most important tools are the punch and the die. The punch is carried upon the extremity of the ram of the press and transmits the power, acting upon the bottom of the cartridge-case, which is inserted in the larger end of the die, the latter being strongly secured to the head of the press opposite the hydraulic cylinder. The die consists of a ring of hardened and tempered steel, the interior having the shape of a truncated cone, the axis of which is in a straight line with that of the punch. The opera-

PUNCHES, DIES, AND TOOLS. 199

tion of drawing is performed by placing a partly drawn case, properly centred, in the larger end of the die, and advancing the punch until it touches the bottom of the cup. The pressure then comes into play, forcing the cup through the small end of the die, thereby reducing the diameter of the cup and the thickness of the walls and increasing the length, a process which involves considerable flow of metal.

During the process of drawing, the cartridge-case is subjected to stresses in general oblique to the surface, represented by P (in the stress diagram in Fig. 361). This stress may be resolved into two, one of which is normal and the other tangential to the surface of contact between the brass and the die, called respectively N and T. If E represents the total pressure exerted by the punch upon the bottom of the cartridge-case, it is clear that equilibrium will exist when the vertical components of the normal and tangential forces are together equal and opposite to the force E. If a is the angle formed by the wall of the die with the axis of the punch, the equation of equilibrium will be: $E - T \cos a - N \sin a = 0$. Thus it will be seen that the forces N and T vary with the magnitude of the angle a. As this increases $\sin a$ increases and $\cos a$ diminishes, and consequently the values of the components T and N also decrease and increase respectively. When $a = 90°$, $5 \sin a = 1$, and $\cos a = 0$, and E is then equal to N. When $a = 0$, then $\sin a = 0$ and $\cos a = 1$, E being equal to T. In the first case this stress would be entirely normal, and in the second case entirely tangential and tensile. The two extreme cases, however, never occur. In practice the flow of the metal is never achieved by simple tensile forces, and compressive forces are always present. The drawing of a 6-inch cartridge-case will now be described in detail.

CUPPING THE SHEETS.

The cupping is divided for convenience into two stages, the first being done with the punch and die illustrated in Fig. 350. Before commencing it is necessary to centre the die relatively to the punch, the breadth of the annulus being measured at three points. The stroke of the punch is then adjusted, so that at the end of each stroke it does not exceed what is necessary to thrust the disk clear through the small end of the die, and so avoid waste of time and power. At the commencement of the stroke an extra length of stroke of from 4 inches to 8 inches is given in addition to the amount actually necessary to clear the die, in order to give the operator time to place the disks upon the die. The die, punch, and disk are then well greased, and the latter is placed upon the upper surface of the die. Water is admitted to the cylinder and the punch advances, driving the disk through the die and out at the smaller end, whence it falls in the form of a cup into a receptacle placed below the press. The maximum pressure attained is 1,000 pounds per square inch, as shown by the gauge attached to the hydraulic cylinder. The cup is then annealed for about 28 minutes at 1,364° F., having a steel clip placed round it. The scale which forms on the surface of the cup is subsequently removed by pickling in lead-lined

FIG. 350.   FIG. 351.

wooden troughs containing dilute sulphuric acid, of a strength of 1 to 4, for a period varying from 8 to 15 minutes according to the strength of the bath. The cups are then washed by immersion in lead-lined wooden troughs, through which runs a stream of water, every trace of acid being quickly removed. The second cupping operation is made in exactly the same manner as the first, except that the punch and die shown, Fig. 351, are substituted for those previously used, the same precautions being observed for centring and lubricating. The maximum hydraulic pressure indicated by the gauge is 1,150 pounds per square inch, while the subsequent annealing lasts 20 minutes at a temperature of 1,202° F. The pickling and washing processes which follow this and all other annealings are as before de-

PUNCHES, DIES, AND TOOLS. 201

scribed. The behavior of the metal during cupping is an efficient test of its quality. The presence of impurities or improper annealing are quickly shown by cracks or a roughened surface.

First Drawing.

Fig. 352 shows the punch and die used in this operation, also the resulting piece. The maximum hydraulic pressure is 1,300

FIG. 352.   FIG. 353.

pounds per square inch. The pieces are then annealed at 1,202° F. for 28 minutes.

Second Drawing.

This is performed with the tools shown in Fig. 353. with a maximum hydraulic pressure of 1,350 pounds per square inch. The subsequent annealing is at 1,202° F. for 26 minutes.

Third Drawing.

This is performed with the tools shown, Fig. 354, with a maximum hydraulic pressure of 1,320 pounds per square inch. The subsequent annealing is at 1,184° F. for 25 minutes.

Fourth Drawing.

Before drawing, the bottom of the piece is flattened preparatory to indenting, which takes place after the fifth drawing, and is necessary for the formation of the primer hole. Flattening is accomplished by pressing the piece between the punch and a flat steel disk supported on the die. The disk is then withdrawn and the drawing proceeds as usual. The tools are shown in Fig. 355. The maximum hydraulic pressure is 1,000 pounds per square inch, and the subsequent annealing is at 1,166° F. for 22 minutes.

Fifth Drawing.

This is the last operation performed in the vertical press. The tools used are shown in Fig. 356, the maximum hydraulic pressure being 700 pounds per square inch. The subsequent annealing is at 1,166° F. for 20 minutes.

Indenting for Primer.

This operation is performed in the vertical 1,000-ton press with the tools shown in Fig. 357. Upon the ram which moves upward is placed a pressure plate, to which is hinged a steel

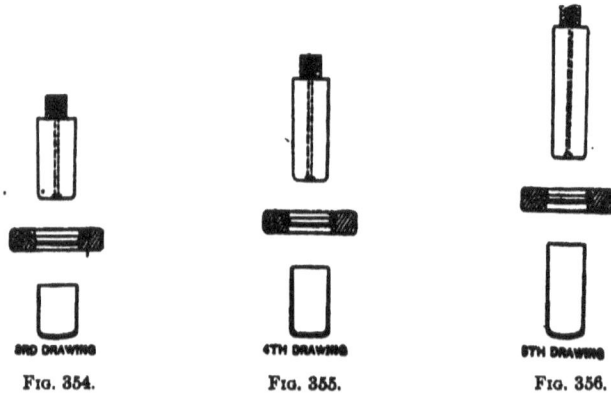

Fig. 354.   Fig. 355.   Fig. 356.

punch-shaped piece having the same external form as the interior of the cartridge-case as it leaves the fifth drawing, and with an indentation at the top. This can be hinged to one side to facilitate the insertion and withdrawal of a cartridge-case. Upon the under side of the upper head of the press is a fixed holder, into which is screwed a flat piece of tempered steel having a small projection in the centre. The object of this is to form, in conjunction with the recess in the punch, the metal boss on the inside of the case, for the primer. The cartridge-case is sub-

PUNCHES, DIES, AND TOOLS.

jected to a pressure of about 314 tons between the two surfaces, with a hydraulic pressure of 2,500 pounds per square inch. No annealing is required after indenting.

Sixth Drawing.

From this operation onward the two larger of the three horizontal presses are used, because the length which the cartridge-cases have now reached does not permit of their manipulation in the shorter-stroke vertical press. The tools used for the sixth drawing are shown in Fig. 358. Up to this point the cartridge-

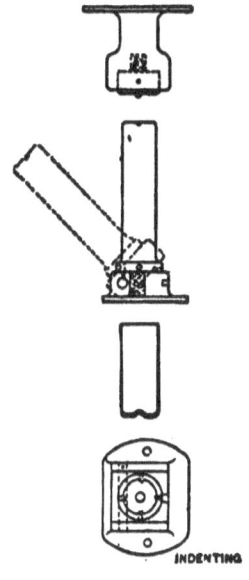

FIG. 357.

cases have been able to strip themselves from the punches by catching on the under side of the dies. Their expansion at the moment when drawing is complete, and when they are relieved from the considerable lateral pressure, prevents their being again drawn up through the die by the retreating punch. But from the sixth drawing onward, the lateral pressure is less, and other means are adopted. Under each die is an attachment containing eight fingers pressed inward toward the axis by springs. Dur-

204   PUNCHES, DIES, AND TOOLS.

ing drawing, they give way before the advancing case, retiring into recesses. But when the end of the case has passed them they spring out and keep the case from following the punch back, the inclination of the recesses in which they move assist-

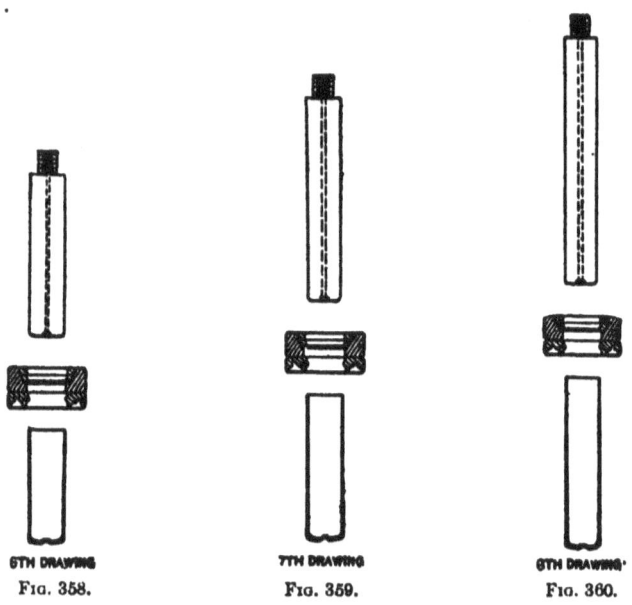

FIG. 358.   FIG. 359.   FIG. 360.

ing this action. The sixth drawing may be performed either on the 18-inch or the 16-inch horizontal press, either having sufficient power. The subsequent annealing is at 1,166° F. for 18 minutes.

Seventh Drawing.

The tools used for this operation are shown in Fig. 359. The subsequent annealing is at 1,166° F. for 15 minutes.

Eighth Drawing.

The tools used for this operation are shown in Fig. 360. The subsequent annealing is at 1,112° F. for 14 minutes.

Ninth Drawing.

The tools used for this operation are shown in Fig. 362.

PUNCHES, DIES, AND TOOLS. 205

The subsequent annealing is at 1,058° "snap," for it is no such thing.

Tenth Drawing.

The tools used are shown in Fig. 362 a. This is the last drawing operation, and the blanks undergo no annealing upon its completion.

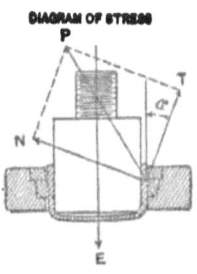

Fig. 361.

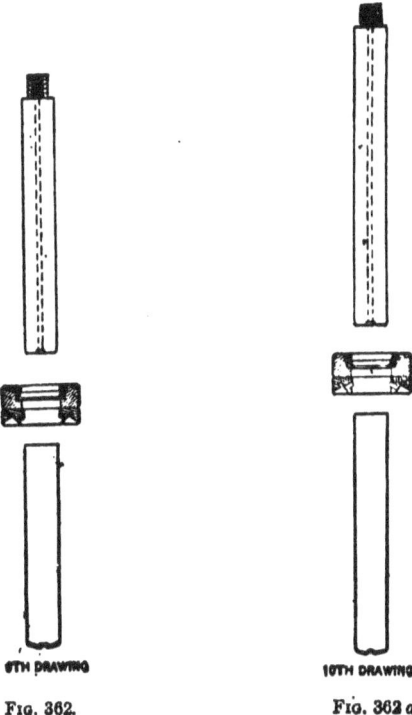

Fig. 362.   Fig. 362 a.

Heading.

The formation of the head of the cartridge-case is one of the most interesting of the operations in the process of manufacture. The total pressure which the head of the cartridge is called upon to stand under fire is enormous. With the 6-inch quick-firing gun used in the Spanish service, for which the cartridge cases are intended, the pressure caused by the explosion is about 17 tons per square inch. Even this pressure is exceeded when test-

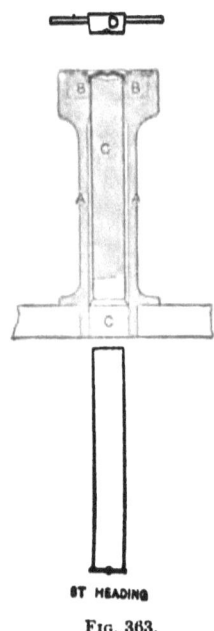

FIG. 363.

ing the guns, which is done with three discharges at a pressure of 20 tons per square inch. When the area of the cartridge-case head is considered, some idea may be formed of the enormous aggregate pressure to which it is subjected. It is essential for the satisfactory working of the guns that no deformation should take place under fire, and it is therefore important that during manufacture the head should be subjected to a pressure two or three times that likely to be experienced in practice. The operation of forming the head is made in the vertical 2,500-ton press in three stages. The tools used for the first stage are shown in Fig. 363. An iron casting, *A*, termed a bolster, is placed upon the ram of the press and serves to support the die-holder and die *B*—which latter imparts the form of the flange to the head. Inside the bolster is fixed a steel stem *C*, over which the cartridge-case is slipped in the condition in which it leaves the tenth drawing. This stem, which must be capable of withstanding an aggregate pressure of 1,650 tons, is of the best hardened steel, and is without question the most delicate of all the tools employed in the process of manufacture. The first heading operation is performed by inserting the cartridge-case between the stem and the bolster. Upon the top is also placed the punch

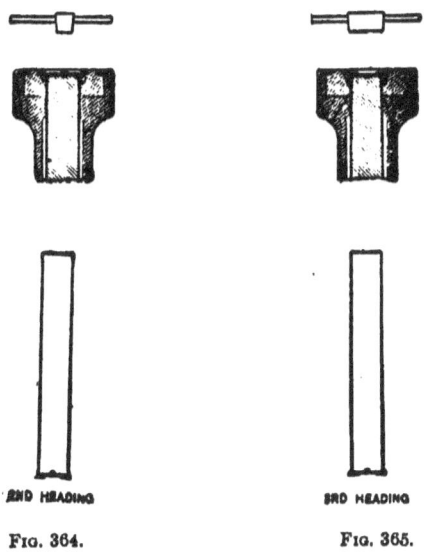

FIG. 364. FIG. 365.

D of hard steel, provided with a central depression, the object of which is to reduce the area of contact over which pressure is exerted on the head of the cartridge-case. The total pressure is 1,600 tons, which leaves the head with a central internal and external projection, and forces the metal outward to form a flange.

In Fig. 364 is shown the second heading operation. This is performed with the same tools as the first, except that a smaller punch, 3 inches diameter, is placed over the cartridge-case, in-

208   *PUNCHES, DIES, AND TOOLS.*

stead of the punch D previously used. A total pressure of 600 tons is exerted, with the result that the outside projection is flattened, and all the metal is driven into the internal boss, thus allowing sufficient metal for the primer holes. Finally, the third heading operation is performed with the tools shown in Fig. 365, a total pressure of 1,650 tons being applied, with the result that the head is rendered flat and shapely.

Tapering.

This operation is for the purpose of giving to the cartridge-case its final external form, enabling it to fit the chamber of the gun, and to be easily inserted and withdrawn. It is performed

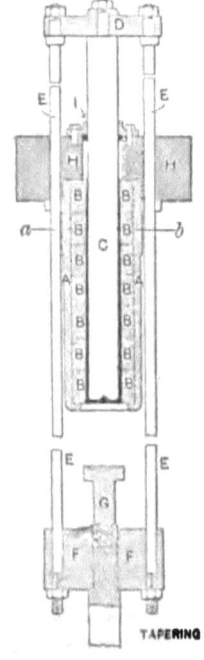

FIG. 366.

in one or the other of the horizontal presses, in order to take advantage of their longer stroke. To the fixed head H of the press, Fig. 366, is bolted the cast-iron bolster A, inside which are placed seven rings of tempered steel B, the internal length

PUNCHES, DIES, AND TOOLS. 209

of which when thus assembled is exactly equal to that of the gun chamber. The cartridge-case is driven into this space by the press, but as it is necessary forcibly to extract it after the operation, the special apparatus shown is made use of. The cylindrical extractor C, having a head shaped to fit the inside of the headed cartridge-case, is connected rigidly with the ram of the press through the crossheads D and F and the tie-rods E, and moves therewith, its position being kept central by the guide I. The punch G, bolted to the ram of the press, forces the cartridge-case in during the forward stroke, while the extractor C forces it out during the return stroke. At Trubia the tapering is divided into two operations with annealing between, to avoid risk of cracking. Before the first tapering the cartridge-case is annealed at 1,040° F. in a small vertical furnace, care being taken to allow the head to remain outside the furnace in the air. It is then placed in the press and forced about half its length into the chamber, the precaution being taken to adjust the stop of the press so as to limit the stroke to half its usual length. On the return stroke, by the aid of a wooden distance-piece inserted between the extractor and the head of the cartridge-case, the latter is forced out. The case is then returned to the vertical annealing furnace, where it is exposed to a temperature of 932° F., care being taken as before not to anneal the head. Tapering is then completed in the press, the cartridge-case being driven completely home into the die chamber.

Other Mechanical Operations.

The remainder of the operations are of a mechanical nature, such as turning the end, the head, the steps in the chamber, the attachment for the primer, cutting to the exact length, etc., none of which involve any features of special technical interest. It may, however, be mentioned that throughout the whole course of manufacture the thickness and diameter of the cartridge-case are carefully checked with calipers and gauges, and particularly for the first two or three cases in each lot, in order to verify the accuracy of the dies and the setting of the tools. The ends of the cases are frequently turned to length between the various drawing operations, since there is a tendency, due either to the irregularities of metal or to uneven annealing, to stretch unequally, leaving ragged edges. It is also of great importance that the thickness of the end of the cartridge-case should be closely checked, and this is performed by limit gauges. Lubrication of the punches and dies is effected by olive oil or soapy water.

SECTION VIII.

The Manufacture and Use of Dies for Drawing Wire and Bar Steel.

MAKING DIES FOR DRAWING WIRE.

THE process known as wire-drawing, as understood in the manufacturing world, is the art of reducing the diameter of metals by successive passages through holes of gradually decreasing size in plates called dies. The gradual decrease gives a corresponding increase in the length of the "wire," the actual ratio varying by constant proportion depending upon the reduction in diameter. The holes in the dies are usually made to conform to one of the numerous wire gauges, or to a system expressed in decimals, depending on the method employed in the mill where they are used. The variation or decrease in successive sizes, which is termed "drafting," is limited by the tensile strength of the metal, and also by the desired amount of hardness which is required in the finished article, and each wire-drawer has his own ideas on the subject. On coarse sizes, say to No. 5 B. and S. gauge, several numbers may be covered by each reduction. From these to about No. 40, one size at a time is usually reduced; and from thence the reduction is made by a difference of $\frac{1}{1000}$ of an inch at a time, and by $\frac{1}{10}$ of a thousandth, when a size nearing $\frac{1}{1000}$ inch diameter is reached. These figures are, of course, dependent on the nature of the metal, and very often much less differences are required.

The dies, of course, are the all-important factors of successful work. They must be of a substance which will not wear quickly or unevenly; they must be capable of severe usage; and lastly, they must be as inexpensive as it is possible to make them and still be durable.

When great accuracy is required in the sizing of wire, the

diamond is by far the best material to use for a die. It is the only substance which can be used for very fine sizes, and they are even used as coarse as No. 12 B. and S. gauge. This seems to be rather a large size, but still the principal objection is the first cost, and as the nature of the material is brittle, many fear to use them for sizes heavier than No. 15. They are without question the most economical in the end, where large quantities of wire are drawn, and even when they become worn, may be reamed out considerably with comparatively small layout.

DIAMOND, SAPPHIRE, AND AGATE DIES.

The substance used is the black diamond or "Bortz," which is found principally in South Africa and Brazil. It is extremely hard and the boring of the hole a slow and tedious operation. Take the case of a piece $\frac{1}{8}$ inch thick. The spindle which does the drilling runs at a speed of about 5,000 revolutions per minute, and, running ten hours a day, will require two or three weeks to get through that distance. When the drilling operation is completed, the stone is set in a brass plate about $1\frac{1}{4}$ inches diameter and $\frac{1}{4}$ inch thick, and stamped with the size, usually expressed in decimals of an inch, and the weight of the stone used.

In a comparative test made as to the relative hardness of common agate and diamond, a piece of the former about $\frac{1}{8}$ inch thick was placed in the drilling machine and operated on in a similar manner. The drill went through that distance in seventeen minutes. Sapphires are also used for dies, and are hard enough for some purposes, while their comparatively low price makes them desirable in some cases; but where durability and accuracy are the prime factors, the diamond stands as the best known substance in the world.

For sizes down to $\frac{3}{4}$ of an inch in diameter, a tool-steel plate with a hole in it of the required size is generally used. If the object is merely quickly to reduce the diameter, this hole does not require accuracy; but if the work is to be finished on that

PUNCHES, DIES, AND TOOLS.

size, it should be carefully polished, as every nick will show a scratch on the work.

CHILLED IRON DIES.

Dies made of chilled iron are usually employed for sizes below ¾ inch down to No. 10 B. and S. gauge, and are also made from close-grain iron, being cast in a cast-iron mould, or "chill," which makes them very hard and dense. The holes are cast conical in form, the large end being at the back where the wire enters. These holes, however, do not usually extend clear through, a thin wall of metal being left on the small end to be punched out and finally reamed to the required size. This reaming is done by means of a steel drill of square section revolving at a very slow speed, giving a finished hole of constant size, ranging from about $\frac{1}{4}$ to $\frac{1}{8}$ of an inch long, depending on the size. These chilled-iron dies are usually made with four to eight holes each about 1¼ inches apart; the thickness of the die varying from 1 inch to 1½ inches, and the length according to the number of holes cast. They are inexpensive at first cost, and, when worn, may be reamed out to larger sizes without impairing their usefulness. Hundreds are used in every mill, and their manufacture is an art by itself, requiring a high order of skill and much patient practice to become proficient in.

STEEL DIES.

Steel dies are also used in large quantities, for sizes ranging from No. 10 to No. 18 gauge. They are made of high-grade tool steel, either from round rods with a single hole, or rectangular bars with several holes. These holes are of conical form also with the small end reamed to size, and may be used indefinitely. The steel is not hardened, so that when the hole becomes enlarged by use, the front of the die is hammered or "upset," and the hole again reamed out, the operation being done by hand with tools of the required size.

THE MAKING OF DRAW PLATES BY CASTING IRON AND STEEL WITH DIAMONDS.

Draw-plate making is no novelty to the tool-maker of wire experience, but the drawing of very fine wire involves some peculiarities of treatment which are suggested by an invention of Frederick Krause, of Jersey City, N. J. The invention relates to the formation of diamond draw plates, and the process is exhibited in the illustrations, Figs. 367 to 371. Fig. 367

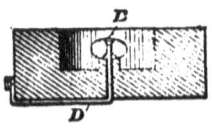

Fig. 367.

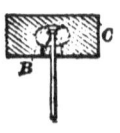

Fig. 368.

shows a mould with a diamond in place; Fig. 359 shows the diamond incased with steel or iron. The method of casting further material about the article is seen in Fig. 369, and Fig. 370 shows the enlargement as cast. A modification of the process is presented in Fig. 371.

In the illustrations a mould *A* is shown with a perforated bottom. A diamond already drilled is placed in the mould and

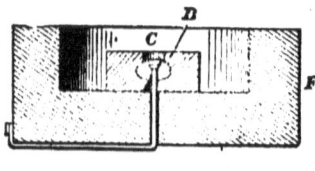

Fig. 369.

a wire *D* extends through the hole in the diamond *B* to hold it securely in the temporary position. The molten steel or iron is then poured into the mould and surrounds the diamond, the latter sitting on a slight elevation above the bottom of the mould. The steel or iron when cold is removed from the mould, as indicated at Fig. 369. At the under side of the stone which rests

PUNCHES, DIES, AND TOOLS. 215

on the seat in the mould, the iron and steel casing or shell C leaves the hole in the diamond B exposed. At the opposite side the casing C is bored out to form an opening communicating with the perforation in the diamond. The wire D which serves to hold the diamond in place when it is cast around with metal, is removed when the casing is bored.

The plate can now be employed for wire drawing, but in order to facilitate the handling of the tool, it may be enlarged by forming a casing E about the casing C. This enlargement can be cast by placing the plate as seen in Fig. 368 in a mould F, and holding it in position with a wire, as already shown. The perforations in the diamond are plugged with graphite, clay, or some similar substance, after the wire D is in place to prevent the metal from flowing into the cavity when the casing is made. The enlargement is bored out to free the opening, as was done in the first instance.

The danger of injuring a diamond during these operations will be apparent to any one who has undertaken to handle these

FIG. 370.   FIG. 371

easily chipped crumbs of carbon for industrial purposes. The inventor does not overlook this or the possibilities involved, and says plainly that the heat during the casting operation may injure the diamond. He advises the wrapping of the diamond with sheet metal. This procedure is indicated as G in Fig. 371, and the retaining wire there illustrated has a head which cannot slip through the opening in the diamond when the shank is passed through the wrapping and stone. During the casting of the part C the wrapping G is supposed to shield the diamond from the molten steel; immediate contact is prevented anyway, and the wrapping becomes alloyed or fused with the casing so that when the draw plate is complete the encased diamond appears as illustrated in Fig. 368.

THE DRAWING OF ROUND AND RECTANGULAR BAR STEEL.

Round and other bar stock is made in the following manner: The stock is point 1 to point 25 steel, and is first pickled and washed, a process which removes all scale from the bars. The rounds are drawn through dies $\frac{1}{16}$ inch smaller in diameter than they are rolled, and other shapes are reduced in similar amount. They are then machined, straightened, hand-straightened, cut to length, covered with oil to prevent rusting, and then stocked or shipped.

PICKLING THE STOCK.

The pickling operation consists of soaking from 10 to 15 tons of stock in a solution of sulphuric acid and water for four or five hours, washing with water and then through a bath of lime-water, the washing being continued, one bar following the other through the tanks. Care is taken to see that nothing sticks to the bars that can mark them while passing through the die. As illustrating the necessity of cleanliness in this respect, if we chalk-mark or tie a hair or a thread around the bar, traces of them are visible on it after passing through the die. It would be hard to guess at the depth of these marks, owing to their extreme shallowness.

MAKING DRAWING DIES OF CAR WHEELS.

Many mills make their dies for round stock out of car wheels, and chill the metal around the draw hole. Fig. 372 shows a 3-inch die. For about $\frac{3}{4}$ inch the hole is parallel, then bell-mouthed to a $\frac{13}{16}$ radius. As the shoulder at a gets dull it is ground or reamed sharp, as represented by the dotted line b. When but about $\frac{1}{8}$ inch of the cylindrical part of the hole is left, the die is enlarged for another size, and so on, as long as the chilled part lasts. A wrought-iron or steel band is shrunk on to materially strengthen the die.

Shafting commercially known as 3-inch is, as nearly as can be made, 2.995 inches in diameter. The cylindrical part of the

PUNCHES, DIES, AND TOOLS. 217

hole in the die is 2.994 inches in diameter. There is no measurable outside expansion of the die, and the inside is not get-at-able. It is probably more than the contraction of the bar; the latter expands some as soon as free from the die. The increased density of the bar, due to drawing, can only be determined approximately by its elongation, owing to its initial roughness and

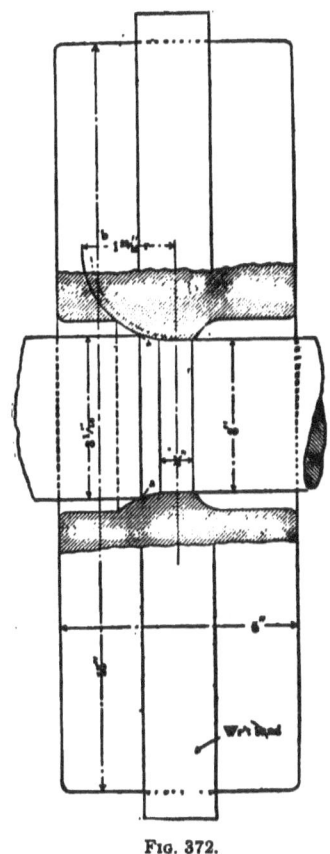

Fig. 372.

irregularities; the bars also do not stretch alike. The well-marked results of the drawing are increased strength and stiffness, and a great decrease in ductility. Occasionally a bar breaks while being pulled through the die, and the fracture, except for color and texture, resembles that of hard cast iron, although before it was put through the drawing machine it would have stretched 25 per cent without breaking. The larger the bar, other things being equal, the more liable it is to break by drawing, although the reduction is the same, $\frac{1}{16}$ inch in all sizes from $\frac{3}{4}$ inch to 5 inches.

The pull required to draw $1\frac{1}{16}$-inch bar through an inch die is 15,000 pounds, and 30,000 pounds to draw a $2\frac{1}{16}$-inch bar through a 2-inch hole, showing that the pull varies as the diameter or periphery, and not as the area of the bar, as some might suppose. Bars are often drawn with a much greater reduction than $\frac{1}{16}$ inch, and the changes due to drawing vary somewhat with the amount of reduction. High carbon steel bars go through the die on the jump, and leave well-defined swells on them where they stop in the die for an instant. Rods $\frac{3}{4}$ inch and below are pulled four at a time, with a reduction of $\frac{1}{32}$ inch. Above $\frac{3}{4}$ inch and up to $1\frac{7}{8}$ inches two bars are drawn at a time. All bars are more or less kinked in drawing, and, although the pull tends to keep them straight, when they fall from the die they assume all kinds of crooks. When pulling several bars at a time some can be felt lagging behind and then again going ahead of the others, athough the grips do not allow much variation in this respect. One peculiarity noticed in drawing the bars through the die is the ease with which they can be rotated. One may grasp an inch rod with the hand tight enough to twist it perceptibly in the die, as far as the grips permit. This is another example wherein longitudinal motion eliminates cylindrical friction. On the other hand, if the rod is rotated in the die, a force of 1 pound will, in time, pull or wear it through. Bars are drawn through the die at about 17 feet per minute, and at this speed requires about 1 horse-power, net, to pull a $1\frac{1}{4}$-inch bar. A stream of black oil lubricates the die and rod during the drawing.

CHILLED DIES VS. TOOL-STEEL DIES.

Chilled dies last about ten times longer than tool-steel dies, and the reason for this is that unlike metals wear better than like surfaces. The only exception known to this rule is the case of cast-iron dies. This metal is less "quarrelsome" than other metals, with itself or with others. The number of feet pulled

PUNCHES, DIES, AND TOOLS. 219

through a 1¼-inch die varies largely, but 4,000 feet may be taken as the average without sharpening.

Fig. 373 shows a die for drawing "flats," or rectangular sections. The frame pieces $A\ B\ C$ are bolted together as shown, A and C having holes at their centres for the bar to pass through. The die proper is formed of the solid cylinders $a\ b\ c\ d$, all having a common diameter of 3¹⁵⁄₁₆ inches, and the bases or ends of $b\ c\ d$ are bored out at right angles to their axes, to a radius of 1¹⁵⁄₁₆ inches, so that the base of one cylinder will fit the cylindrical part of another, and when arranged as in the figure will make a bar $2 \times \frac{1}{2}$ inches. The length of cylinder c determines the thickness of the bar; the length of the others is immaterial. It will

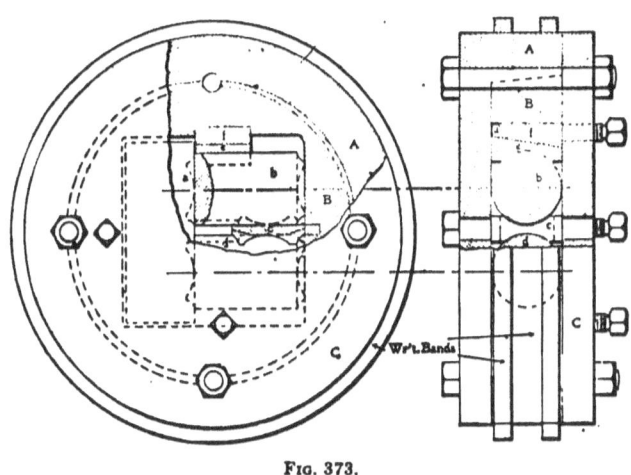

FIG. 373.

be noticed that cylinder a may easily have twenty or more wearing places, while $b\ d$ each have sixteen, and c eight wearing spots without grinding. When the die gets dull on one side, it is faced about, and the bar is drawn through from the other side. The die is adjusted with set-screws acting as wedges and "dutchmen" f and e, respectively, behind each cylinder, these not shown except at cylinder b. The work turned out of this die is first class, the surfaces are true and smooth, and the corners neat and sharp. It will be seen that by suitably grooving a and c cylinders, hexagon, octagon, or oval bars may be drawn through this die, but not with so many advantages as bars having a rectangular section. This tool might approximately be called a universal die.

The machine for drawing the bars is constructed of two horizontal columns made of five T-rails, each disposed to admit a carriage between them. One end of this double column receives the die and supports an idle sprocket wheel and shaft, while the driving sprocket wheel and shaft are located at the other end of the column. These sprocket wheels carry a sprocket chain made of links 6 x ¾ x 10-inch centres, 2-inch rivets. The carriage is provided with a hook to engage the chain to pull with, and lets go when there is no tension on the chain. The carriage is also provided with wedge dies, taper 1 in 6, which grip the reduced end of the bar and pull it through the drawing die and let go of it, owing to general shake and recoil, when the bar leaves the drawing die.

SECTION IX.

Pens, Pins, and Needles; Their Evolution and Manufacture.

PROCESSES FOR GOLD-PEN MAKING.

THE gold pen of to-day has been brought to its present degree of perfection by the American manufacturer, and the industry from its inception has been characterized by the use of American methods. For the production of the gold pen a high degree of skill is necessary, and only experts are employed in the different departments.

The gold used in making the pens is obtained from the United States Assay Office. It is then melted and alloyed to about 16 carats fine, and rolled into a long narrow ribbon from which pen blanks or flat plates in the shape of a pen, but considerably thicker than the finished pen, are cut by means of punch and die in a power press. The blunt nib of the blank is notched or recessed at the end to receive the iridium that forms the exceedingly hard point which all good gold pens possess. The iridium is coated with a cream of borax ground in water, and laid in the notch formed in the end of the blank. It is then secured by a process of sweating, which is nothing more or less than melting the gold of which the pen is formed so that it unites solidly with the iridium. The blank is then passed between rollers of peculiar form to give the gradually diminishing thickness from the point backward. The rolls have a small cavity in which the extreme end of the iridium-pointed nib is placed, to prevent injury to the iridium. After rolling, the nib of every pen is stiffened and rendered spongy by hammering. This is the most important process in the manufacture of the pens, as the elasticity of the nib depends entirely upon this operation. The pen is then trimmed by a press similar to that which is used for cutting out the blanks, or by automatic machinery. When the blank has been trimmed; the name of the manufacturer and the number of the pen is stamped on it. The pen is then given its convex surface in a suitable press, the blank being pressed between a concave die beneath and a convex punch above. Quite a little force is necessary to bring the pen to the required convexity, and when this operation is completed, two jaws approach the blank and press it up on the opposite edges, thus giving the pen its final shape. The next step is to cut the iridium into two points by holding it on the edge of a very thin copper disk, which is charged with fine emery and oil and revolves at a high speed. The nib is then slit by a machine and the slit cleared by means of a fine circular saw. After slitting, the nibs are brought together by hammering, and the pen is burnished on the inside in a concave form and on the outside in a convex form. This is necessary in order to give the pen a uniform surface and greater elasticity. These nibs are then set by the fingers alone, after which operation the pen is ground by a lathe with a thin steel disk and a copper cylinder, both charged with fine emery and oil. After the grinding is done the pen is polished upon a buff wheel, which completes the process of manufacture. Before the pen is placed upon the market, however, it is given a thorough inspection to see that it possesses the proper elasticity, fineness, and weight, then passed to an inspector who tests it and weighs it.

MANUFACTURE OF STEEL PENS.

The real inventor of the steel pen is unknown to posterity. France, England, and the United States each have claimants for the honor, and it is difficult to decide to whom it belongs. Arnoux, a French mechanic, made metallic pens with side slits in 1750. Samuel Harrison, an Englishman, made a steel pen for Dr. Priestly in 1780. Peregrine Williamson, a native of New York, while engaged as a jeweler in Baltimore, made steel pens in that city in 1800. He met with signal success and produced a very good article.

The first manufacture of steel pens by mechanical appliances was in England during the third decade of the nineteenth century, and the names associated with it were John Mitchell,

PUNCHES, DIES, AND TOOLS.

Joseph Gillot, and Josiah Mason, each doing something toward perfecting the processes of manufacture by mechanical means. At the period when these men commenced operations the pens in use were very crude specimens, made from a piece of steel formed into a tube, and filed into the shape of a pen, by hand, the joint of the two edges forming the slit. By degrees a press was contrived to do the cutting, bending, and marking; and machinery was devised for cleaning and polishing. Experiments were made with the object of securing the best possible quality of steel, and by the year 1860, when the manufacture of steel pens was first begun in this country, the article had been brought to a considerable degree of perfection.

The pens in use half a century ago were mostly fine-pointed, and while they gave satisfaction in certain lines of penmanship, some objections were made to them for business and rapid writing. Since that time there has been a gradual improvement in the material used and the process of manufacture, and the fine-pointed pens have given way to some extent to the stub and other blunt-pointed pens.

The many prefixes, such as Peruvian, Damascus, Amalgam, and Silver, used to describe the pen, are but fancy names and do not indicate the quality of the article. The material used for all kinds is cast steel of the best quality, imported from England or Sweden. The very best variety is that made from Swedish iron, which has in its granular structure a peculiar density and compactness not found in the iron ores of this country. This is the reason for the failure up to the present of home manufacturers being able to successfully produce steel for pens.

PRELIMINARY PROCESSES.

The steel used in the industry is received in sheets varying in length, width, and thickness. These sheets are cut by the manufacturer into strips of convenient width, and are packed in an oblong iron box, which is placed with the open top downward in another box of the same material, the interstices being then filled up with a composition in order to exclude the air. The boxes are placed in a furnace, gradually heated until they are a dull red in appearance, and then gradually cooled. In this process the strips become covered with bits of small scale. To remove this roughness they are immersed in a bath of diluted sulphuric acid, which loosens the scales, and they are then placed in wooden barrels containing water and broken pebbles. These barrels are revolved until the whole of the scaly substance is removed and the strips are of a silver-gray appearance. The strips are then taken to the rolling mill, where they are passed between successive rolls until reduced to the required gauge, the more common thickness being the one-hundred-and-sixtieth part of an inch. This operation requires considerable care and skill, as the variation of one-thousandth part of an inch in the thickness of the strip would seriously affect the flexibility of the pens. The strips are now three times their original length and have a bright surface.

FORMING AND SHAPING THE PEN.

In these preliminary processes the labor is performed by men and boys, but the process of forming and shaping the pen, that begin at this point, are carried on by women and girls, who are more adapted to the work. The cutting of the blanks is accomplished by a blanking punch and die of the usual construction. This die is set in a bolster and is perforated by a hole the exact shape of the blank; and a punch, also the exact shape of the blank, is attached to the face of the press ram. The operator with her left hand introduces one of the strips of steel at the back of the press and pulls the handle toward her with the right hand. This causes the ram to descend, driving the punch into the die, thus perforating the strip of steel with a scissor-like cut, and making a blank which falls through the opening in the die into a drawer below. The operator then pulls the strip of steel toward her until it is stopped by a small projecting stop-pin in the face of the die, and the operation just described is repeated, and again repeated until the whole of one side of the strip is perforated, when the strip is reversed and the other side treated in a similar manner. In the operation of cutting, a small V-shaped indentation is formed in the blanks upon the upper edge

PUNCHES, DIES, AND TOOLS. 225

of that part inserted in the holder, which may be found upon careful examination, and which plays an important part in the succeeding processes, as it enables the operator to distinguish between the smooth and rough sides of the blank.

The next process, called marking, is done by a stamp. The precise mark required is put upon a piece of steel, which is placed in a hammer of the stamp. The stamp is operated by foot. The operator takes a handful of blanks with her left hand, and by a dexterous motion makes a little train of them between the thumb and forefinger, presenting the first in the most ready position to be passed to the other hand. By the right hand the blank is placed, with the point toward the worker, in a guide upon the bed of the stamp, where the hammer falls upon it and makes the impression of the name cut upon the punch. So skilful are the operators in this process that they can stamp between 30,000 and 35,000 pens a day. Should the impression be made unusually large, the marking process is deferred until later in the course of construction.

The next process in the manufacture is "piercing," which produces the elasticity desired and causes the ink to attach itself to the pen. A piercing punch and die are located in a screw press, and an ingenious arrangement of guides is fastened thereto. The operator then places the blank in its proper position and so manipulates the press as to cause the ram to descend, driving the punch into the die. In order to soften the blanks, so that they can be properly shaped, they are put through a process of annealing. In this process the blanks are freed from the dust and grease that have become attached to them, and are carefully placed in round iron pots, which are again enclosed in large ones, covered over with charcoal dust to prevent the entrance of gases and put into the furnace, where they are heated to a dull red and then allowed to cool off gradually.

After this process is concluded the blanks are soft and pliable and readily assume the various shapes into which they are made by the next process, called "raising." In this operation a punch and die are again brought into use. The punch is fitted into a contrivance fixed in the bottom of the press ram; the die is placed in a bolster, a cylindrical piece of steel attached to the bottom of the press, with a groove cut for the reception of the die. Four pieces of steel called guides are fixed to the bolster in such a position that the operator is enabled to slide the blank into the die, where it is held by the guides until the punch descends, forcing the blank into the die, and giving the pen its shape. The blanks are then placed in thin layers in round pans with lids and go through the process of hardening.

HARDENING THE PENS.

In the operation of hardening the pens, the pans in which they are placed are put in the furnace for a period varying from twenty to thirty minutes, at the end of which time they come to a bright red heat. The pans are then taken from the furnace and their contents thrown into a large bucket immersed in a tank of oil. This bucket is perforated, and when lifted from the tank retains the hardened pens while the oil drains off into the tank again. The pens are then placed in a perforated cylinder, which is set in motion and drains off the remainder of the oil. At this stage of the manufacture the pens are very greasy and as brittle as glass. To remove the grease adhering to them they are again placed in perforated buckets and immersed in a tank of boiling soda water.

TEMPERING THE PENS.

After the thorough cleaning the pens are put into an iron cylinder, which is kept revolving over a charcoal fire until they are tempered to the degree required. This process is regulated according to the color shown by the pens, which indicates the varying temperature of the metal. After this operation the pens are black and rough at the point. To remedy these defects the pens are subjected to the process known as "scouring," which consists in dipping the pens in a bath of diluted sulphuric acid, which removes all extraneous substances acquired in the hardening and tempering processes. Great care is exercised in this operation, as the acid is likely to injure the steel. The pens are then placed in iron barrels with a quantity of water and a mate-

PUNCHES, DIES, AND TOOLS.

rial composed of annealing pots broken and ground fine enough to pass through a fine riddle. The barrels are set in motion, which is continued for a period varying from five to eight hours. At the end of this time the pens are placed in barrels with dry pot for about the same period, after which they are put into other barrels, together with a quantity of dry sawdust. They are then ground between the centre piece and the point. This is done by girls with the aid of a "bob," or "glazer," a circular piece of alder wood about 10½ inches in diameter and ½ inch in width. Around this a piece of leather is stretched and dressed with emery. A spindle is driven through the centre and the two ends are supported in bearings. The mechanism thus arranged is set in motion by means of a leather belt, and the operator, holding a pen firmly, grinds off, with a light touch, a portion of the surface.

SLITTING THE PENS.

The last and by far the most important operation performed by mechanical means in the manufacture of steel pens is that of "slitting." The tools used for this purpose are two oblong pieces of steel called cutters, which are about 1½ inches long, ⅜ inch thick, and 1¼ inches wide. The edges of these cutters are equal in delicacy to the cutting edges of a razor. One of the cutters is fixed in a press with a pair of guides screwed on either side, and the other cutter is held by a bolster, having attached to it a small tool called a rest or table. The operator places the pen upon the table, pushes the point up toward the guide, and, by operating the machine, makes the upper cutter descend and meet the lower one, thus slitting the pen.

At this stage of the process of manufacture the outer edge of each point is smooth, but the inside edges are sharp and rough. To remedy this defect the pens are again put into iron barrels with pounded pot, and kept revolving for five or six hours, when they are removed and polished in sawdust. The pens are then colored by being placed in a copper or iron cylinder which revolves over a coke fire until the requisite tint is obtained. If the pens are to be lacquered they are placed in a solution of shellac dissolved in alcohol. This solution is afterward drained off and the pens are placed in iron cylinders that are kept revolving until the pens are dry. The pens are then scattered upon iron trays and heated in an oven until the lacquer is diffused equally over the whole surface of the pens. The lacquer gives the pens a glossy appearance and prevents rust; and when the pens have cooled they are complete as far as manufacturing processes are concerned. Before they are offered to the public, however, they are given a very careful inspection, to see that no inferior ones are put on the market.

PROCESSES OF PIN MANUFACTURE.

The old process of pin manufacture by manual labor was very slow and tedious, since each pin passed through the hands of fourteen to eighteen individuals. The modern pin is made in the United States by the improved automatic Atwood and Fowler machines. The process of pin manufacture by modern machines may be briefly described as follows: Coils of wire are placed upon a reel, whence the wire is drawn automatically by a pair of pincers between fixed studs that straighten it. A pin length is then seized by a pair of lateral jaws, from which a portion of the wire is left projecting, when a snaphead die advances and partially shapes the head. The blank is then released and pushed forward about one-twentieth of an inch, when the head is given another squeeze by the same die. By this repetition of the motion the head is completed and the blank is cut off the wire in the length desired. About one-eighth of an inch of wire is required to make the pin head. If the attempt were made to upset this head in one operation the wire would be more likely to double up than to thicken as desired.

These headed blanks then drop into a receptacle and arrange themselves in the line of a slot formed by two inclined and bevel-edged bars. The opening between the bars is just large enough to permit the shank of the pin to fall through, so that the pins are suspended in a row along the slot. When the blanks reach the lower end of the inclined bar in their suspended position they are seized between two parts of the machine and passed along, rotating as they move, in front of a cylindrical

PUNCHES, DIES, AND TOOLS.

cutter, with sharp teeth on its surface, that points the pins. They are then thrown from the machine properly shaped, and if they are brass pins they are cleaned by being boiled in weak, sour beer. After they are cleaned they are coated with tin. This is done by placing alternate layers of pins and grain tin in a copper pan and adding water along with bitartrate of potash. Heat applied to this produces a solution of tin which is deposited on the surface of the pins. The pins are then taken from this solution and brightened by being shaken in a revolving barrel of bran or sawdust. Lastly, the operation of "papering" takes place. This process is performed now by an automatic papering machine something in the following manner. The pins to be stuck are placed in a hopper, in connection with which a steel plate is used, with longitudinal slits in it corresponding to the number of pins which form a row in the paper. The pins in the hopper are stirred up by a comb-like tool, the shanks drop through slits in the steel plate, and the pins are suspended by their heads. Long narrow sheets of paper are presented by the operator to the action of the machine, by which two raised folds are crimped, and the row of pins collected in the slit steel plate is then, by being subjected to the same action, pressed through the two crimped folds. These operations are repeated until the requisite rows of pins are stuck in each paper.

THE MAKING OF NEEDLES.

The making of needles was without doubt one of the first arts practised by man, and dates back to the remote period when man first evidenced a desire to cover his form with clothing. Remains of ancient civilized and uncivilized nations bear evidence of the use of the needle made of various materials. Excellent specimens made of fishbone, horse's bone, and bronze have been found in the caves near Brunequel, France, and on the sites of ancient lake dwellings of Central Europe. In Egyptian and Scandinavian tombs bronze needles have been found varying from $2\frac{1}{2}$ to 8 inches in length.

The first introduction that Europe had to the steel needle came with the Moors at the time of the Saracen invasion, but it is not certain that these people were the inventors, because the Chinese lay claim to having used steel needles from time immemorial.

In its original shape the needle was in appearance like the modern shoemaker's awl, which was used for merely perforating materials meant to be fastened together along their edges, so that they could be laced together by hand. As the use of this needle involved two operations, it was soon discarded for a needle with a circular depression near the blunt end for holding the thread, and thus did away with the lacing operation. Since this needle, though it did well enough for coarse work, was inadequate for finer work, the needle with the eye was introduced.

Since the introduction of the steel needle the model has remained the same, and progress in the art of needle-making has been confined to devices for perfecting the material used and the methods of construction. In the early days of needle manufacture, when the trade was practised at home or in small shops, the materials and devices used were very crude. After the manufacture of the needle was started in plants provided with conveniences and facilities for its production, improvements were slowly introduced in performing the different operations.

The most notable improvements prior to 1870 may be summarized as follows: Drill-eyed needles were first made in 1826 and were followed two years later by the burnishing machine, by means of which the eye secures its beautiful finish. In 1840 the process of hardening in oil succeeded the former method of hardening in water, in which a large proportion of the needles became crooked, so that their straightening involved considerable time and expense. The stamp to impress the print and groove and the press with a punch to pierce the eye, though suggested as early as 1800, were not in general use until 1830, and by 1886 were superseded by an automatic machine. In 1839 a simple method was invented by a Mr. Morrell for polishing many thousands of needles simultaneously, and in 1869 a machine was brought out by a Mr. Lake for doing many of the operations previously performed by hand. The more recent improvements have been made in devices for heating and ventilating, and for getting rid of injurious dust which rises from the emery wheels in the grinding process.

The process of manufacturing the common hand-sewing needle, as carried on in Germany, France, and England, differs from that in use in the United States.

Needle-making as an industry was put on a permanent basis in the United States shortly after 1852, when the peculiar kind of needles used in machinery was introduced. As the sewing-machine is essentially an American invention, and the most important feature of the invention of the machine was the needle constructed by Elias Howe for the making of the lock stitch, it was very natural that this part of the sewing-machine should be manufactured in this country. It is estimated that from 6 to 8 per cent of all the operative labor involved in the construction of the sewing-machine is employed in making the needle. With the successful manufacture of the different varieties of sewing-machine needles began the manufacture of needles for knitting-machines. As the demand for sewing- and knitting-machines increased there was a corresponding demand for the needles used in these machines, and the industry developed rapidly.

The needles made are of various lengths and patterns to suit the requirements of the different sewing-machines. Besides those differing generically, such as straight and curved, or specifically, such as long, short, round-pointed, and chisel-pointed, there are many peculiar patented needles for use in particular sewing-machines. Among the endless varieties of sewing-machine needles the most prominent is the common needle used in the household sewing-machine. This needle has the eye at the pointed end, with a long groove on one side and a short groove on the opposite side, and is used in connection with a shuttle or other device for carrying a second thread, which is passed through a loop of the thread in the needle, thus forming the double lock stitch. The purpose of the grooves is to protect the thread from wearing or tearing during the operation of the machine.

In addition to the common household sewing-machine needles there are needles for use in sewing leather, including many varieties to suit the various machines. Some of these needles, in distinction from the common sewing-machine needles, have a hook instead of an eye. The material to be sewed is perforated with an awl, and the thread is then pulled through by the hook. In most leather sewing-machines, however, the needle itself perforates the material and pulls the thread through. In sewing cloth only the needle with a round point is used; but for sewing leather there are points of various shapes, known as twist, reverse twist, wedge, cross, chisel, reverse chisel, and diamond. A very interesting needle, used in the manufacture of boots and shoes, is that of the Goodyear welting machine. This needle is a segment of a circle in shape and puts the welts upon boots and shoes with remarkable alacrity and accuracy.

MAKING STEEL SPRING AND LATCH NEEDLES

The steel spring and latch needles used in making hosiery and stockinet work are extensively manufactured in the United States. The former is constructed by reducing the working end on a taper to an approximate point, and then bending the reduced portion over upon itself so as to form an open loop, a groove having been previously made in the needle so as to come opposite the point. In the operation of the needle the point stands out at the proper time for the yarn to be taken, which is carried through to form the stitch. As the forward motion continues the point is depressed into the groove by coming in contact with the mechanism arranged for the purpose, and thus the passage through the loop is secured without catching. The latch needle has, instead of the spring barb, a short rigid hook, which is formed by tapering the working end to an approximate point and bending it in combination with the latch. The latch is contained in a groove milled in the body of the needle and is pivoted upon a rivet which passes through the wall of the groove. As the latch, the walls between which it is riveted, and the diameter of the rivet are extremely delicate, each part being but one-hundredth part of an inch thick, great care and skill must necessarily be exercised in manufacturing the needle. The purpose of the latch is to aid in forming and casting off the stitch by preventing the yarn from being caught under the hook except at the proper time.

When the sewing-machine needle was first made here the processes of manufacture were similar to those employed in England in making the common hand-sewing needle, and required a great deal of manual labor. The reducing of the shank to the required size and putting in of the grooves on the sides of the needle were accomplished by stamping between dies. By this method the superabundant material was thrown out at each side as a fin, cut off by hand-shears, and later removed by means of a die and punch in a press in a trimming operation, after which the needles were rounded up and pointed by filing. Gradually these operations were replaced by rolling, grinding, turning, and milling, and finally machinery was invented to do the work entirely automatically.

MANUFACTURE OF THE SEWING-MACHINE NEEDLE.

In the course of the manufacture of the sewing-machine needle it passes through the following states: Blank, reduced blank, reduced and pointed blank, grooved, eye punched, hardened and tempered, hand burr dressed, brass brushed, eye polished, first inspection, hand straightened, finish pointed, and finished. There are two methods in use for the manufacture of the modern sewing-machine needle. In most respects the processes are similar, but they differ in the manner of forming the blade. In one method the blade is formed by cutting the blank down to its required size, and in the other method the wire is cut into short pieces about one-third the required length of the needle when finished, and then by a process known as cold-swaging these are brought to the proper length.

As the modern machinery used in the first process mentioned is largely of private designs, the manufacture cannot be described in detail, but it may be fairly inferred from the following method used a few years ago:

METHODS OF PRODUCTION.

At that time the needle was made from the best quality of crucible steel wire, which was received in coils, and after being straightened by means of an automatic machine was fed into a machine devised for forming the large head or end of the needle, and cut off into blanks of the required length. The blanks were then sent to machines, three in number, for roughing, dressing, and smoothing. The first two worked with coarse and fine emery wheels, respectively, and the third with an emery belt. Into these machines the blanks were fed from a hopper on to a grooved endless travelling carrier, which exposed to the action of the emery wheel that portion of the blank which was to be reduced in diameter from the shank of the needle. The portion not reduced was that designed to be placed in the end of the needle bar of the sewing-machine. As the needles passed the emery wheel they were rotated by a pair of reciprocating plates, so that they were equally ground on all sides. After the process was completed by the emery belt in the third machine, the needles were passed to a machine where the two grooves on the sides of the needle were made by two circular saws past which the blank was fed automatically. The saws were pressed in against the needles and then withdrawn at such times as would give the required depth and contour to the groove. The eye was then punched by a belt-driven punching machine, after which the needles were heated to a cherry-red in a reverberatory furnace with a charcoal fire, taken out and immersed in whale oil. They were then placed in sheet-iron pans suspended from arms of a revolving shaft, and tempered in an oven heated by the surplus heat from the furnace. Next, the needles were cleaned on an emery cloth, being held in bunches of about twenty between the finger and thumb and rotated while being pressed upon the cloth. They were then taken, with the grooves upward, by a flat-jawed tongs carrying seventy at a time, and held against a scratch brush of brass wire, which revolved 8,000 times a minute, to polish the grooves. The brush of brass wire was soon replaced with a bristle brush, which finished the polishing of the grooves. While yet held in the clamps these nee-

PUNCHES, DIES, AND TOOLS. 235

dles were threaded in gangs on cotton thread, which was covered with oil and emery, and then drawn back and forth in various slanting positions so that the polishing powder would act on all parts of the eye. When removed from the thread the needles were cleaned by a revolving hair-brush, and the eyes, points, and blades inspected. Imperfect ones were thrown aside and the good ones sent to the hand straightener, who rolled them on a surfacing anvil at the level with his eye, who detected any curvature and corrected it by a tap of a small hammer. The final operations were finish pointing, which was done on a fine emery wheel, and finish polishing, done by a revolving hair-brush charged with crocus and alcohol.

COLD-SWAGING PROCESS OF POINTING NEEDLES.

In the second method of manufacture the wire is fed into a machine called a straightener and cutter, which straightens the wire and cuts it off into blanks about one-third the length required for the finished needle. The blanks are then placed in a small iron cylinder rotated in such a manner as to keep the blanks in constant friction, and thus remove burrs, scale, and dirt. They are then ready for the cold-swaging machine. The blanks are placed in a hopper from which they are taken automatically, one at a time, and their ends are presented to the action of a set or revolving sectional steel dies. By the constant opening and closing of these dies while in rotation the ends of the blanks are compressed and drawn out to form the blades. After swaging the blank is stamped in order to identify it. In the process of swaging there results a slight variation in the length of the needles, and they are trimmed by the clipping and straightening machines. The prominent feature of these machines is the arrangement of the screw-feed for simultaneously carrying the needles across, so that the ends of the shanks are aligned against a fence, and forward, so that the points are presented to a cutter which trims all to a uniform size or length. After passing the cutter each needle is struck by a die that stamps upon the shank the descriptive number. The other processes involved in this method of needle manufacture are similar to those described in the first method.

Since the invention of these automatic machines for the different processes, the mechanism employed has been so combined as to effect a transfer of the blank from one operation to the next without the intervention of hand labor. In such combination of machinery there has been marked development during the past fifteen years, and the industry has fully kept up with the progress of the other wire-working processes.

SECTION X.

Punches, Dies, and Processes for Making Hydraulic Packing Leathers, Together with Tools for Paint and Chemical Tablets.

EFFICIENCY OF LEATHER PACKING.

WHERE high-pressure water is employed for power transmission, one of the most important and expensive items is the packing of rams, etc., so that they may slide backward and forward and yet be water-tight. The higher the pressure the more important this question becomes; in fact, in many cases the limit to the pressure depends on the efficiency of the packing. There are many kinds of packing in use, each having his own special points, but there are many places in which the well-known self-acting leather packing is by far the best, if, indeed, it is not indispensable. Its chief disadvantage is, however, its initial cost, which, by the side of other packings, is high. Nowadays, these leathers can be bought ready made, but there are many firms possessing hydraulic installations where the engineer in charge is expected to make his own leathers, and, where this is possible, it is the better and cheaper method.

THREE KINDS OF HYDRAULIC LEATHERS.

In the following, descriptions are given of the apparatus needed, the method of making, and a few hints for the successful working of the three kinds of self-acting leather packing; for, although apparently a simple operation, the making of good hydraulic leather needs considerable practice and care. The three kinds of leathers are shown in Fig. 375.

To all who have studied hydraulic machinery it is well known that the depth D of the leather has practically no effect on its water-tightness and friction, as the joint is made at one place only in its depth, *i.e.*, the point *A* in each case. This fact is clearly shown in Fig. 374, which shows sections of the three kinds of leathers which are worn out. As will be noticed later, the part *A* which shows most wear is also the part which receives most stress during manufacture. Also the deeper the leather the greater the stress at this point, and the more care needed in the manufacture to retain an even thickness throughout the finished leather, so that it is evidently advisable, and

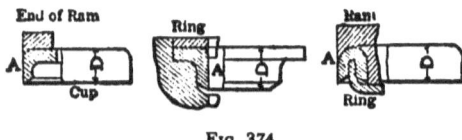

Fig. 374.

certainly cheaper, to keep the depth of the leather as small as possible.

A leather should be exactly made to fit the recess made for it, except in the depth, and, where possible, should be fitted with a guard ring for it to bed against, rounded out to exactly the same curve as the leather, Fig. 374.

In the case of the U leather the guard ring is very neces-

Fig. 375.

sary, and should have holes drilled in it to allow the water to get to the leather. The distorted shape of the U leather shown in Fig. 375 is due to the fact that no guard ring is fitted. Attention to these details results in the leather keeping its shape much longer than it would otherwise, and consequently it will last longer; also, the leather opens out more quickly when the pressure is first put on.

PUNCHES, DIES, AND TOOLS.

THE SELECTION AND PREPARATION OF LEATHER FOR WORKING UP.

Leather specially prepared for the purpose is sold by curriers. It is softer than ordinary leather, but should be of the very best quality; a piece from "middles" is best. The procedure is then as follows: Select a piece of sound leather the required diameter, free from "knots"; look especially for "pin pricks," which open out and show better when the leather is wet and stretched. If one of these fine holes is found, lay the piece aside for a smaller size, because the hole is sure to open out during manufacture, and when finished the leather will be porous. When a sound piece has been selected, pare off with a knife all soft part on the flesh side, and make the leather blank and even thickness all over, and from $\frac{1}{16}$ to $\frac{1}{8}$ inch according to the size of the leather. Be careful to measure from the moulds *exactly* what the thickness is, as great trouble is experienced in "blocking" the leather, if too thick. Now soak the blank well in water, working it between the thumb and fingers until soft and pliable all over; rub tallow well into it, and it is then ready for the moulds. Great diversity of opinion exists as to which is the better side of the hide for wearing purposes, whether the smooth hair side or the rough flesh side. The hair side is closer grained, and therefore more waterproof than the flesh side. Thus, if the hair side be put inside, and the flesh side for the wearing side, the leather will be waterproof almost as long as it lasts. On the other hand, if the hair side is but for the wearing surface, the leather will wear better and has a much neater appearance. It is always best, however, to put the smooth side inside, and let the rough side take the wear.

MAKING OF CUP LEATHERS.

These are not often used for rams exceeding three inches diameter, and should vary in depth from $\frac{3}{8}$ inch for 1 inch diameter and less, up to about $\frac{5}{8}$ inch for 3-inch leather. Place the prepared blank smooth side upward, centrally on the die, and punch centrally above, as shown in Fig. 376. Then force the punch in, gradually and at intervals, until the leather is just level with the bottom of the die; trim off the top edge so that it is just level with the die top. Now force the leather a bit fur-

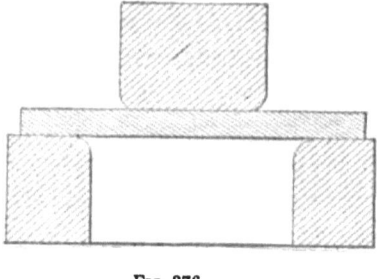

Fig. 376.

ther through the die, and insert ring, as in Fig. 378. This saves the ring from spreading the leather at the round *A*, Fig. 378. Put the plate on the bottom and force the ring a little way in.

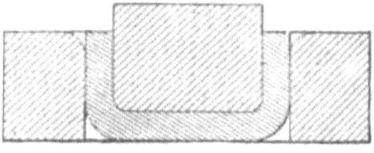

Fig. 377

Draw or knock the die down on the plate, and clamp it in this position (Fig. 378). Remove the clamped mould and leather and put them in some warm place, such as a boiler-room or the

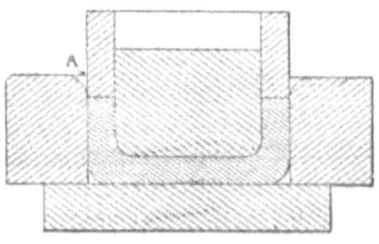

Fig. 378.

top of a steam-engine cylinder for small sizes, and leave to dry. This usually takes from a few hours to a day, according to the size of the leather. Take care that the place is not too hot, or it will seriously deteriorate the leather. When taken from the mould the leather should have set quite hard; in fact, the better the leather the harder will it set. The bevel edge can now be put on with an ordinary knife, and smoothed round with fine glass paper. The leather is then finished. This bevel should be at about 30°, and it is very important that it should continue right up to the edge of the leather, leaving a sharp edge all the way round. Table 1 gives the diameter of blanks from which to block the leather.

TABLE 1—*Blanks for Cup Leathers.*

	In.	In.	In.	In.	In.	In.	In.
Outside diameter of finished leathers..........	3	$2\frac{3}{4}$	2	$1\frac{1}{2}$	$1\frac{1}{4}$	1	$\frac{3}{4}$
Diameter of blank.......	$4\frac{1}{4}$	$3\frac{7}{8}$	$3\frac{1}{8}$	$2\frac{1}{2}$	$2\frac{1}{8}$	$1\frac{7}{8}$	$1\frac{3}{8}$

MAKING OF HAT LEATHERS

Like cup leathers, these seldom are used for rams exceeding 3 inches in diameter. The total depth should vary from $\frac{1}{2}$ to $\frac{3}{4}$ inch. The various stages in the manufacture are as follows: First, with the rough side upward (Fig. 379), force into the dies

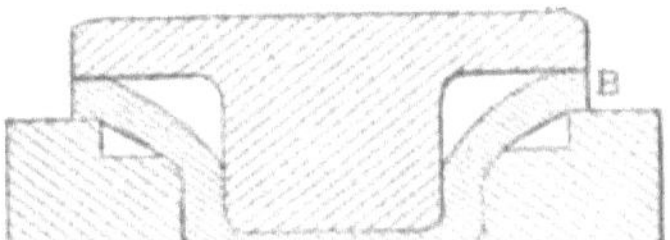

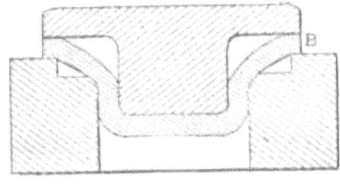

FIG. 379.

as much as possible of the part *B*; the remainder may be sheared off by the sharp edge. Insert the ring as in Fig. 380 and punch the bottom out. Force the ring in a bit, and clamp in this position (Fig. 381). It can then be dried and finished off the same as the cup leather. The same ring that was used

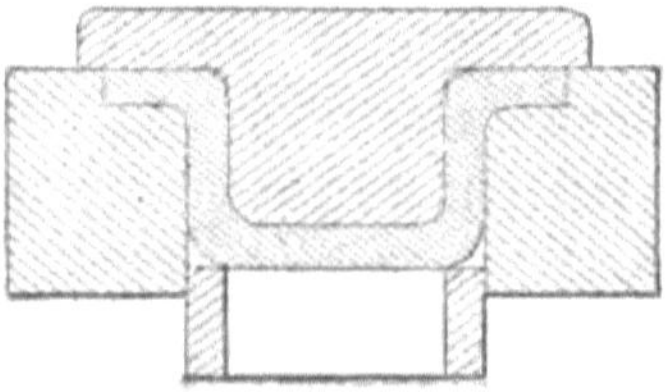

FIG. 380.

for the cup leather will answer. Table 2 gives the diameter of the blanks from which to make the leathers.

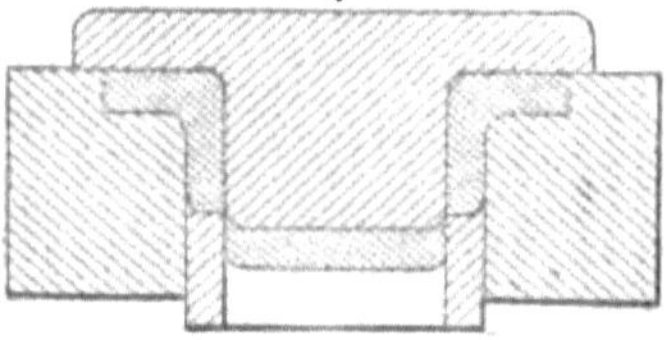

FIG. 381.

TABLE 2—Blanks for Hat Leathers.

	In.	In.	In.	In.	In.	In.	In.	In.
Inside diameter of finished leathers..	2 3/4	2 1/4	1 3/4	1 1/2	1	1	3/4	1/2
Outside diameter of finished leathers..	3 3/4	3 5/8	2 7/8	2 5/8	2 1/8	1 5/8	1 9/16	1 9/16
Diameter of blank..	5	4 5/8	4	3 5/8	3	2 3/8	2 1/8	2

MAKING OF U-LEATHERS.

This is by far the most common form of leather. It is superior as a packing to the cup or hat leather, and also has less friction, being more supple; but it is more expensive and needs more care in its manufacture, and hence, where possible, the hat or cup leather is used. The U-leather can be used for any

PUNCHES, DIES, AND TOOLS.

size ram from 1 inch diameter upward. The depth of U-leathers is as follows:

For Rams up to 1.......... ½ inch diameter, ½ inch deep.
For Rams above 1.......... ½ inch and up to 3 inch diameter, ⅝ inch deep.
For Rams above 3.......... inch and up to 20 inch diameter, ¾ inch deep.
For Rams above 20......... inch from 1 to 1, ½ inch deep.

The manufacture is explained by the sketches, Fig. 382, showing the first position, with the smooth side of the leather inside. Leave it in this position from about half an hour to a few hours, according to the size of the leather; then invert, put in the cen-

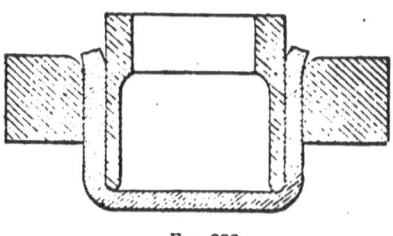

FIG. 382.

tre block, and force the middle portion out again, as shown in Fig. 383. Let it dry in this position. Then take it out of the mould and scribe a line on the side all way round, with a scribing block, showing the depth of the leather (Fig. 384). Now

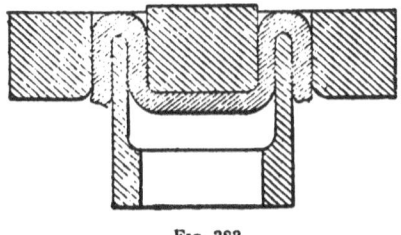

FIG. 383.

fix in a vise, and, with an ordinary wood saw, saw off by this mark, and finish flat by rubbing on a piece of glass paper laid on a flat surface. The bevel edge can now be put on and finished off with sandpaper as before. The following plan is recommended if a large number are to be made. When the leather has dried, take it out of the mould and fasten it to a wooden face-plate of a small lathe. The required depth is then soon cut off with a wood-turning chisel, and the bevel put on and

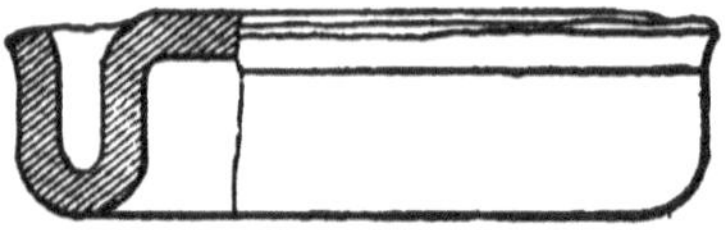

FIG. 384.

polished. The burnished appearance which some makers' leathers have is put on in the lathe by means of a hardwood stick and soft soap; but, of course, this only improves the appearance and not the quality of the packing. Table 3 gives the diameter of the blanks from which to make the U-leathers.

	In.	In.	In.	In.	In.	In.	In.	In.
Inside diameter of finished leathers..	18	14½	11	8½	8	6	4½	3¾
Outside diameter of finished leathers.	19¾	16	12½	10	9½	7½	6	4¾
Diameter of blank..	25¾	20	16½	14	13½	11	9¼	7½

	In.	In.	In.	In.
Inside diameter of finished leathers...........	2¾	2	1½	1
Outside diameter of finished leathers.........	3¾	3⅛	2½	2
Diameter of blank.........................	6½	5¼	4⅛	3½

PRACTICE TO ADOPT FOR WORKING HYDRAULIC LEATHERS.

In all operations great care should be taken that all dies and blanks are central, so that no part is unduly stretched. Always use a disk of leather, and never attempt to block a leather from a washer, because the middle cut out of the finished leather can always be used for smaller sizes, so that no leather need be wasted. With regard to exerting the pressure, the practice of cutting a hole in the centre of the blank and then drawing the punch down with one central nut and bolt is not recommended, as the leather invariably draws away from the hole, and in many cases tears; and, again, the hole is cut often bigger than that needed in the finished article. If a great many leathers are to be made, a small hydraulic press is very useful, but usually the quantity to be made does not warrant such an expenditure. For the smaller sizes up to about 2 inches the ordinary bench vise is all that is needed. For the larger sizes a bench fitted with two slots (at right angles to each other) to take 1-inch tee-headed bolts are all that are needed; the clamps, of course, are used in the usual way. This simple tackle has been used for U-leathers 26 inches in diameter, and $1\frac{3}{4}$ inches deep.

RE-LEATHERING A MACHINE.

Always keep the full number of spare leathers in stock, and as soon as one is used, make another to replace it. This is economy, because a leather when made up does not deteriorate like the raw hide. It also saves much inconvenience if a leather gives way unexpectedly. Before placing a leather in a machine, take the leather and rub it well with good dubbin of the best tallow, until it is soft; neatsfoot oil is very good, but it is expensive. If it refuses to soften and still remains hard, put it in warm water for a few minutes; but if this is done, it must be put into the machine immediately, or it will swell out of shape and have to go into the mould again. The same applies when cleaning a machine, and if a leather is taken out wet it should be at once put into a mould, for if it be allowed to dry out of the machine it will be found impossible to get it into the machine again. Always measure the depth of the recess which is to contain the leather, and also the depth of the leather, to see that there is plenty of play depthways, for if a leather is nipped in the depth it prevents it from opening out under the water pressure. Also be careful that the bevel is on. This seems a small point, but it is very important and often overlooked. The leather will not work well unless the bevel is on and correct; in machines in which leathers without the bevel have been put, much trouble and delay were caused. When the bevel was put on the machines worked properly.

The only difficulty experienced in inserting a new leather is in the case of a U-leather, when the mouth of the cylinder is not fitted with a gland, but has simply a recess turned in it. If the leather is of large diameter its own springiness allows it to be pushed in, if done gradually and with care. If of small diameter, as much as possible must be pushed into the recess, and then with a hammer shaft the remaining portion is knocked straight down. It will appear as in Fig. 385, and can then be easily

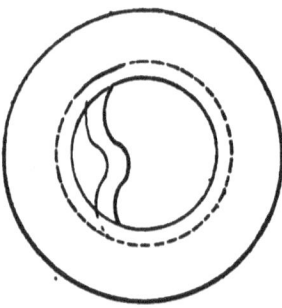

Fig. 385.

pressed into the recess; but the leather must not be too soft, or this method will make it very much out of shape.

When in use keep the ram bright and well lubricated. If possible, the part of the cylinder that the leather works over, or the part of the ram that works over the leather, should be sheathed with brass or, as in more modern practice, copper; the leather lasts much longer and needs less lubrication. When working against gray-iron and the lubrication is poor, the leather seems actually to tear away small portions of the iron from the surface. Even when in slow motion, when the pressure is high, as in a flanging or stamping press, this is quite apparent, for if a well-worn leather is removed and allowed to dry, small specks of iron, quite visible to the naked eye, are seen to be embedded

PUNCHES, DIES, AND TOOLS.

in the working face of the leather. Finally, always use water as clean and free from grit as possible.

MAKING PUNCHES AND DIES FOR CHEMICAL TABLETS.

The tools described and illustrated in the following were used in a multiple press, Richards patent, for making a tablet like Fig. 386, four at a time. Fig. 386 contained 40 grains of powder and was made, as shown, quasi-hexagonal in shape, two corners only being flat. This was done for three reasons: First, it had to pass through the neck of a bottle, which it could not

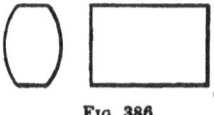

Fig. 386.

do if it were a perfect hexagon, the neck of the bottle not being large enough; second, to assist in releasing it from the top of the punch; third, to strengthen the punch, which was made like Fig. 387. The job required eight punches like Fig. 387, four of them longer than the other four; a die like Fig. 391; two punch-holders like Fig. 388, and a head ring like Fig. 389. The

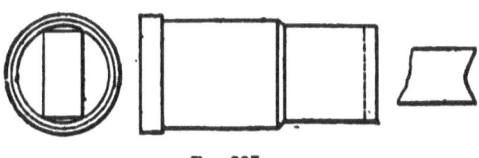

Fig. 387.

die had to be as hard as possible without cracking, and the punches hardened and drawn to a light straw, as the powdered medicine is very hard on dies and punches; they had to be a working fit and highly polished.

The first thing done was to hunt for a piece for a temporary jig. A piece was found like Fig. 390, which was then faced

248   PUNCHES, DIES, AND TOOLS.

off on both sides; then laid off for the tapped holes *a*, *b*, and *c*, and the casting clamped to the face-plate—using tapped holes *a* and *c*—and the circle, the diameter of the die laid out and bored to one inch in depth to fit the die; the balance of the hole was finished to fit the small diameter of the holder, Fig. 389. Next

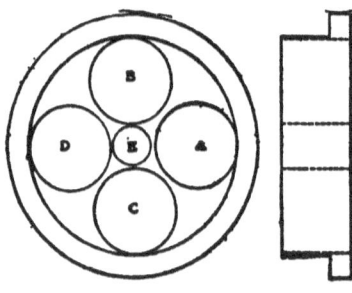

Fig. 388.

a circle, of as large diameter as possible, was put on the face-plate, and four equidistant points laid out on it; it was then put on the lathe, and the blanks for Figs. 388, 391, and 389 faced off on both sides, holding them tightly by screw *d*.

A ⅜-inch hole was then put in the centre of Figs. 388 and

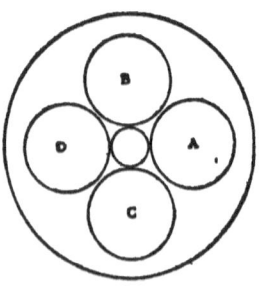

Fig. 389.

389. Fig. 389 was put in holder, Fig. 390, and a tram was made. After putting one end of the tram in a light punch mark previously put in the lathe bed and the other end in one of the four divisions in the face-plate, a line was drawn at the centre height of Fig. 389 with a pointed tool in the tool-holder; then

the face-plate was turned to the next division in the face-plate and another line scribed, until there were four lines on Fig. 389. The last line drawn from Fig. 389 to the holder, Fig. 390, was as shown at *X*, then the holder, Fig. 390, was moved 4$\tfrac{11}{16}$ inches, which was the centre distance between the holes *A*, *B*, *C*, *D*, from the centre *E* of the punch-holders, Fig. 388 and the

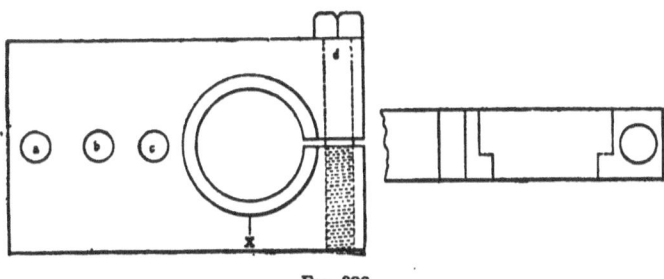

Fig. 390.

head ring, Fig. 389, using *a* and *b* tapped holes, and the first hole was put in and reamed. Then screw *d* was loosened and the head ring turned until the next line came on the line *X* and the next hole was put in, etc., until the entire number had been finished. Holders, Fig. 388, were then finished each with a hole,

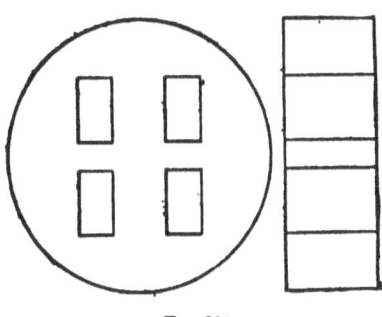

Fig. 391.

and a plug made to fit the hole in the two punch-holders and the head ring; a plug was also made to fit the $\tfrac{3}{8}$-inch hole in their centres. After this one of the punch-holders and the head ring were plugged together and put in the jig, Fig. 390, with one of the lines on Fig. 389 in line with the line *X* on the jig. The next hole was then bored in the punch-holder, screw *d* was loosened, the holder finished like the head ring. Next, each piece was put on a ⅜-inch mandrel and finished on the outside. The jig was removed from the lathe and the die, Fig. 391, put in one side and a punch-holder in the other and clamped together with *d*. A piece was then bored .368 and turned to fit the holes in the punch-holder and hardened a little. The die was then drilled with a .368 drill. Next, a broach like Fig. 392 was made, .006 smaller than the punches and a tit turned to fit the

FIG. 392.

hole in the die. Then the round end of the broach was put in the die and scribed for the punches, after which the die was drilled and the stock cut and filed so that the broach would enter. Then a small piece of steel was laid off and filed and broached to fit the broach, after which it was turned to fit the punch-holder. The die and punch-holder were then tightened together in the jig, the broach guide was put in the punch-holder, and the die was broached and filed a little at a time

PUNCHES, DIES, AND TOOLS. 251

until finished. A little trouble was experienced after hardening, but it did not amount to much.

The punches were cut off so that two could be made out of one piece, four sets, one large and one small, like Fig. 387. The punches were faced and turned straight $\frac{1}{64}$ over size, and then finish turned to micrometer measurement. The punches were then cut apart by milling. They were then fitted to the die and put in the punch-holder which fit in a nut and was clamped to the plunger. They were then put in the lathe and faced to the same length. Next, the bottom punches were taken and put in the head ring and punch-holder, and after getting them in the plunger the die was slipped on to keep the punches from turning. They were then faced. It was not possible to use the same method for facing the top punches because they did not come through the die. Now the top punches were all the same length, likewise the bottom punches. Next, an end mill was made the shape of the top and bottom of the tablet, and the end of the punches were milled, after which they were polished and tempered, and then polished again.

The job required nineteen days to complete.

THE MAKING OF PAINT TABLETS ON PUNCH PRESSES.

New fields are constantly being found for the use of the power press. Though originally designed for the working of sheet metal, they are now being used with great profit in a large number of other industries, one of the latest of which is the manufacture of water-color paints such as children use. The presses designed and used for this purpose are entirely automatic. Before describing the work and the manner of producing it, it will be well to go back a little to the preparation of the material which is to be worked up.

After the color has been thoroughly mixed and has reached the consistency of putty, it is taken from the vat and rolled out like piecrust until it is about a quarter of an inch thick. The material is then cut into long strips, about $1\frac{1}{2}$ inches wide, by means of a hand-roll like a common rolling-pin, but having thin, round cutters at stated intervals. The next operation is done with a similar roll but with cutters only about three-quarters of an inch apart. This roll is moved across the strips and makes small cakes about 1½ or ¾ inches. These cakes are allowed some time to dry.

Then comes the work of the power press. The semi-hard cakes are placed on the dial over the pockets, which are regularly spaced around the dial. The cakes are inclined to be a little irregular, and for this reason it is necessary to force them into the dial pockets; this is done by one of three punches, this one being located at the extreme left of the ram. The die is located below the dial directly under the forming punch in the middle of the ram which, passing through the pockets, pushes the cakes into the die and exerts enough pressure on them to *condense the moisture, densify* the material, give each cake a uniform appearance, and at the same time *emboss* a suitable design.

After this operation the tendency of the cake is to stick in the die, and to overcome this the bottom of the die is removable. By means of the cam-actuated knockout with which the press is equipped, and which is situated below the index-plate indexing mechanism, the bottom of the die is raised, as the punch ascends, and forces the formed cake back again into the dial pocket. The third punch, which is at the right of the ram, is set on an arm considerably in front of the centre of the slide, and is used to push the finished cake through the pocket into a receptacle below the press bed to receive it. The dial feed is operated in the usual way. The presses used for this work will make 36,000 paint cakes per day of ten hours, and the weight of the machines completely equipped is 800 pounds.

SECTION XI.

Drawing, Re-drawing, Reducing, Flanging, Forming, Reversing, and Cupping Processes; Punches, and Dies for Circular and Rectangular Sheet-Metal Articles.

A "QUINTUPLE" COMBINATION DIE FOR PRODUCING FIVE DRAWN AND EMBOSSED TIN SHELLS AT ONE OPERATION.

In Fig. 393 are shown front views of the upper and lower sections of what may be called a "quintuple combination die." The tools were used for the production of five shells, 1¼ inches

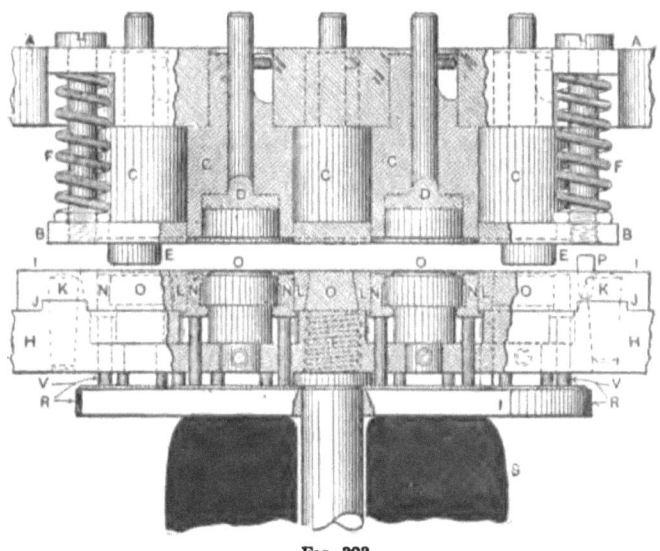

Fig. 393.

in diameter by 1/16 inch deep, from decorated sheet tin at one operation, the work done including blanking, drawing, and embossing. The tools should be properly called "a gang of combination dies." The principle of construction carried out in the tools is essentially the same as that found in the usual single-acting combination drawing die, and no attempt will be made here to give a detailed description of the making of the various parts, as the proper practice is illustrated and described fully in other pages of this work. A general description of the assembling of the parts and the work accomplished will suffice for all purposes, and in connection with the engravings, point out to the practical man, how similar tools may be adapted for the production of work of the same nature in like or even greater quantities. We can see no reason why the principle may not be further extended in the tools to produce a greater number of

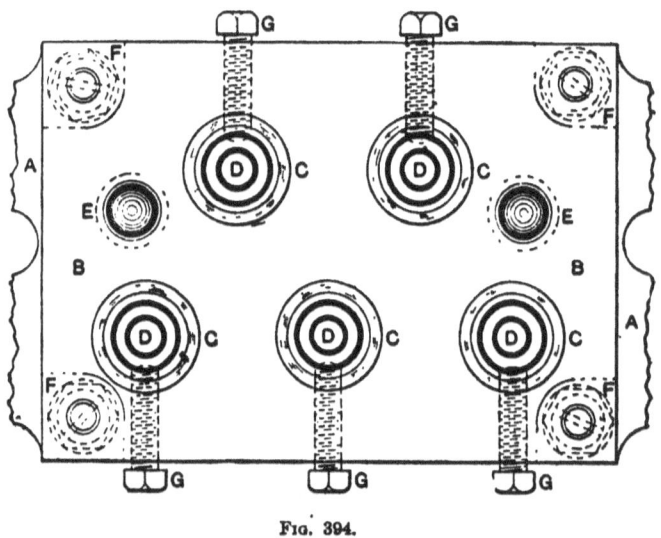

FIG. 394.

shells than those shown here, if a press of suitable dimensions and of sufficient strength can be procured.

As stated before, in Fig. 393 is a front view showing, in conjunction with other things, sectional views of the first row of punches and dies, so that their construction may be clearly understood. Fig. 394 is a plan view of the punch, or upper section of the tools; Fig. 395, a plan of the die sections; Figs. 396 to 400, details of the smaller working parts of the punch section;

PUNCHES, DIES, AND TOOLS. 255

and Figs. 401 to 402, details of the smaller or working parts of the die section.

By taking Fig. 393 as the main reference drawings, and referring occasionally to Figs. 395 and 394, the construction of the tools will be more easily understood. The positions of the details in the finished tools, Figs. 393, 394, and 395, may be determined by comparing the reference letters, which are identical in all drawings.

The lower or die section consists of, first, the main die plate $H\ H$, in which are located the five drawing and embossing

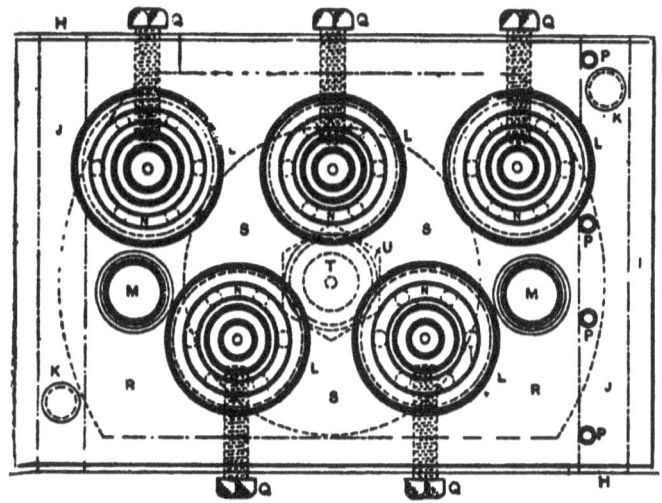

FIG. 395.

punches O, O, O, O, O, the five blank holder-rings N, N, N, N, N, the thirty spring-barrel tension pins (six to each die), the spring-barrel stud T, and five set-screws Q, Q, Q, Q, Q, for fastening the drawing and embossing punches. Then we have the upper die plate $I\ I$, in which are contained the five cutting dies L, L, L, L, L, the two punch-locating stud-bushings $M\ M$, and the four gauge pins P, P, P, P. The two die plates are of machine steel, and are located in alignment with each other by the two taper-hardened dowel pins $K\ K$, and fastened by bolts at the ends, which are not shown. R is the upper spring-buffer plate of cast iron, and *S* the rubber buffer. The machining and finishing of the seats for the various dies in the die plates necessitated some very accurate and careful work in order to insure the perfect alignment of the working parts.

In the upper or punch section, Fig. 393, we have, first, the holder, in which are contained the five combination cutting punches and drawing dies *C, C, C, C, C*, with their combined strippers and embossing pads *D, D, D, D, D*, the spring stripper *B B* for the stock, and the two punch-locating studs *E E*, while *G, G, G, G, G*, Fig. 308, indicates the set-screws for fast-

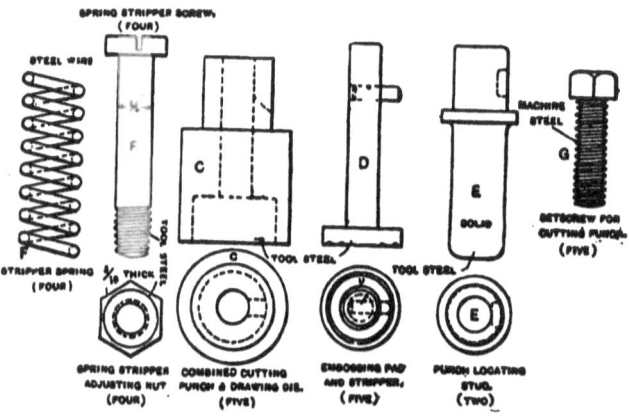

Figs. 396 to 400.

ening the punches within the holder. The holder is cast iron, while the stripper plate *B B* is machine steel. The combined cutting punches and drawing dies are of tool steel and were hardened, drawn, and carefully lapped and then ground, as were also the five combined strippers and embossing pads, *C, C*, etc. The locating studs were also accurately made.

It will be noticed that the two die plates *H H* and *I I* have their perfect alignment further assured by locating them by the tongues and grooves. The five cutting dies were forced into taper holes in the plate *J J*, while the five blank-holder rings *N, N, N, N, N* were made so as to press up against the shoulders of the drawing punches *O, O*, etc., thus eliminating their tendency to work the cutting dies loose. The thirty spring-buffer

PUNCHES, DIES, AND TOOLS. 257

tension pins were of drill rod, and were accurately finished to size in length, being finished to within a limit of variation of one-thousandth inch. Insignificant as they appear, these tension pins play a most important part in the proper working of the dies. Unless they are finished accurately, so as to be the same length, there is no possibility of drawing the metal uniformly, free from wrinkles, or without marring the decorations, through unequal pressure which will be communicated to the blank holder rings by the inaccurately made tension pins. The manner in which the tools are used and the shells produced will be understood from the following description.

Both sections are set up in the press, which is of the inclinable type, and equipped with a two-bar knockout to slide, so as

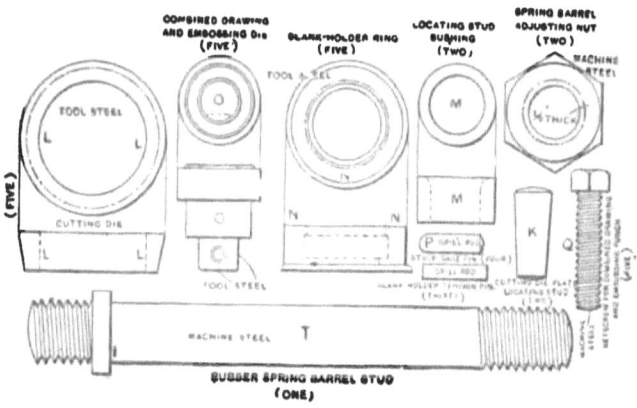

Figs. 401 to 410.

to actuate the stripping and embossing pads D, D, etc., in the punches. The punch section is fastened to the face of the press ram by bolts, and the die section to the press bolster by large cap screws. We might state that the press in which the tools were used was very powerfully geared and weighed 3,500 pounds. It had a standard stroke of 2 inches, a maximum stroke of $3\frac{1}{2}$ inches, and a circular opening in the bolster of 13 inches diameter. Both sections of the die were located in alignment by lowering the press slide and entering the two locating studs $E\ E$ into the bushings $M\ M$. Then the stroke was adjusted, the

17

spring-buffer pressure regulated, and the work was ready to proceed.

The press being inclined, the metal is fed from right to left, one edge resting against the gauge pins P, P, P, P, and the end extending to the edge of the back row of three dies. The press, being stepped, the punch section descends and the stripper $B\ B$ flattens and holds the stock securely on the face of $J\ J$, while the blanks are being cut into the cutting dies N, N, etc., and drawn up into the cutting punch and embossed. As the slide rises the stripper $B\ B$ sheds the stock from the punches, and the double-bar knockout in the press slide actuates the five stripping and embossing pads D, D, etc., causing them to descend and strip the finished shells from the punches when they drop to the die face and fall off at the back by gravity.

When it is realized that the product of the tools shown here was equal to five times the work of a single combination die, their output per day of ten hours can be imagined, and the saving in labor and expense appreciated. Is it any wonder then, that, not counting the cost of material, circular decorated and embossed shells of the dimensions here given can be furnished the consumer at the rate of 2 cents per hundred, and in some instances, when the quantity ordered is well up in the millions, at even a lower rate?

SET OF PRESS TOOLS FOR DRAWING AN ODD-SHAPED CUP.

The set of tools described in the following and illustrated, possesses many points that will be of interest to any one engaged in this class of work, and the experience had in producing the punch by which the final operation was performed will prove of assistance to any one having a similar punch to make or a similar cup to draw. The engravings, Figs. 411 to 414, show progressive operations from start to finish by which the piece, Fig. 414, was produced. This is a corrugated, conical cup, drawn from a round blank of soft brass $\frac{1}{16}$ inch thick and $2\frac{1}{16}$ inches in diameter. The corrugations project externally and internally from six equally spaced, square grooves which converge radially

PUNCHES, DIES, AND TOOLS. 259

and disappear within ¼ inch of the bottom. The specifications in this case required that there should be developed on the outside a distinct shoulder at the base of the conical part, and that this shoulder should be formed on the corrugations only. On the inside no shoulder should be visible, but the formed grooves be made to disappear uninterruptedly near the bottom. It will be seen that the pressure in the last drawing has to be much

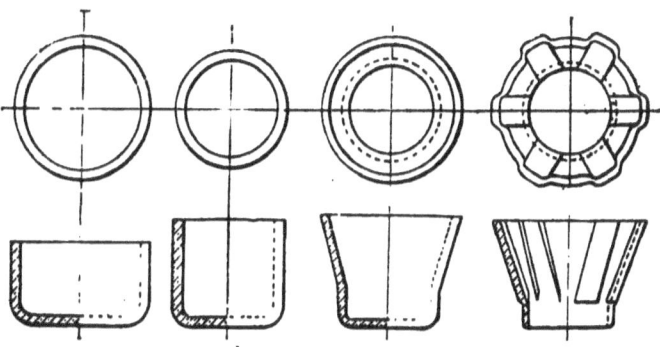

FIGS. 411 to 414.

greater than ordinary in order to accomplish these results, for the stock at the point where the straight and conical portions met was pressed out quite thin in order to develop the shoulder.

The dies and the shanks of all of the punches used were made of a uniform size so that the two holders, one for the punches and one for the dies, were all that were required. The punches were secured by a set-screw and the dies were seated in the holder and held fast by four set-screws, equally spaced around the side, with their points set into the circumference of the dies. To make the change from one operation to the other it was only necessary to loosen the screws, remove the tools in use, and substitute the ones next in the set. Each drawing operation was followed by a careful annealing in order to insure the equal flow of the metal and to minimize any possibility of cracking.

The main problem was to make the finishing punch as cheaply as possible and have it stand up under the severe work required of it. It was first made in sections by turning it to shape and then dovetailing to receive the elevated pieces which were made separately and forced into the dovetails up against a shoulder. Then the end was drilled and tapped and the straight tip screwed on. The parts having been carefully fitted were marked, removed, and hardened, and then replaced as before. The punch made in this way had been used for a short time when the elevated pieces began to chip and crack, where the strain was greatest, and often one hundred or more cups would be run through before the defect was noticed. For this reason the hardened sections were replaced by soft ones, but after a short run they would flatten down and make continual repairing necessary. To overcome the trouble the punch was finally made as shown in Fig. 415, from a solid piece of tool steel. This was milled, chipped, and filed to a finish, and after hardening was drawn to a light straw temper, after which no trouble was experienced.

The first and second drawing dies were made from tool-steel disks 5 inches in diameter and 2 inches thick, with holes to draw the cups as shown in Figs. 411 and 412. These dies were of the plain push-through type with a sharp edge on the under side to strip the work from the punch which was made sligthly tapering and of a diameter equal to the inside of the cup. In this case the punch was made two thicknesses of stock plus .008 inch less than the hole in the die. Previous to the last operation it is general to make the punches more than twice the thickness of stock smaller than the die and to taper them $\frac{1}{8}$ inch per foot to assist in stripping the work, for in most cases the exact sizing of drawn work is of little importance up to the last operation. Consequently tapering the punch and making it below size will in no way interfere with the finishing of the work which is all done in the last operation by a punch and die that must be made of suitable dimensions to meet the requirements. The rule usually adopted is to make the punch smaller and tapering for short cups of thick metal and nearer to size for thin stock, for the latter is more apt to develop body wrinkles if not properly pressed out in the die. The die for the third operation is made the size of the finished die but with plain walls.

PUNCHES, DIES, AND TOOLS.   261

The die for producing the finished work is shown in Fig. 416. The part A was first turned and bored in the lathe and then fastened on a special arbor fitted to the spindle of the dividing head of a milling machine. Six $\frac{1}{4}$-inch holes were drilled and reamed, as at $B B$, the piece being indexed so that they would be accurately spaced. A special angle iron was made for holding this piece at the angle of the sides of the die, and two holes were carefully drilled in its face to correspond with a dia-

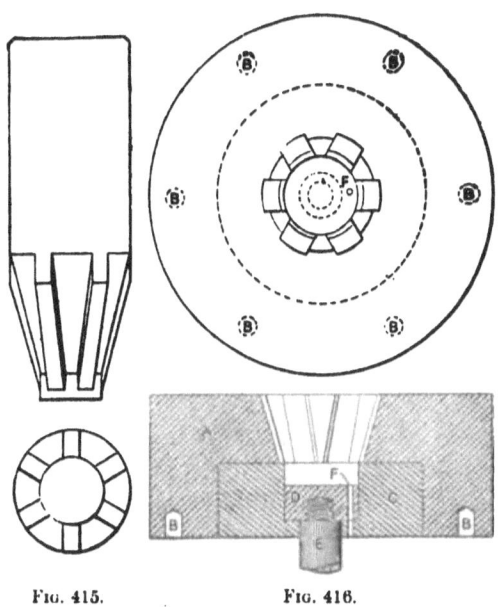

Fig. 415.   Fig. 416.

metrically opposite pair of drilled holes in the back of the die. In these holes were driven $\frac{1}{4}$-inch guide pins which projected $\frac{1}{16}$ inch above the face of the angle iron. The angle iron was then bolted to the knee of the shaper and the die located on its face, by the pins, and strapped firm and true. The six drilled holes provided a means for accurately locating the die for planing the grooves, which was done with a formed tool screwed on the end of an extension-shaper tool of the kind usually used for internal work. After shaping, the grooves were filed and polished dead smooth and the die hardened. The piece C was turned and ground to a light driving fit in the back of the die and both

262 PUNCHES, DIES, AND TOOLS.

pieces surfaced on the under side. D is a steel pad and E a stud for stripping the finished work. They are operated by the press ram on its upward stroke. The vent F is for the escape of air. Following the drawing of the cup the bottom was punched out in a plain cutting die.

DRAWING AND FORMING TOOLS FOR DOUBLE-ACTING PRESS.

The engraving, Fig. 417, shows a combination punch and drawing or forming die for producing the cup shown in position in the die. The cup is made in a double-acting press in one operation.

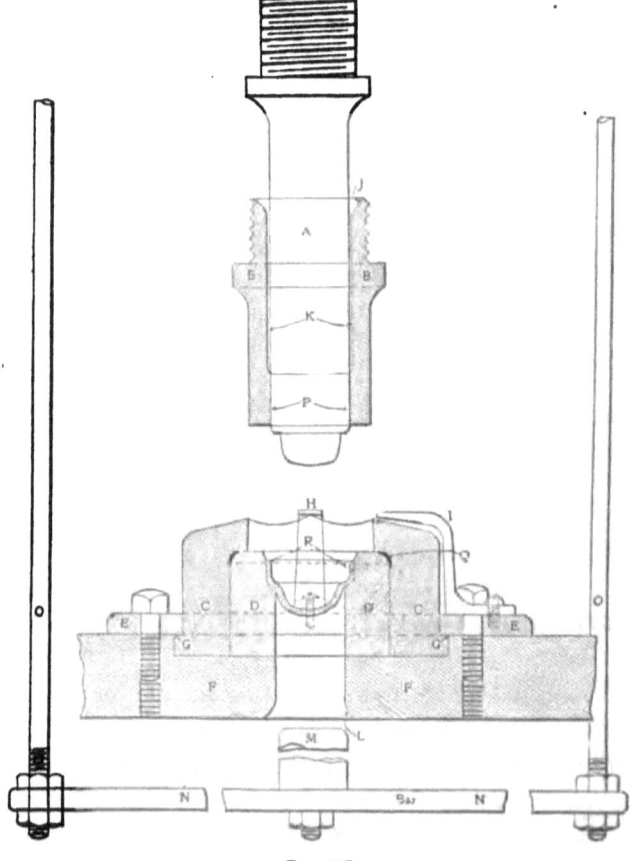

FIG. 417.

PUNCHES, DIES, AND TOOLS.

A is the forming punch threaded at its upper end to screw into the upper or drawing head of the press. Its lower end is shaped to form or draw the stock to the shape required. *B*, the cutting punch, is threaded to screw into the lower head of the press. It is rounded at *J* to clear punch *A*, recessed at *K* to avoid grinding the entire length inside. It is ground outside of the required diameter and at *P* to a neat sliding fit on punch *A*.

C is the cutting die made the size of the blank, tapered inside about 2°, and made about $\frac{1}{32}$ inch shearing on the top. The corner is cut in at *Q* to avoid grinding; the shoulder at *G* provides for clamping. This die is ground inside only. *D*, the forming die, fits in *C* and is held by it. *E* is a boiler-iron clamping ring bolted to plate *F*, and clamping dies *C* and *D*. *F*, a cast-iron plate or holder, is secured to the bed of the press. *H* is an end stop and *I* a side gauge for feeding the stock. *H* extends up over the back of the die as *I* does over the side.

M is a stripper attached to cross-bar *N*, and rods *O O* connect this to the upper head of the press. The threads on the rods are made long enough to allow for adjusting the stripper.

The stock is punched by *B*, forming a round disk, the punch then continuing down, pressing the blank on die *D* and holding it while punch *P* draws and forms it down into the die *D*.

COMBINATION BLANKING AND FORMING DIE.

Fig. 418 shows a combination die and punch for blanking, forming, and punching, complete at one operation, the piece shown at the right. It is of galvanized iron, $4\frac{3}{4}$ inch diameter with $2\frac{1}{8}$-inch hole, three equidistant ears and a half-round ring rib on one side.

The lower part, or die, Fig. 418, comprises the outside cutting die 1, the inner forming block 2, the inner cutting die 3, and the knockout arrangement 5, 6,' 7.

The face of the punch first comes in contact with the cutting edge of the die, cutting the blank. The punch then forces the blank on to the forming block 2, the latter, and also block 4, being pressed downward until they rest on the bottom disk 12.

264   PUNCHES, DIES, AND TOOLS.

The pressure between the face of the punch and the forming block raises the circular rib on the blank, and the inner punch 3 makes the hole in the middle. Then the punch travels upward, releasing the blanks, which are thrown or ejected upward

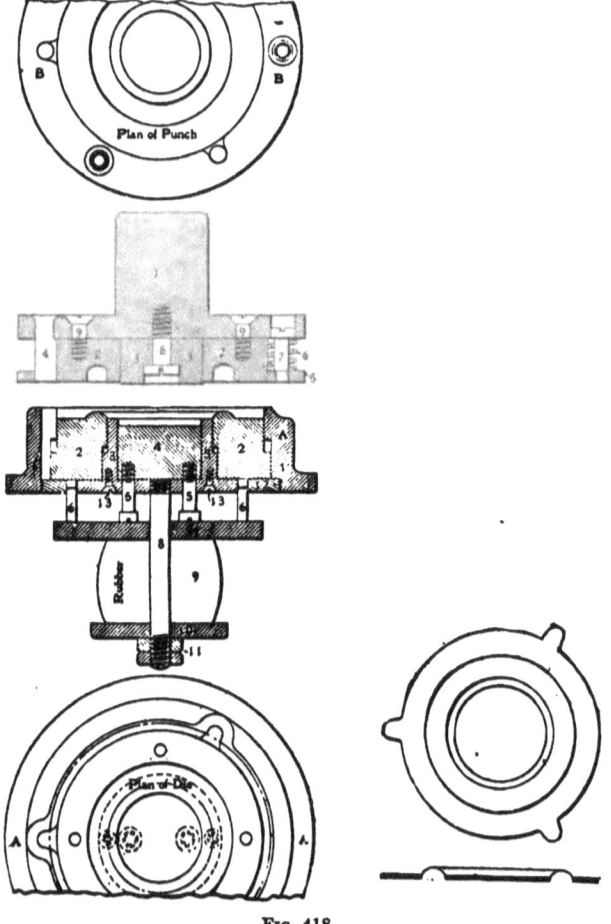

Fig. 418.

by the spring pressure on the forming block 4. The stripper 5 on the punch removes the "waste" from the punch.

In making the different parts of the lower member, the die 1 may be made of a solid block of tool steel about 7 inches round by 2 inches thick. It can also be made of a forging of tool and machinery steel. It is machined all over; the inner sides are

PUNCHES, DIES, AND TOOLS.

filed out to about $\frac{1}{8}$ inch in diameter larger than the finished blank, to allow for the draw on the "rib"; the cutting edge is made straight down, without any clearance; the three recesses for the projections on the blank or ears are cut to shape. About $\frac{3}{4}$ inch below the face the die is recessed on the inside $\frac{1}{2}$ inch larger in diameter, forming a square shoulder. At the bottom the die is again recessed about $\frac{3}{8}$ inch larger in diameter and $\frac{3}{8}$ or $\frac{7}{16}$ inch deep. This is threaded and the round disk 12 that comes up against the shoulder and flush with the bottom of the die is fitted into it. Four $\frac{1}{4}$-inch holes are drilled in it for the four steel pins, 6.

The inner forming block 2 is best made of tool steel, left soft, the face turned to the shape of the inner side of the blank, and the outside to fit in the die so that it will work up and down nicely without any shake, the straight part of the face being $\frac{1}{4}$ inch below the cutting edge or face of the die when the block rests on the bottom, and flush when the block is pushed up to the inner shoulder of the die. The hole in the centre of the forming block is bored out to size of the inner circle of the round rib. About $\frac{3}{4}$ inch from the top it is recessed $\frac{1}{8}$ inch on each side, forming a square shoulder, the outside of the inner cutting die 3 fitting nicely into it, so that the forming block 2 will work up and down. The inner cutting die 3 is made of tool steel hardened and ground all over, with four drilled and tapped $\frac{1}{4}$-inch holes in the bottom for $\frac{1}{4}$-inch flat-head screws 13, fastening it to the bottom plate 12; the top of the inner die comes up flush or level with the flat part of the inner forming block 2 when it rests on the bottom. The knockout block 4 is made of machinery steel, fitting in the inner die 3 so that it will work freely up and down, and when it rests on the bottom to be $\frac{1}{4}$ inch below the face of the inner die. It is drilled and tapped at the bottom for two $\frac{3}{8}$-inch steel screws 5 fitting snugly into it but loosely in the bottom plate. These screws are left long enough so that the distance between the heads of the screws and the bottom of the plate is to equal the distance between the face of the knockout block 4, when it rests on the bottom, and the face of the die 1. The four steel pins 6 are $\frac{1}{4}$ inch shorter than the distance from the top of the disk 12 and the top of the plate 7, which is a round machinery-steel plate ⅜ inch thick with a ⅝-inch hole in the centre fitting over the central stud 8⅝ inches in diameter, screwed into plate 12. A rubber spring barrel 9, about 3 x 3 inches, held up by a washer 10 and two ⅝-inch lock-nuts, 11, complete the making of the bottom part or die.

Referring to the punch section, for the punch-holder 1, a machinery-steel forging is best, but it may be made of cast iron, which will of course have to be heavier. Fastened to it is the round punch 2, of tool steel hardened and ground all over, the outside fitting nicely into the die. The face is recessed about ¼ inch deeper than the height of rib in the blank, to allow for grinding the face of the punch; the width of recess to correspond to the width of the rib in the blank. The hole in the centre of the punch is large enough to receive the inner punch 3 which is fastened to the shank with a screw 8 in the centre. The inner punch is left about $\frac{1}{16}$ inch longer, and projecting above the punch. It is made a nice fit in the inner die, which must be about $\frac{1}{16}$ inch smaller than the diameter of the hole in the finished blank to allow for drawing the rib.

The three smaller punches 4, for punching the ears of the blank, are of tool steel and fitted by turning or grinding to the circle of the outside of the punch. The ends are rounded, fit a round hole in the shank, and are slightly riveted on the back to keep from pulling out. The outsides fit the recesses in the die. The stripper T, on the outside of the punch, is made of ¼-inch flat machinery steel, and is held in place, to come slightly above the face of the punch, by three screws 7. Three steel springs, 6, supply the tension to the stripper. In sharpening the punch and die in the course of use, the inner punch and die are packed to keep the proper position.

DIE FOR RE-DRAWING LARGE SHELLS.

Fig. 419 shows the die for the third and fourth operations in the drawing of a large shell. The die for the third operation only is here shown. For the fourth operation the die and stripper are changed. All the parts here shown are round, so that

PUNCHES, DIES, AND TOOLS. 267

dies, gauges, and strippers for other articles may be fitted to them.

A is the shell after the second operation, in position for the third operation. *B* is a cast-iron or steel gauge or guide. Cast

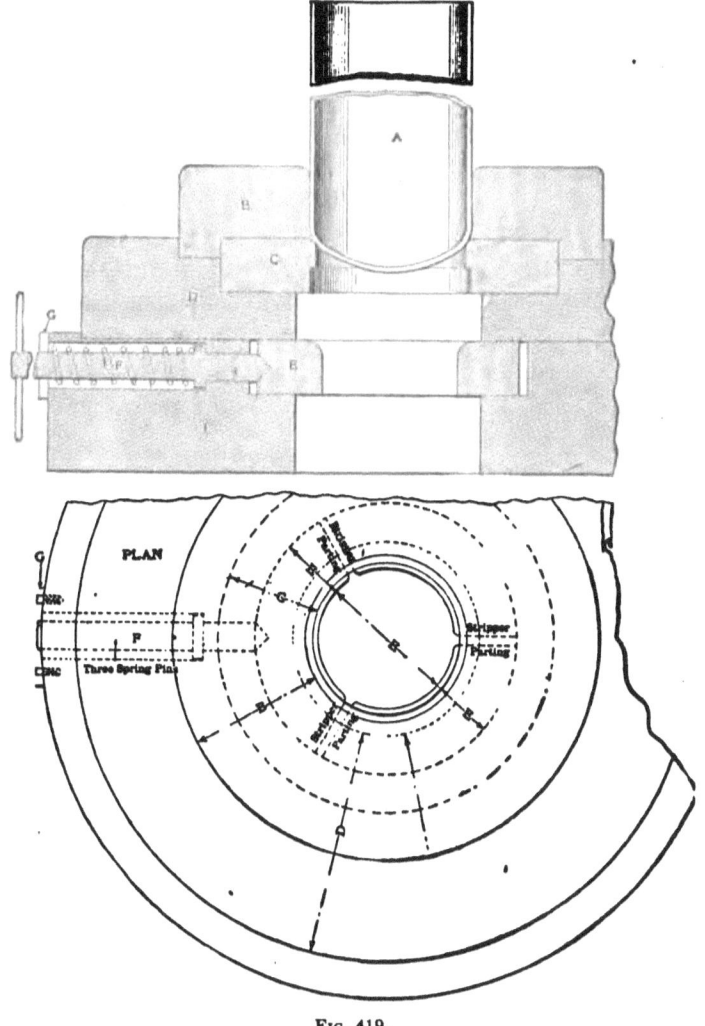

Fig. 419.

iron was used, which answered the purpose very well and was easy to work. *C* is a steel die; the upper or cutting corner is rounded, and after hardening is polished as smooth as possible.

Too much care cannot be given to the finish of the parts coming in contact with the shell in drawing, as any roughness will show in the finished tube. The die is recessed, as shown, to do away with friction as much as possible. D is a cast-iron holder in which the die C and gauge B are fitted. This is to let into bolster I about ½ inch. E is the stripper, the upper corner of which is rounded and the lower corner left square. It is made of tool steel and tempered as hard as possible. The hole is bored the same size as that in the die. The piece is then cut into three equal parts. Care must be taken not to cut away more than will allow the parts to close just enough to do the stripping, or, in other words, to hug round the punch after the shell passes through.

On the down stroke of the punch the stripper is forced open, allowing the shell to pass through, after which the stripper is closed against the punch by the springs acting against pins F; the stripper is in position to strip on the return stroke of the punch.

It is well to call the reader's attention to a part in the design that was encountered and which gave considerable trouble. The outside of the stripper was made nearly the same diameter as the die, which necessitated a large recess in I, thereby causing the reduction of the support under die C and casing D, causing D to bind in and prevent E from working freely. Another important point is the rounding off and polishing of the inside corners of the stripper after the splitting; if not rounded, marks will show on the finished shell. The die is secured to the press bed by bolts passing through holes drilled in D and I. The punch for the third and fourth operations, owing to its small diameter, is made in one piece. In the punch for the first and second operations the diameter is much larger, and it would require a much larger piece of steel, costing more to replace when worn.

DRAWING DIE WITH BLANK-HOLDER ATTACHMENT.

Fig. 420 shows a die for the first operation in drawing a shell from No. 14 gauge sheet metal. The diameter of the blank was 5¼ inches.

The drawing die A is made of tool steel, and has two impor-

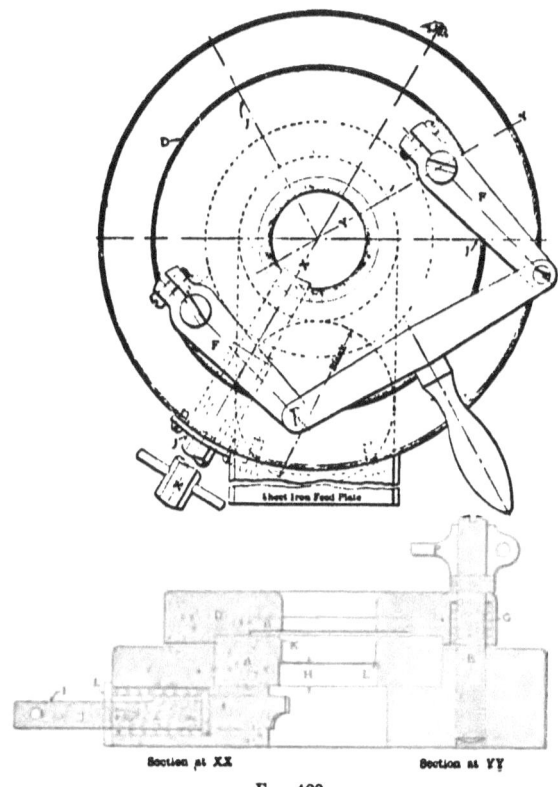

Fig. 420.

tant features at K and H. The rounding at K is ½ inch; if made smaller, the bottom will tear out, and if larger, wrinkles will form and tear the metal at the top. For this size blank the smallest punch that can be used with success for the first operation is three inches; smaller will tear the bottom out of the metal. H provides clearance and also serves as a stripper.

D is a cast-iron guide and clamp plate. *B* is a hardened steel ring inserted in *D*. This is the diameter of the sheet-metal blank. *E E* are clamp bolts operated by levers *F F*. *G* is one of the springs to keep *D* up while the stock is being fed in under the punch. The purpose of *D* is to hold the blank down while drawing; if this is not done the metal will wrinkle and make bad work. *B* is inserted to overcome the hard wear of drawing. *F* is secured to *E* by a split clamp; the reason for this is that after *D* is clamped on the blank, *F F* are set and made fast in a convenient position for the right hand of the operator, leaving the left hand free to feed the blanks on the sheet-iron feed plate.

This drawing was done on a large Stiles geared press, which runs continuously.

COMBINED CUTTING, DRAWING, AND KNURLING DIE.

Figs. 421 to 426 show a set of cutting and drawing tools used in a double-acting power press for cutting and drawing a so-called knurled cover, in one operation; also a die bed or bolster

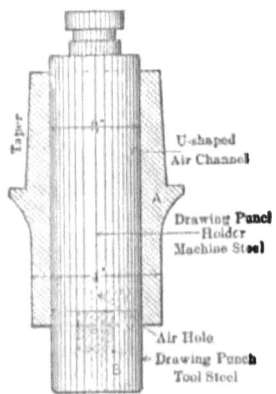

FIG. 421.

which is a quick time-saver, and is used in connection with this set of tools. These tools cut and draw a brass cover 3-inches in diameter and ½ inch deep; thickness of metal 0.050 inch—which

PUNCHES, DIES, AND TOOLS. 271

is used on a cylinder-shaped box of metal, the outside of the cover being ribbed or milled for convenience in screwing on or off the box, as shown in Fig. 426.

Fig. 421 shows the cutting punch A, which is 4 inches in diameter at the cutting part. It has a taper shank, and is made

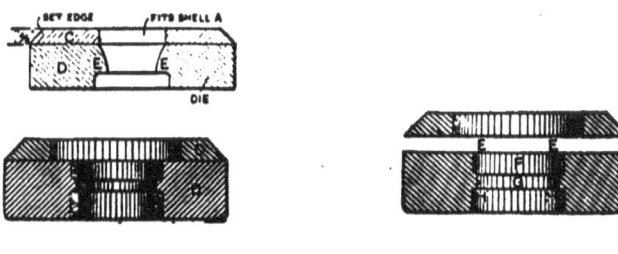

FIG. 422.　　　　　FIG. 423.

a driving fit in the punch-holder, Fig. 425, which is held in the press by means of two ⅜-inch screws.

The drawing punch, B, is made in two parts. The drawing part is made of tool steel, and is 3¼ inches long and has a 1-inch hole tapped in the end where it screws on the punch-holder. The punch is hardened and ground, and is lapped smooth to size. The air hole shown, not only makes it easier to draw a shell or cover, but enables the cover to be more easily stripped

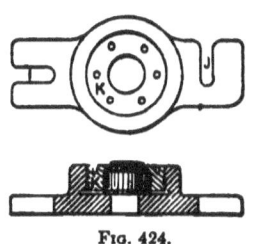

FIG. 424.

from the punch. The drawing-punch holder is used for different sizes of drawing punches, and is held in the press by the yoke key, or as some die-makers call it, a U-shaped key, which fits the nipple on the upper end of the holder.

The die is made in two parts, namely, the cutting part and the drawing part. The cutting part is made of tool steel, and is hardened and ground on the top and bottom; it is bevelled on

272   PUNCHES, DIES, AND TOOLS.

the side to correspond with the bevel on the threaded collar shown in Fig. 424 at K. The 4-inch hole is ground and has very little clearance, for the reason that if too much clearance is given it will cause the blank to shift from side to side after it is cut, thereby causing the cover to be uneven in its length.

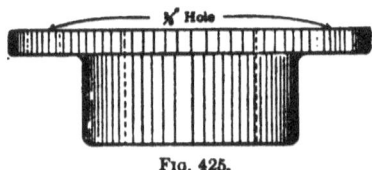

FIG. 425.

The drawing part is made as shown in Fig. 423. F and G are the parts that draw up the ribbed cover after the blank is cut. The mouth of the die is rounded, as shown at $E\ E$, thereby making it easier to force the blank through the die. The diameter of F is made 0.020 inch larger than G, and it draws up the cover before it is ribbed by G. The diameter of G is made a trifle smaller than the required size, and the cover expands slightly after having passed through the die. The corners $H\ H$ are made sharp and strip the cover from the drawing punch on the upward stroke of the press. Part G is ribbed or milled with the knurl shown in Fig. 426.

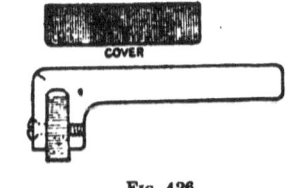

FIG. 426.

Fig. 424 shows the bolster for holding the die. The die is held in position by the threaded collar K, which is screwed down upon it by the aid of a spanner wrench having two pins which engage in the holes shown. This bolster differs from the one most generally used inasmuch as it has the slots for the screws running at right angles with each other. When taking the bolster out of, or placing it in the press, the screws need never be any more than just loosened up, while with the other

PUNCHES, DIES, AND TOOLS. 273

style one screw must usually be taken out entirely before the bolster can be removed. When placing this bolster in the press, the slot I is engaged with the screw on the left, after which the bed is swung to the right until the slot J engages the screw.

The arrangement for guiding the metal, and the stripping of it from the cutting punch, are not shown, but are similar to those described in other pages of this chapter.

THE BLANKING AND DRAWING OF RECTANGULAR SHELLS.

The method of constructing a "double-action" punch and die for drawing oval and square work, and described in the following, has numerous good features.

Figs. 427, 428, and 429 show the die bed E, planed from front to back for the lower or drawing die D, and from right to

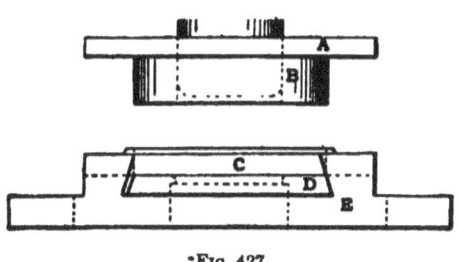

Fig. 427.

left for the upper or blanking die C. The dies are both planed to an angle of 10° from the vertical, and keyed into the bolster. This arrangement allows the die to be made of separate pieces easily worked, admits of grinding the faces of both dies after hardening, the removal for repairs of either one of the dies, without disturbing the other, and also gives a means of adjustment between the two dies, in relation to the drawing edge to the cutting edge, that is not easily attainable in any other manner. The blank-holder and blanking punch B is screwed and dowelled to the plate A in the usual manner.

Figs. 431, 432, and 433 show outlines of the successive operations, eight in all, which were necessary to draw a rectangular shell $1\frac{3}{4}$ inches by $\frac{7}{8}$ inch to finish $1\frac{1}{4}$ inches in depth as per

18

274   PUNCHES, DIES, AND TOOLS.

Fig. 426. The finished shell is shown in Fig. 426. The specification called for a shell of the above dimensions, to be made of sheet copper, the corners to be full and sharp; at the bottom to

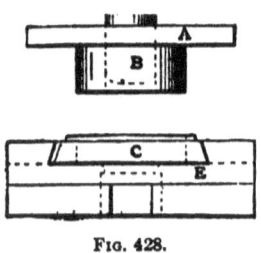

FIG. 428.

a radius of 1/16 inch was allowable. The shell was to gauge No. 26 B. and S., .0859 inch after finishing.

Fig. 429 is a plan of the blanking and drawing die for the

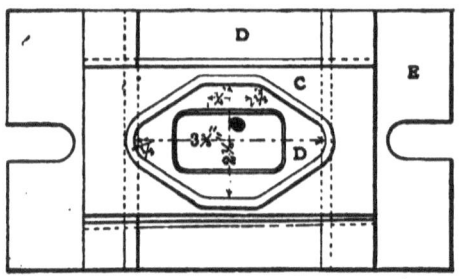

FIG. 429.

first operation, and gives the outline and dimensions of the blank. The subsequent operations were performed in dies similar in all respects—except as to shape—to the dies used in the re-drawing of ordinary round work.

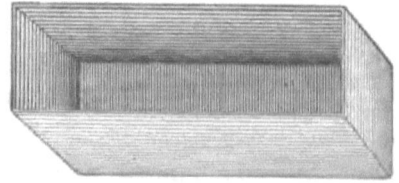

FIG. 430.

The following table gives the dimensions of the drawing dies and shows the reduction in operations. The data will be of

PUNCHES, DIES, AND TOOLS.

value to all mechanics who may happen to have a similar job of metal drawing to accomplish.

1st op. from blank,	2.375	x	1.1875 inch	radii	.3125
2d "	"	"	2.250	x	1.0625	"	"	.250
3d "	"	"	2.125	x	.9375	"	"	.1875
4th "	"	"	2.0625	x	.875	"	"	.156
5th "	"	"	1.9375	x	.750	"	"	.0945
6th "	"	"	1.875	x	.687	"	"	.0625
7th "	"	"	1.812	x	.625	"	"	.0315
8th "	"	"	1.750	x	.562	"	"	none

The metal used was No. 24 B. and S. gauge, .020 inch, and was reduced two gauges in passing through the various opera-

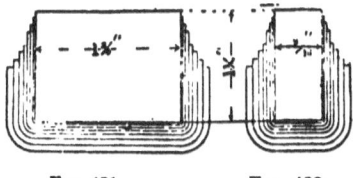

FIG. 431.　　　FIG. 432.　　　FIG. 433.

tions. An annealing was necessary after each drawing except the seventh, the last drawing being done without annealing to give required stiffness to the shell.

PUNCH AND DIE FOR DRAWING A TIN FERRULE.

In Fig. 434 is a sectional view of a punch and die for forming a tin ferrule at one stroke of the press. Fig. 435 is a plan or face of the punch, and Fig. 436 an outline in section of the ferrule to be made.

The main punch C is of soft tool steel, and has the stem c turned integral with it. E is the small punch with the stem, e, driven into the main punch and slightly riveted. B is the cast-iron body of the die, and has lugs for bolting to the press. K is the die proper, or cutting edge, and is of hardened tool steel, ground inside, on the bottom, and outside. J, the die centre for forming the ferrule over, serves also for the small cutting edge and is of hardened tool steel lapped inside. H is the drawing ring which is pressed up tightly against the face of the punch by

276   PUNCHES, DIES, AND TOOLS.

a large spring—not shown—acting on the three pins I. This ring H also pushes the ferrule up off J after it is formed. To

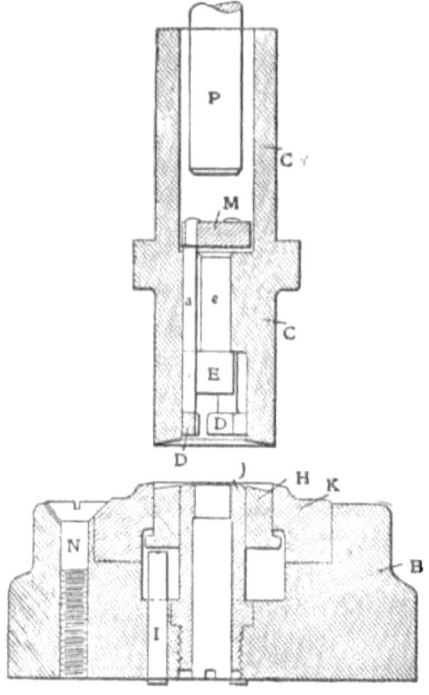

Fig. 434.

extract the finished ferrule from the punch C, three pins, d, are used, which have heads on their lower ends fitting the annular space between punches C and E, and are driven tightly into a

Fig. 435.   Fig. 436.

disk M, which strikes the stationary knockout P. The holes for pins d are drilled somewhat larger than the space between punches, as shown in Fig. 2, and the pins d are flattened on the inner side.

DRAWING FLANGED CUPS.

In the designing of dies for drawing up plain cups by the reducing process no rule can be given whereby the number and

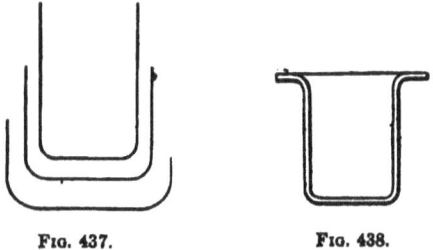

FIG. 437.   FIG. 438.

diameters of the different dies to use for any one single cup can be determined except approximately. Practice and long expe-

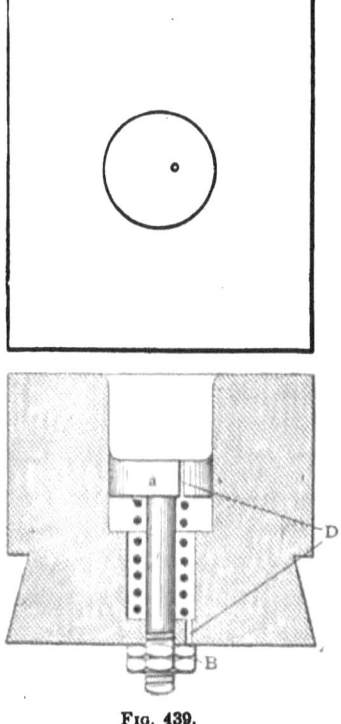

FIG. 439.

rience chiefly govern the making of the dies and the eliminating of the unnecessary operations. Smooth polish, lubrication, and

278 PUNCHES, DIES, AND TOOLS.

not too long a surface bearing in the depth of the dies are points essential to secure.

Fig. 437 shows the external outlines of three drawing operations to produce a cup $1\frac{5}{8}$ inch long, $1\frac{3}{8}$ inch diameter, and with a flange diameter of $1\frac{13}{16}$ inches, from soft brass blank $1\frac{7}{8}$ inches diameter and $\frac{3}{32}$ inch thick. Fig. 438 is a section of the finished cup.

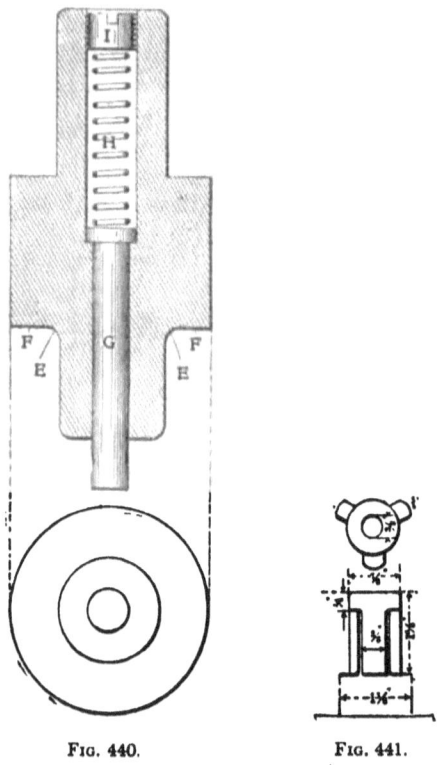

Fig. 440.   Fig. 441.

The drawing dies are plain round ones, with square shoulders on the under side to strip the work from the punch as it is pushed through. The punches for the first two operations taper .002 inch in 1 inch; the third one is perfectly straight. All are provided with air holes to facilitate the stripping of the cups. Annealing is necessary after each operation. Following the drawing operations is the squaring off the ends in the lathe to the desired length to form the flange, and the flanging follows.

PUNCHES, DIES, AND TOOLS.

Place the cup in the die, Fig. 439, which is made .002 inch larger than the last of the three drawing dies, and as it is forced down by the punch, Fig. 440, the knockout *A* in the die descends and brings up on the bottom; meanwhile the flange is being turned out by the radius in the punch *E* and flattened out by the shoulder *F*, when it comes in contact with the face of the die. As soon as the punch is elevated clear of the die, the work is raised sufficiently to be removed by the operator, by the knockout *A*, Fig. 439, moved upward by the ⅛-inch coil spring *C*, which is regulated by the jam-nuts *B*. *D* indicates holes for the air to escape. The spring pin *G*, Fig. 439, is below the end of the punch far enough to shed the work and to press firmly upon the bottom of the shell while the flanging is being accomplished.

BLANKING, DRAWING, AND HOLE-CUTTING DIE WITH POSITIVE KNOCKOUT.

Figs. 442 to 445 illustrate a somewhat novel blanking, drawing, and hole-cutting die for producing in one operation a bracket made of IX tin. Fig. 441 shows the shape, with dimensions, of the bracket produced.

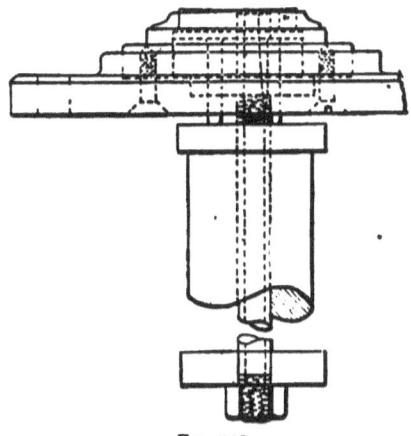

FIG. 442.

Fig. 443 is a section at *H*, in Fig. 445, through the punch, and the new and original feature with which the positive knockout is arranged. The central punch *B*, which is made of tool

280 PUNCHES, DIES, AND TOOLS.

steel, is held in position by the hardened tool-steel key C, the knockout or stripper D being bored to a working fit to receive the central punch B, and turned to fit nicely at E to guide the same, but a loose fit at F, assuring a free working fit for stripper or knockout. The key C is a driving fit in the stem of the punch

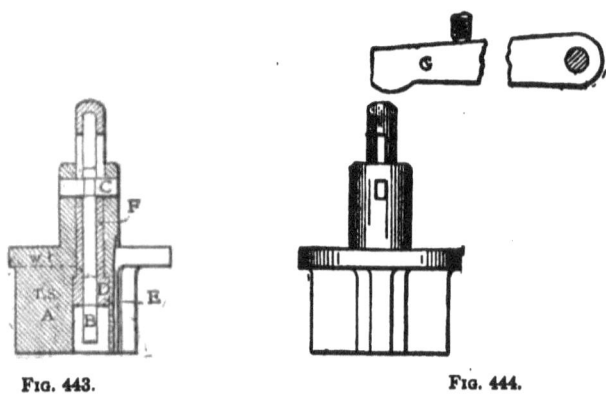

FIG. 443.   FIG. 444.

and free in the slot of the centre punch B and stripper or knockout D. The slot in D is made long enough to permit its descent to the face of the punch. On the upward stroke the ascent is continued until the knockout lever is struck.

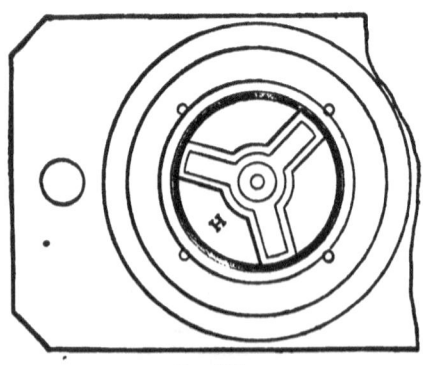

FIG. 445.

These brackets were made of different heights, from ¼ to 1¼ inches, so it was necessary to leave the centre punch B, as shown in Fig. 443, long enough to punch the lowest bracket.

To all practical die and tool-makers it is well known that a spring in any die or punch is objectionable—that is, if there is

PUNCHES, DIES, AND TOOLS. 281

any positive movement known that can be attached, and especially does this hold good in a combination die where there is a long draw and no room for springs. Another objectionable feature of a spring in this case was that it would bend the blank or the three legs, *i i i*, Fig. 443, if the spring was strong enough to extract the blank, and should the spring fail to act and the operator not see it, there might be two blanks in the punch.

Fig. 445 is a plan of the die laid out to leave as little scrap as possible. Fig. 442 is a front elevation of the die.

The spring rubber J was attached to a tube instead of a rod, to allow the scrap of the centre hole to pass through. It was necessary in this case to use a long rubber. It will be found generally that better results can be had by using short lengths of rubber for a spring than by one long piece, as each piece expands independently of the other.

DIE FOR RAPIDLY BLANKING AND DRAWING SMALL SHELLS.

Figs. 446 and 447 illustrate a gang combination die for cutting and drawing eight shells at a time. The tools were used in a No. 20 Bliss press.

For the die a crucible tool-steel disk was used. After being turned to 6 inches diameter by 1 inch thick, it was laid out for the eight cutting edges, each hole carefully centred with a "wiggler" or centre indicator on the face-plate of the lathe and bored to about .005 less than the size of the blank, which allowed a little for grinding after being hardened. After the eight holes had been bored in succession, four $\frac{1}{4}$-inch holes were drilled and countersunk for flat-head screws. Two more $\frac{1}{4}$-inch holes for dowel pins and another for a gauge pin were bored also, after which the die was hardened. All the holes, with the exception of the eight cutting edges, were plugged up with fire-clay before hardening. After the hardening and tempering, the die was ground flat on both top and bottom, and the side or wall was rounded up. Before doing anything more with the die, a punch plate or holder was prepared, of machinery steel, a duplicate of the die as to diameter, thickness, and the holes for screw and dowel pins. A cast-iron bolster for the die was then machined and bored to the exact diameter and depth of the die, drilled for the dowel pins, and tapped for the flat-head screws with which to secure the die. After the die was secured in the bolster with the screws and dowel pins, it was fastened to the face-plate of the lathe and adjusted so that one of the eight cutting edges ran perfectly true, and it was then ground to the exact required size of the blank. The *die* was then removed from the bolster with-

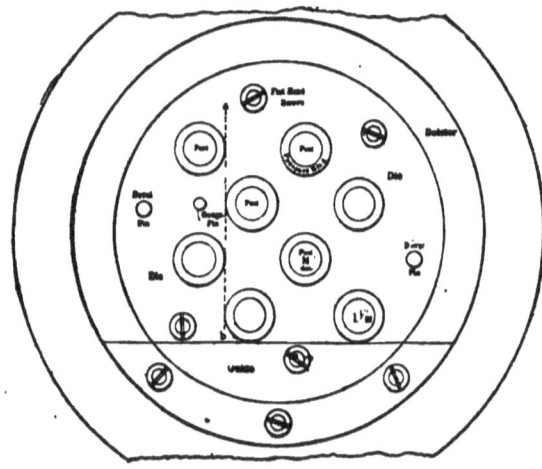

FIG. 446.

out shifting or disturbing the position of the latter on the face-plate, and the bolster was then bored or recessed to receive the combination post or centre, as shown in the drawings. After this had been accomplished, the *punch-holder*, with the face down, was secured in the same manner and by the same means as the die, and this was bored out to receive the shoulder piece. The bolster was then removed from the face-plate, and the punch-holder was then removed from the same. The die was replaced and the other seven cutting edges were ground; the bolster was recessed for the posts, and the punch-holder was bored in succession, as described. By doing all of these parts in this way a perfect alignment of all the punches was secured and everything fitted perfectly.

A combination post or centre for each recess was then made of tool steel and turned the exact diameter of the shell wanted.

PUNCHES, DIES, AND TOOLS. 283

The lower end of it fitted the recess in the bolster and was fastened to the bolster by three small screws. Three pressure pins, also made of tool steel, extended through the base of each post and through the bolster to act in connection with the pressure rings shown in the drawings. The centre of each recess was tapped for a ⅜-inch stud, to which rubber pads and plates were attached in the usual combination fashion, one for each post. The punches were also made of tool steel, and after being finished, were hardened in oil and ground to .001 inch diameter smaller than the cutting edge of the die.

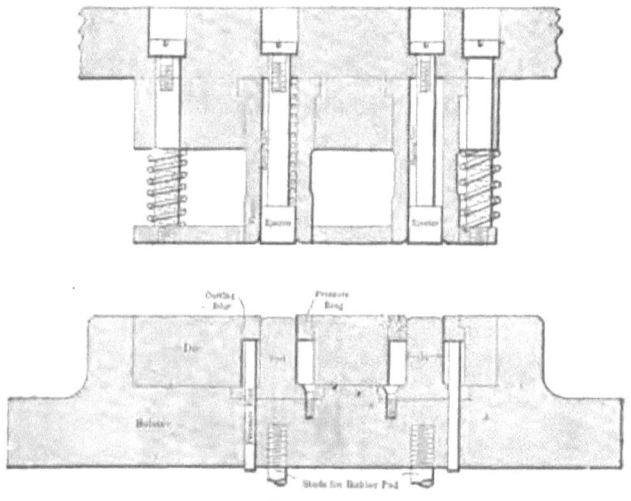

FIG. 447.

Each punch was equipped with a spring-actuated knockout or ejector, and was fastened from the back of the punch-block with button-head screws, as shown. The punch-block was made of machinery steel and was fastened to the lugs of the press ram instead of using the regular 1½-inch shank. A spring stripper, to strip the stock from the punches after being cut, was fastened and held in position, similar to the ejectors in the punch, by long button-head screws, which extended through the punch-holder and punch-block. A guide for the stock, a strip of steel about ⅛ inch thick, was attached to the bolster on the side nearest the operator (marked "Front").

The press in operation was inclined, which facilitated the

discharge of the shells, there being no obstruction to retard the same after being drawn and ejected. These shells were made in this die at the rate of 120,000 per day. It will be noticed from the position of the gauge pin (which, by the way, was bent to bring the end from which the strip was gauged almost to the edge of the nearest cutting edge, or to the dotted line *a b*, Fig. 446) that the first stroke of the press cuts and draws only six shells and that every stroke thereafter makes eight; the feed being alternately one space, three spaces, one space, three spaces, and so on. This die has been in almost constant operation since its finish, as the demand for the shells was tremendous.

SET OF DRAWING AND FORMING DIES FOR MAKING TIN NOZZLES.

In Figs. 448, 449, 450, and 451 are shown a set of dies for the production of nozzles for tin cans of large sizes used to ship liquids. The dies are of the combination type, used in single-

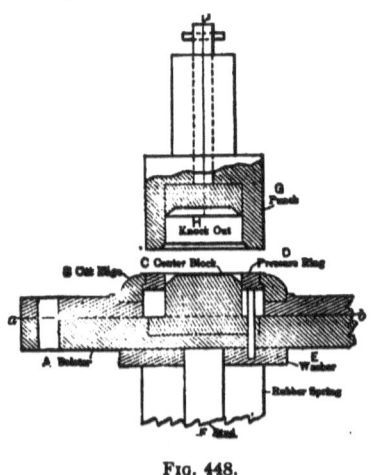

FIG. 448.

action presses and do from one to three operations at one stroke of the press. From 12,000 to 15,000 pieces of finished work can be turned out per day from these dies according to the speed of the operator. Figs. 452 to 455 illustrate the work.

The first die, Fig. 448, is composed of five principal parts:

PUNCHES, DIES, AND TOOLS.   285

A is a gray-iron casting bolster plate made to separate at the line *a b* so the die can be readily taken apart for repairs. *B* is the "cut edge" set into the top plate and held down by three flat-head screws (not shown). *C* is the centre block set into the lower plate of the bolster and also held in place by flat-head screws, not shown. *D* is the pressure-ring or blank-holder which rests on three pins (one shown) which in turn are supported by the washer *E*, which rests on the rubber spring surrounding the stud *F*, and held in place by another washer and nut (not shown) with which to regulate the pressure while draw-

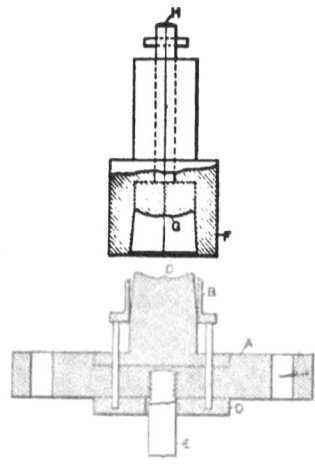

FIG. 449.

ing the shell. *G* is the punch and drawing die combined, the outside diameter of which is fitted to the cut edge *B*. The inside diameter is fitted to the centre block *C* plus twice the thickness of metal. *H* is a forming pad made to fit the top of the centre block *C*. It forms the top of the shell at the end of the stroke and also serves as a knockout for the shells.

In operation the tools are set into an inclined press. The punch coming in contact with the cut edge *B*, cuts the blank, which is held by the pressure-ring *D* against the end of the punch *G*, but as punch *G* continues down, the blank is drawn over the centre block *C*; and, as the punch ascends, the stem *I*, which is at the top of the punch shank, comes into contact with a bar in the press, thus pushing the pad *H* down, and the shell represented in Fig. 408 slides off back of the press.

Fig. 450 shows the second operation or re-drawing tools. *A* is a bolster plate, *B* is the drawing ring, supported by pins and a rubber spring, the same as in Fig. 448. The centre block in this die is tapered and the punch *F* is also bored out tapering to fit it. The pad in punch *F* is of peculiar shape, as will be noticed and will be explained further on. The shell is placed

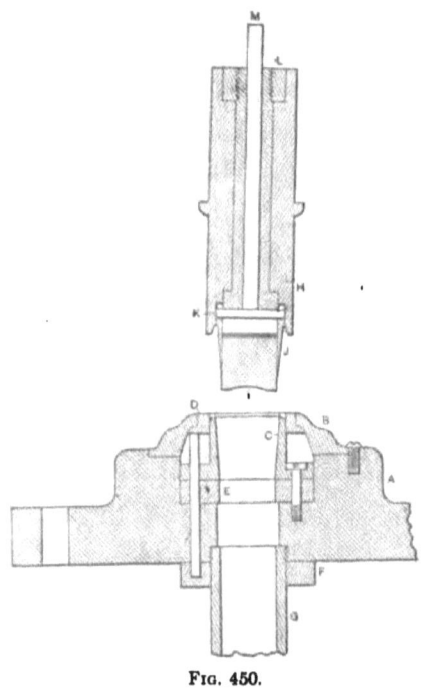

FIG. 450.

on the drawing ring, and the punch, as it descends, draws it down and compresses it to the shape of the centre block *C*. The shell is knocked out on the up-stroke by the stem *H*, same as in the first operation, and the drawn piece looks like Fig. 453.

Fig. 449 illustrated the tools for the third operation, which really consists of three operations. *A* is the bolster plate of the die; *B*, the trimming die; *C*, the centre block; *D*, the drawing ring; *E*, the lower die; *F*, washer; *G*, a tube through which the bottom of nozzle passes after being punched out. These bottoms

PUNCHES, DIES, AND TOOLS. 287

are used for roofing shells for fastening tar paper in place on roofs, etc., so that in the process we really make two articles at once. These tools are used in an inclined press. As the punch comes down, punch I cuts the bottom out, and at the same time punch H trims the lap edge; as it continues on down it presses the shell over the edge of the centre block C. As the punch ascends the knockout bar comes in contact with the pin M, carrying the stripper J down by the cross-pin K and ejecting the nozzle in the shape of C, Fig. 454.

Fig. 451 represents the tools for the fourth and finishing operation. It consists of a simple punch and die, yet much depends on these tools, for the nozzles all have to be of an exact

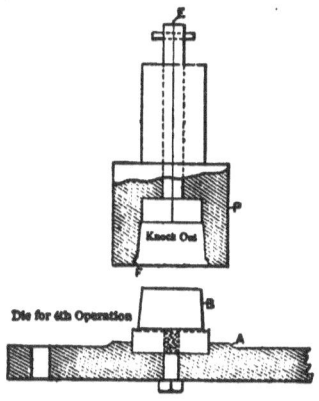

FIG. 451.

size on the finished edge to receive a sealing cap, and this cap when closed on has to be watertight. The die consists of a bolster-plate A and a die-block B, made of tool steel, hardened and tempered and ground to size. The punch is also hardened and tempered and ground to gauge. The tools are set in the press, the nozzle is slipped on the die-block. The punch D in coming down comes in contact with the shoulder F on the inside of the punch. As the punch continues to descend, this edge is curled over and pressed down to the shape of D, Fig. 455. As the punch rises, the shell is knocked out by the knockout-stem, same as in all the other dies.

There are many mechanics who seem to feel that it does not require any *good* tools to produce a tin can, whereas tools for such work require the utmost accuracy in points of construction; besides, tin itself is a very peculiar metal to handle in drawing dies.

CUPPING TOOLS FOR A DOUBLE-ACTING CAM PRESS.

In the following a description and illustration are given of a set of cupping tools used in connection with one of the Waterbury Farrel Foundry & Machine Company's presses of the double-action cam type, and equipped with the Bates automatic roll feed. In Fig. 456, A is the cutting or blanking punch, and B is the cupping or drawing punch. The blanking punch A has a straight shank S', which is threaded at one end and is held securely in the punch-holder by a threaded ring or collar.

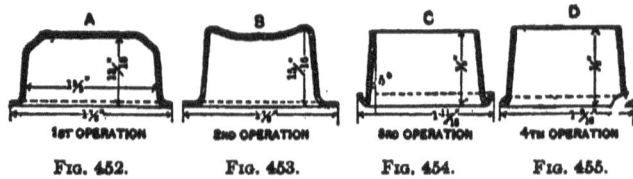

Fig. 452.   Fig. 453.   Fig. 454.   Fig. 455.

The drawing punch B is secured to the "gate" of the press by a yoke key at the shoulder S. In making the blanking and drawing dies we construct each one independently of the other, for the reason that the bottom drawing die D has to be replaced much oftener than the top or blanking die C.

In this kind of double-action press tools the punch A cuts the blank and holds it firmly on the top of the die D, and the punch B descends and pushes the blank through the die D, and, as the gate with the punches ascends, the cup or cylindrical shell is stripped from the punch B at the edge M, which we commonly call the "shedder."

The bottom or drawing die D has a free movement sideways and therefore is self-centring, but is held down in the cast-iron die bed DB by the blanking die C being screwed upon it by the threaded ring R. NN are holes for tightening the ring with a spanner wrench. The punch A is left soft for the convenience of refitting the cutting edge to the blanking die C. The die D

PUNCHES, DIES, AND TOOLS.

is hardened, tempered, and ground, while the "draw" E is carefully lapped to size. The punch B is hardened and ground, but the temper is not drawn. The blanking die C is hardened and tempered and the cutting edge is internally ground ($1\frac{1}{2}°$ taper) to size.

It is very essential that both the blanking and drawing dies

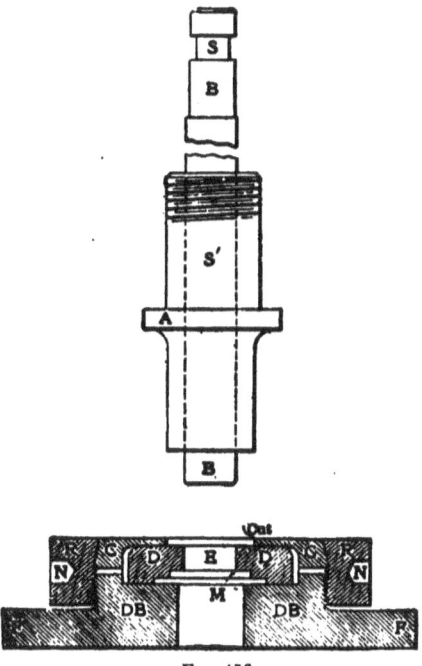

Fig. 456.

should be hardened and ground to a true and level bearing, or else the shells or cups will be "crimped." This set of tools, as described, cut and drew $3\frac{1}{4}$ tons of grass shells, from No. 26 metal, B. and S. gauge, in three days with one setting of the tools. Of course, it cannot be denied that the first cost of the tools was quite a little more than that of those of the ordinary construction, but this extra cost was more than offset by the surprising results.

19

ECONOMIC CONSTRUCTION OF RE-DRAWING DIES.

A great many die-makers construct re-drawing dies solid, as shown in Fig. 457, and fasten them in a bolster having four screws which are screwed up against the outside taper part of the die. Another method, however, is to make the re-drawing

FIG. 457.　　　　　　FIGS. 458, 459.

die in two parts, Figs. 458, 459, of which the set-edge is made to fit the shell *A*, before it is re-drawn. *D* is the re-drawing die that draws up the shell *B* from shell *A*.

The "set-edges" are made in various sizes, and are used in connection with different sizes of drawing die. It sometimes happens that one "set-edge" is used with as many as ten differ-

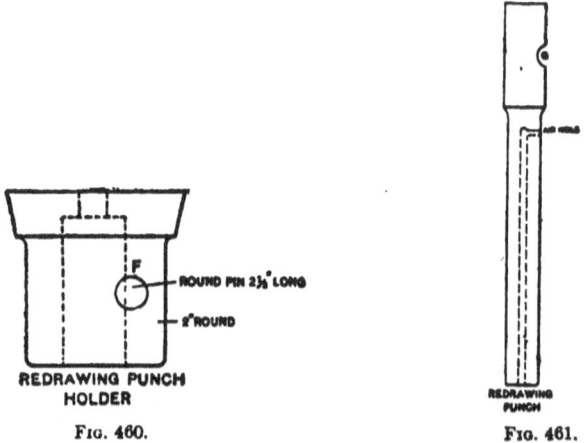

FIG. 460.　　　　　　FIG. 461.

ent drawing dies. It is a well-known fact that when a re-drawing die is worn out, it is shrunk, or, in other words, re-hardened and is ground and lapped back to size. In making the die solid, as shown in Fig. 457, the hole for the set-edge also requires grinding after the die is shrunk. It can, therefore, very readily

PUNCHES, DIES, AND TOOLS. 291

be seen that time and steel are saved by making re-drawing dies in two parts.

Perhaps a word or two will not be amiss at this time in regard to holding the drawing punch in the ram of the press. In many instances the set-screw and check-nut method has been done away with, and another method adopted in its place with better results. Little explanation will be necessary to make this method clear, as the cuts show about all that is required, Figs. 460 and 461.

The gray cast-iron punch-holder fits the dovetail channel in the ram of the press and is held in position by a key (not shown). The round pin shown at F engages in the half-round groove G in the drawing punch, when the punch is placed in the punch-holder, thereby preventing the punch from pulling out when the shell is being stripped from the punch by the sharp stripping edge shown at E. While the drawing punch as shown may appear to be made somewhat longer than is necessary for drawing up the shell B, Fig. 459, it is made long so that the punch can be used also in drawing up various lengths of shells of the same inside diameter.

PUNCH AND DIE FOR REDUCING BRASS TUBING.

In the following are presented sketches and information concerning a punch and die for reducing the ends of brass tubing, as developed by practical experience. The means shown for reducing the tubing consist of a die and punch with a combination of self-contained mechanical movements that can be used only in a press where the ram is at a considerable distance above the bed, and which, moreover, has an unusually long stroke; the travel of the ram in this case being sufficient to permit the easy removal of work $3\frac{1}{4}$ inches long. The tubing used in this die has an outside diameter of .578 inch and is reduced for 1 inch on the end to .471 inch. The walls of the reduced part are slightly thickened and the density of the metal is somewhat increased by the operation, the length remaining unchanged. The work from a die of this kind is uniform as to outside diameter and length, a variation being noticeable on the inside of the reduced part.

292   PUNCHES, DIES, AND TOOLS.

In the accompanying drawings, Fig. 462 is a plan and section of the die and Fig. 463 a sectional of punch. *A* is the cast-iron bed, finished all over and with a way planed longitudinally to accommodate the hardened and ground slides *B*. The pieces before hardening are temporarily dowelled together and indicated on the face-plate of a lathe, the centre of the hole being located where the ends of the slides come together, and hole *C* is machined and polished to size. The slides, after the lathe work is done, are made as hard as possible by heating and dipping in

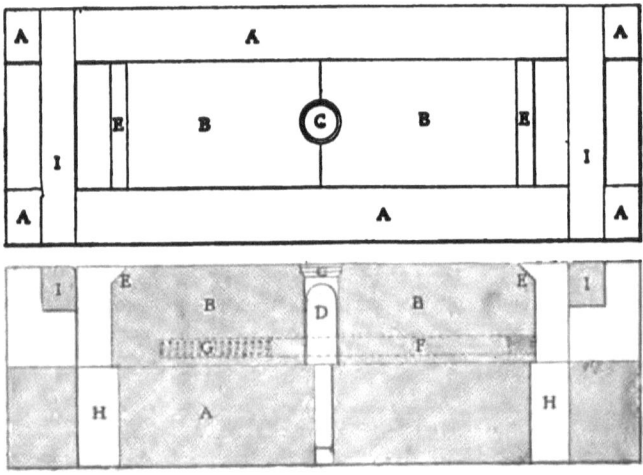

FIG. 462.

a salt-water bath and not drawing them at all. The inner faces of the slides are touched up on the surface-grinding machine to remove any imperfections caused by the hardening and to insure their coming together perfectly, more particularly near the edges of the hole *C*, as the mark of any opening here would plainly be visible on the finished work. The pieces are again dowelled together and the reduced part of the hole *C* is lapped to the required size. From beneath, it is enlarged .002 inch above the size to within $\frac{1}{8}$ inch of its top, to avoid surface friction when the tubing is forced down, as it was found to be impossible to make the die work satisfactorily with a long straight hole. *D* is a stud hardened and driven into base *A*. The space

PUNCHES, DIES, AND TOOLS.

between the wall of the hole C and the stud is the thickness of the tubing plus .004 inch.

The outer ends of slides B are bevelled at corners E 45°.

The construction of the remaining portions of the die and punch, together with their functions when assembled in the com-

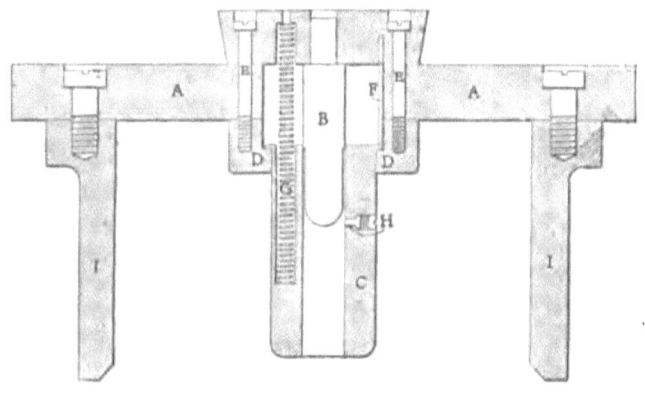

Fig. 463.

pleted tools, and also the operation of the die and punch when reducing the tubing, requires no further description, as the drawings are sufficiently plain to make all necessary points perfectly clear.

INSIDE-OUT SHELL DRAWING.

Fig. 464 illustrates a method for use in a double-action press in drawing a shell and reducing at the same time, as employed in producing aluminum shells in one operation. The same arrangement is used also in the rapid production of catsup bottle caps and other shells made of tin, etc., that cannot be drawn full depth in proportion to diameter without reducing to size. The principles involved are the combining of the usual combination punch and die and the double-action inside-out reduction. The aluminum shells, of which W illustrates the shape, are made of No. 28 gauge, G. P. 12 (or half hard), and they are $\frac{3}{4}$ inch

294   PUNCHES, DIES, AND TOOLS.

diameter by $\tfrac{9}{16}$ inch long, the blank being $1\tfrac{7}{16}$ inch diameter, and they are made from the sheet in one operation.

The die A is what is called a ring die, a combination forging, and the cutting edge of which is hardened and drawn and ground and the body left soft, which enables the screwing in of the combination centre B of tool steel, hardened, and to insert the pressure ring C, which is operated by pins D in connection with a rubber pad and washers and a hollow stud screwed into B (not shown), which allows the shell after reducing to slip

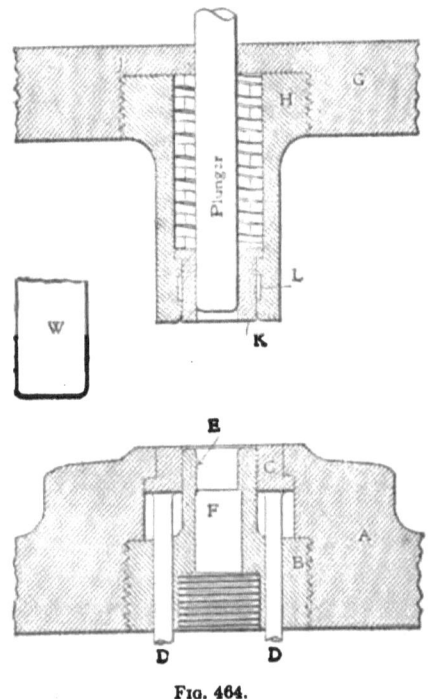

Fig. 464.

through. The combination centre E is bored out (after being turned to draw the first shell) the exact size of the shell; the reducing edge is rounded off as at E, and it has a stripping edge at F.

The punch-block is made of cast iron and has the steel punch H screwed into bottom at J, which bottom acts as a backing for the spring which operates the pressure ring K. The cutting

PUNCHES, DIES, AND TOOLS.   295

punch H is made of tool steel and is bored out to draw the blanks. The boring was a loose fit, or, in other words, instead of boring the punch to take just the thickness of metal, it was given considerably in excess.

The assembling, operation, and accomplishment of the work of the tools can be clearly understood from the illustrations; therefore no further description will be necessary.

REVERSING FORMED SHEET-METAL SHELLS.

The reversing of sheet-metal shells after drawing in order to accomplish something that the ordinary methods would not produce presents many advantages and should be familiar to all die-makers.

Take, for example, the double shell which is used for the

FIG. 465.

elbows of automobile and other cheap horns which require the depth of each shell to be just one-half plus $\frac{1}{16}$ inch greater than the diameter. It is impossible to draw a shell to the shape of Fig. 465, and the tool-maker not acquainted with the reversing method would draw and spin a shell similar to Fig. 466 and

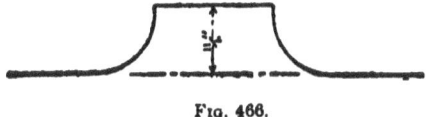

FIG. 466.

then re-spin the outer portion on a chuck of the same shape as the finished article is to be, as in Fig. 465. This, while practical when made of brass, or any other metal that can be annealed, is not practical when made of tin, and is simplified by reversing in the following described manner:

A shell is drawn and spun on a chuck, as Fig. 468, which, as can be seen, throws all the stock up the opposite way, and then, using the spin chuck, Fig. 467, as a drop die, with a lead "force," the stock reverses something as in Fig. 469. Care

FIG. 467.

should be taken in making the chuck that the shape of same is such that there will be just a trifle less stock than is necessary to reach the bottom, as the stock in the reversing stretches, and that the shell clears itself and does not lap or wrinkle before reaching the bottom.

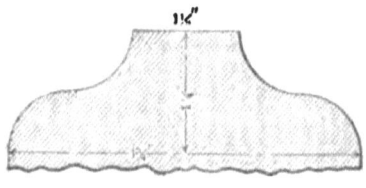

FIG. 468.

Until recently an article, also made of tin, and called by the makers a "cuspidor-hopper," used to be drawn, sunk, and then roller-spun internally to smooth out the wrinkles. Internal spinning is anything but pleasant work, and is to be avoided

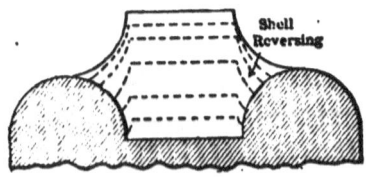

FIG. 469.

whenever possible, and where it comes to spinning out the heavy wrinkles above referred to, and then having the same plated, is a pretty poor job. Anyway, this job was so difficult to perform that it was finally decided to reverse the shell, and then

PUNCHES, DIES, AND TOOLS. 297

see what the result would be. The shell was spun like Fig. 470, and using the old internal spin chuck as a drop die, Fig. 471, it reversed the shell. For simplicity and cleanliness in reversing it beat anything before seen. The wood spin chuck was duplicated in steel and a flange was put on the old internal spin chuck for a bead to bottom on, using the same as a drop die. Reversing the shell in this fashion not only cheapened the cost of manufacture by throwing out four operations, but overcame the diffi-

FIG. 470.

culty in spinning out the wrinkles and the disagreeable job of internal spinning, which is always costly.

One of the advantages of reversing, especially in tin, is that the upper or spun surface is invariably the part that is to be plated or polished; and if the spin chuck is made properly, so

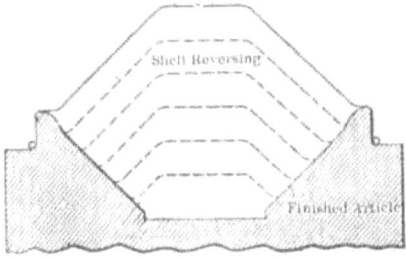

FIG. 471.

that there is always a little less stock than is absolutely necessary, the stock, when reversing, will stretch instead of wrinkle and will always leave the upper surface smooth and clean.

Reversing is not restricted to the articles above mentioned, but is applicable to anything of a similar nature, and the object in selecting these two is to show the comparative difference in shape and the final results the same in both.

DRAWING A CENTRAL HUB IN HEAVY SHEET-METAL BLANK.

The description and illustration of a method and tools for making the stamping shown in Fig. 472 will be of interest to all who encounter difficulties in producing work of a similar nature.

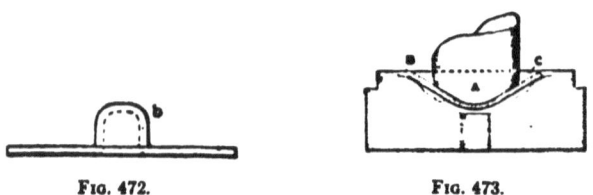

Fig. 472.   Fig. 473.

Afterward the rim was drawn up in the same direction as the hub, in dies of the usual drawing-type construction; however,

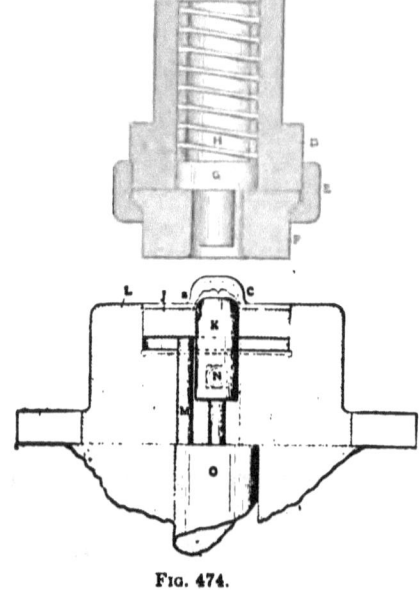

Fig. 474.

here we only describe and illustrate the tools for producing and finishing the hub. The outside diameter of the article when

PUNCHES, DIES, AND TOOLS. 299

finished was about 2.65 inches. The orders on this job were to produce a shell in which the thickness of metal throughout the entire stamping *must* be No. 12, Bir. gauge, or .109 inch. An excess in thickness of the back of the stamping of .015 inch was finally allowed, as there was no press stiff enough to hold the metal to gauge. The main difficulty encountered was to draw the small boss in the centre of Fig. 472 without stretching the metal. After experimenting for some time with the usual method of drawing and re-drawing without success, it was de-

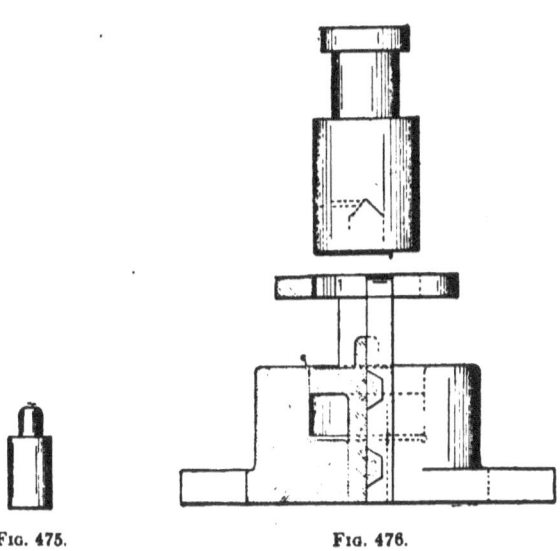

Fig. 475.   Fig. 476.

cided to launch out into new fields, and the desired result was accomplished in the following manner:

The cutting of the blank, which is No. 12 Bir. gauge sheet steel, needs no comment, and is passed on to the second operation shown in Fig. 473. This figure is a vertical section of die B and work C, and a partial axial section and elevation of punch A. The blank is placed in the recess at the top of die B, and the punch A is brought down until the diameter of C is two or three thousandths of an inch less than the push-out J, Fig. 474. The depth of C must be sufficient to give stock enough to form the boss, Fig. 472, after the process of upsetting, the first operation of which is shown by C, Fig. 474, and the last one by Fig. 472. A and B are soft steel hardened in bone dust, are easily worked, and give good satisfaction.

Fig. 474 represents the punches and dies for performing the operation of "upsetting." D is the die-holder, and fits the ram of the press. E is a ring for holding the die F central against the face of D. L is the punch holder, recessed at the top, in which the "push-out" J slides. A hole bored in the bottom of this recess holds the punch K, secured by the screw N. This punch passes up through J and works in conjunction with F. O is the stem of some form of positive knockout, and, through M, actuates J.

This is a reversal of the usual order of things, and is necessary from the fact that if one puts C, Fig. 474, through another

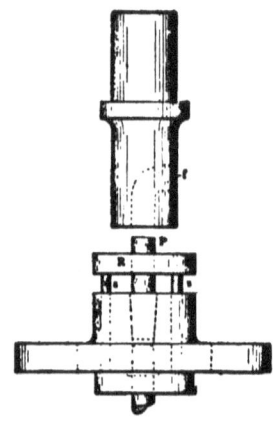

FIG. 477.

"upsetting" operation, in dies made in the usual form, the metal in the. boss "flows" out into the flange, and when one reaches the last operation, the metal is stretched and of no use. By the method shown, C, Fig. 473, is placed on J, Fig. 474, convex side up; F is brought down, and as C is surrounded, the metal flows out into the flange in a slight degree only, and, with a press suitable for the job, is forced into the corner a, Fig. 474. There are five operations or re-drawing operations, the dies and

PUNCHES, DIES, AND TOOLS. 301

punches of which are interchangeable as regards the fit in holders. Good material should be used in this set of dies, and a press of not less than 125 tons capacity. D should be hardened and ground on the face next to F, and should fit the ram of the press

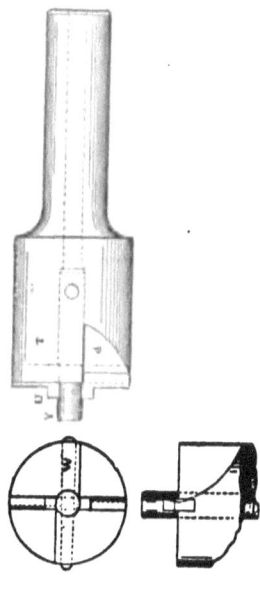

FIG. 478.

at both end and shoulder. The recess L should also be hardened and ground.

Fig. 475 shows the punch which goes with Fig. 472. Fig. 476 illustrates the punch and die for the sizing operation. Fig. 477 illustrates the tools for punching out the hole in the hub; and Fig. 478 shows the trimming cutter used for finishing the rim of hub.

SECTION XII.

Beading, Wiring, Curling, and Seaming Punches and Dies for Closing and Seaming Metal Parts.

AN EXPANSION PUNCH FOR BEADING.

FOR the beading, wiring, curling, and seaming of finished press work various methods are used. In this section we illustrate a few of the most approved methods and describe the construction and application of the tools used for attaining the desired results. Modifications of the design can be adopted for

FIG. 479.

the attainment of results in a large variety of work of both round and square shapes.

The punch and die illustrated in Figs. 479, 480, and 481 were used for throwing out a semicircular bead in a shell of the shape shown in position in the die in Fig. 481.

First make the punch shell A, then make the inner punch B. Sweat A and B together with soft solder and saw cuts into A on the milling machine, using B as an arbor and cutting through A and a little into B, so as to prevent fins forming on

PUNCHES, DIES, AND TOOLS. 303

the inside of the cuts. Then melt apart and turn B down small enough so that A can contract when in use, at the same time removing the saw marks in B.

As there is not a beading die for this punch, consequently the bead cannot be sharply defined or very regular, besides the diameter of the work is increased. To do a really good job, a beading die is also necessary, as shown in Figs. 480 and 481, which shows the complete tool with the punch A contracted. The punch and die constructed as shown allows no chance for

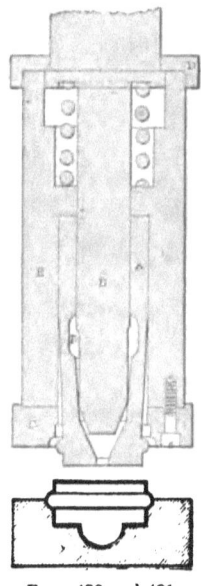

FIGS. 480 and 481.

the sections to get out of order. A and B are the same as in Fig. 479, while E is a soft shell into which A is driven and carries the upper beading die C. Punch A and die C come down together, being forced ahead by a spiral spring at the top, which must be stiff enough to hold the die C down while the bead is being formed. A and C bank at the same time, while B continues to descend until it has expanded A sufficiently to make a good bead.

The recess F is bored in A to give the sections sufficient springiness and should be spring-tempered at this point; all

304 PUNCHES, DIES, AND TOOLS.

other hardened parts to be drawn only to a good yellow color. Only the working parts or end of punch *B* need be hardened. *D* is a soft ring to hold the shell *E* against the tension of the spring.

A SIMPLE WIRING DIE AND ITS WORK.

Formerly it was the practice to make wiring dies for railroad torpedoes different from that shown in Fig. 482, that is with no

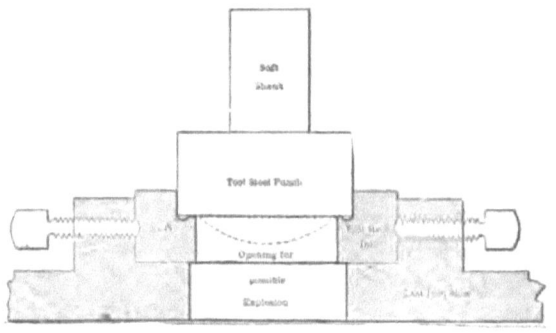

Fig. 482.

drop at *A*, simply a square shoulder with a slight fillet; and the top blank was expected to crawl under the lower one and be

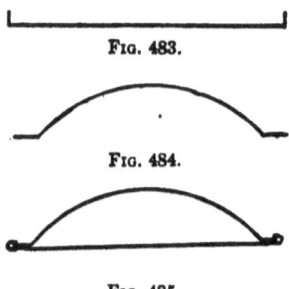

Fig. 483.

Fig. 484.

Fig. 485.

flattened enough to hold the lower blank and also the explosive, as at *B*, Fig. 486. Afterward, however, the tools were changed

PUNCHES, DIES, AND TOOLS. 305

like *A* with splendid results. The metal used in making these torpedoes is light tin, and the opening at *A* is one thickness of metal. The working parts were hardened and tempered. Fig.

FIG. 486.

483 shows the top blank before wiring, Fig. 484 the lower blank before wiring, and Fig. 485 the parts as wired and assembled in the die, Fig. 482.

EXPANDING A DOUBLE BEAD IN A BRASS CUP.

In manufacturing the brass cup, Fig. 487, press tools of rather unusual construction were used. They are shown in

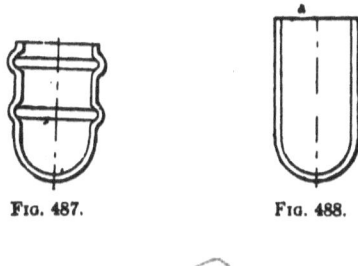

FIG. 487.   FIG. 488.

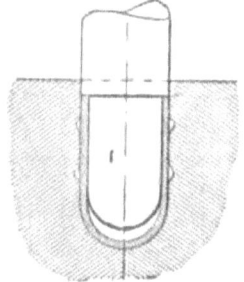

FIG. 489.

Figs. 489 and 490. The job was done in two operations: First the shell, Fig. 488, was drawn in the ordinary manner, and after

20

being trimmed on the edge a, it was finished in the expanding die shown in Fig. 490. In order to extract the shell after it had been expanded, it was necessary that the die should be made in halves, and two pieces of die steel were planed, dowelled, and clamped together and finished, with the grooved hole equally

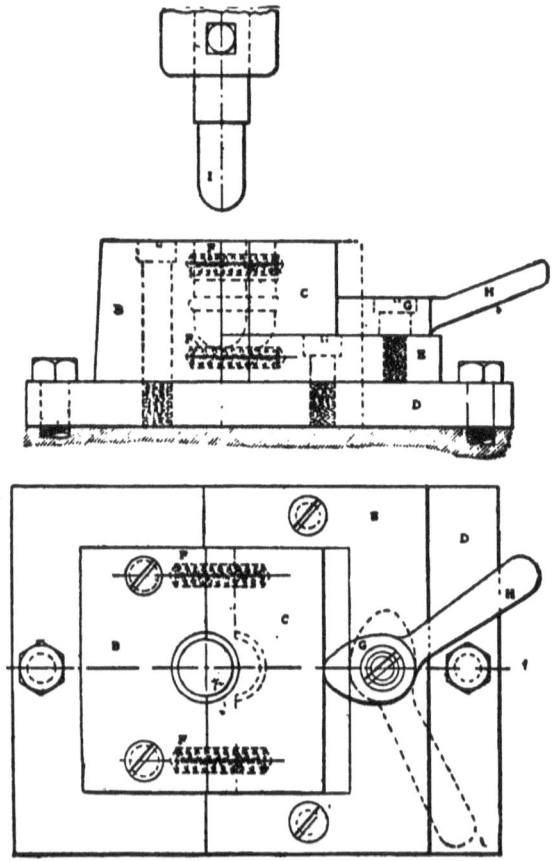

Fig. 490.

halved between them. One block, B, was screwed down to the gray-iron plate D, while the other, C, was free to slide upon it, guided by the machine-steel block E. Placed between die blocks B and C, and retained in position by holes in the same, were four strong coiled springs F. Working on a fulcrum pin G, which screwed into E, was the machine-steel cam H, which was

PUNCHES, DIES, AND TOOLS.

conveniently placed to open and close the die against the pressure of springs *F*. The method of using was briefly as follows: The shell having been placed in the die, which, of course, was closed, the tool *I* (the diameters of which corresponded exactly to the inner and outer diameters of the shell) descended to the position shown in Fig. 489 before doing any actual work. It will be readily seen that by reason of the tool being an exact fit in the shell and the shell bottoming in the die, when the tool continued to the finish of its stroke the flow of metal was bound to force the shell outward into the grooves. The tool was withdrawn, the cam thrown back, and the finished shell extracted.

A PUNCHING AND HALF-WIRING OPERATION.

In the following is presented a method for half-wiring a circular shell. The shell was not completely curled around, but to simply half a circle, to hold an insulation in the part curled, which, in this case, happened to be the bottom of the shell. There

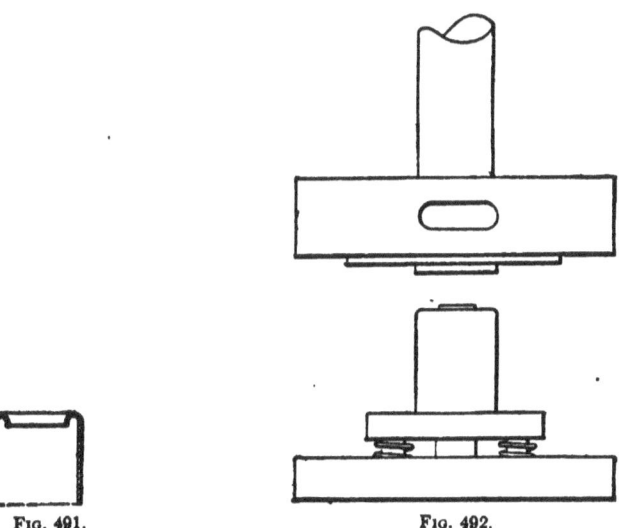

Fig. 491.   Fig. 492.

were two operations performed at once, which was a great saving of time in the production of the shell. The two operations were punching out the bottom of the shell, and forming the half-circle as the press reached the end of its stroke.

308 PUNCHES, DIES, AND TOOLS.

The construction of the punch and die is as follows: Fig. 491 shows the shell as finished, Fig. 492 shows a front view of the tools complete, and Fig. 493 illustrates a sectional view of same. *A* is the punch to punch the bottom out of the shell. It is riveted into *B*, which is the base and fits into the shoe of the press. *C* is the former, with a radius at the top to suit the curve re-

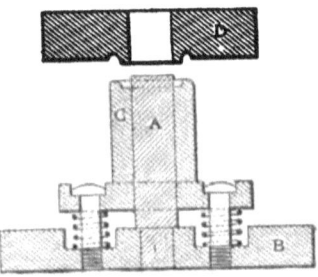

FIG. 493.

quired. It is held up from the base to steady the shell while punch *A* does its work. *D* is the die, and it is also the former which comes down and forms the shell over *C*, and when the die is down to its full stroke punch *A* has pushed the punching of the bottom of the shell out through the slot in the die-holder. Die *S*, after it has formed the shell, has pushed the metal away from punch *A* enough to allow the shell being lifted off easily by the operator.

DETAILS OF HORNING AND SEAMING OPERATION.

In Figs. 494, 495, 496, and 497 are shown details of a progressive seaming set of dies for round work, such as tin pails,

FIG. 494

cans, etc. By one handling of the work, the edges are first upset and finally interlocked and closed together by two strokes of the press.

PUNCHES, DIES, AND TOOLS. 309

Fig. 495 shows a front view of the punch and horn as they are placed in the press. The punch A, by the way, represents a good job of planer work. The cast-iron form holds three pieces of tool steel B, D, E. The pointed edges of D and E are planed one thickness of metal smaller than the grooves in piece B. The function of piece B is to slide in the direction of the arrow to dotted line F. The lower surface striking the top of the horn, it is pushed against springs at C, bending the edges

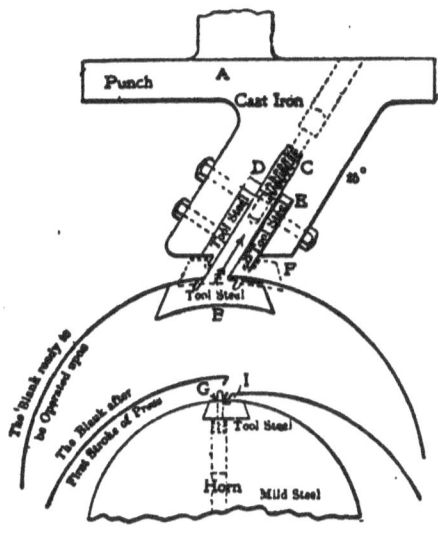

FIGS. 495 and 496.

just as shown in the blank, Fig. 496, after the first stroke of the press. As B leaves the horn on the upward stroke the springs C force B to its normal position again, being held there by fillister-head screws shown more plainly in Fig. 497.

To complete the seam, the edges of the blank are placed together and held directly on the horn and against spring set-edges G and positive stop-gauge H, Fig. 497. The press is then tripped and the under side of B flattens the edges of the blank like the finished seam shown in Fig. 494. This form of seaming die, used in conjunction with the horn, is made for work about 12 inches long. For any longer lengths the horns are supported from beneath by a removable brace; that is, supported by a brace that is instantly and easily swung to one side to admit placing the blank on the horn. There are two forms of seams

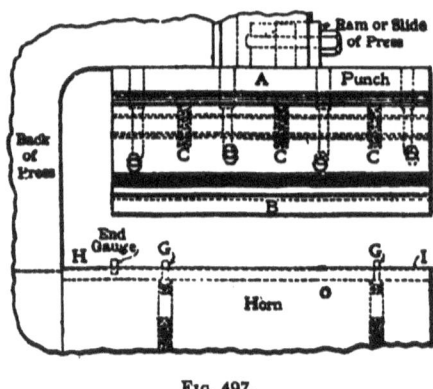

Fig. 497.

used for work of the class mentioned; one turned out and one turned in. The method here described is for the turned-in type; the outward seam is made by planing the groove I in B, and the steel in the horn is left plain.

TOOLS FOR TAPERING HOLLOW SCREW TOPS.

Computing scales, such as are used in the retail trade, are frequently provided with turn-tables, for convenience in turning them, so they can be operated from either side or end of counter. To avoid marring the counter upon which they are placed, rubber-tipped, adjustable foot-screws are provided, which are used also as levelling screws.

The tips were formerly cemented in, but they came loose frequently, so a coaxing or swedging operation was resorted to. The tools shown in Fig. 498 were made to be used on the bench as no press was available.

In Fig. 499 are shown plan and side views of tip, moulded to just fit recess in head of screw, as shown in Fig. 500, and in which they were formerly cemented. In Fig. 501 is shown the

PUNCHES, DIES, AND TOOLS. 311

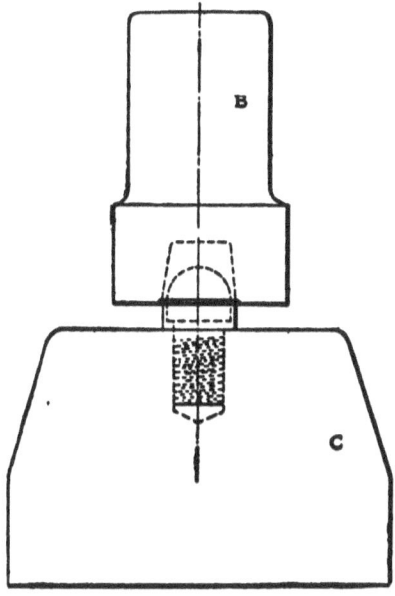

Fig. 498.

Fig. 499.

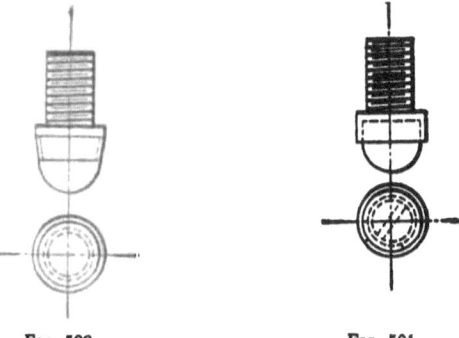

Fig. 500.   Fig. 501.

coaxer *B*, of tool steel, having a taper hole large enough to just allow the head of the screw, with the tip inserted, to enter, and a cast-iron die-block *C* having a hole into which the body of screw fits loosely. *D* is a bench plate on which *C* rests. A light hammer blow on upper end of *B* gives the shape shown in Fig. 501.

SECTION XIII.

Jewelry Die-Making, Eyeglass-Lens and Medal Dies—The Construction of Spoon-Making Punches and Dies.

THE WORKING OF GOLD-FILLED MATERIAL.

THE secret of making good mountings of gold-filled stock lies in so working the material that the gold—which is comparatively a very thin skin covering a brass or filling—shall as far as possible be kept completely covering the filling and free from cracks or rents of any kind which may expose even slightly the base metal. These little bare spots will soon be turned either black or green from the perspiration which, from many persons, is very corrosive and readily attacks these little evidences of

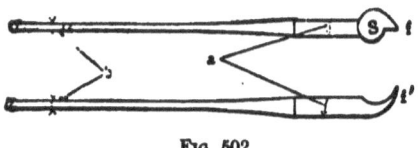

FIG. 502.

carelessness, ignorance, oversight—call it what you will—that allows a defective piece to leave the factory.

In making "bows," or, as they are commonly called, "temples," a portion of which is shown in Fig. 502, the operations are as follows: The gold-filled wire, which is usually purchased from people making a specialty of this material, comes the diameter of a, Fig. 502, and the end which curls behind the ears of the wearer is swaged down to the diameter of b. Of course, swaging also reduces the thickness of the gold, but it is nowhere broken or bruised by the process. Solid gold, German silver, red metal, steel, etc., are all reduced in these swaging machines, some hand feed and some automatic feed. In the case of all

314 PUNCHES, DIES, AND TOOLS.

solid metals, the ends of the temples, which hinge in the "end piece" soldered to the "rims" or "eye wires," are flattened to the proper thickness shown by the dotted outline, Fig. 502, f, and a cutting punch and die trim them afterward to the shape shown by the full lines in the same figure. This trimming cannot be done in the case of gold-filled stock, as this would expose the filling all around the edge of this flattened portion, and the die we are about to describe is for the purpose of flattening this portion of the temple to shape with the gold covering intact all around, and at the same time show the use of a little trick which makes it possible to form up the metal to the point f at the end of the temple.

The temple, after swaging, is pinched off in a die, shown in the lower portion of Fig. 503, to about the shape shown at f',

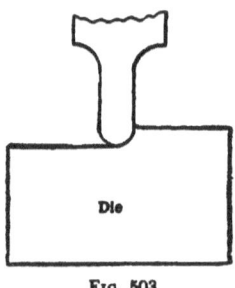

FIG. 503.

same figure. The gold being more ductile than the filling, leaving it flat and thin and neatly covered—the two thicknesses of gold seem to unite at the point. Cutting off these ends would expose the filling. The temples are now flattened to the desired shape in a die, the description of which will be of interest to those engaged at this branch of tool-making.

The die, etc., Figs. 504 and 505, consists of a bed k whose bottom is planed to rest on the bed of a press. It has a hole bored from the top with a flat seat for the plunger j, and is counterbored to receive die D, which is fastened by two screws $h\ h$. A hole to receive m is bored through k at right angles to the front and parallel with the bottom of the bed. The axis of this hole is on a level with the flat seat for j. A tool-steel piece

PUNCHES, DIES, AND TOOLS. 315

m, with its upper side milled out, as shown in the section, has a flat level with the seat for *j*, and is fitted to turn easily in this hole. The end of *m* toward the observer has a lever *l* fitted to it, as shown, and a pin, not shown, prevents *l* from turning on *m*. The left end of *l* has an eye to which is attached a link *'l*, which in use is attached to the ram at its upper end. The die *D* is worked out to the shape shown at *s*, Fig. 502, which shape represents also a section of punch *P* and ejector *i*. *i* is made of such length that when plunger *j* is at its upward limit of travel, ejector *i* will come nearly to the top of *D*. The die *D* is very slightly enlarged at the top, so that when the flattened temple

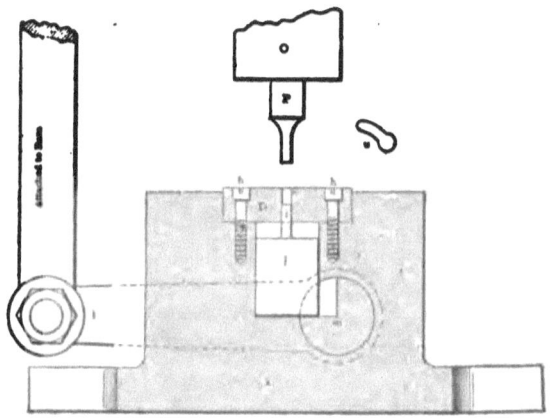

FIG. 504.

is raised to this point by the ejector *i* it will be loose enough to be removed easily. Ejector *i* is a close but easy fit in the smooth hole of die *D*. To secure best results in the finished temples the hole in the die *D* must be very smooth of interior finish.

In making these dies, whose shape must be duplicated, about the following procedure takes place: The die *D*, which is sawed off a bar already turned to the proper diameter, is faced on both sides, and an approximation of the size and shape required is worked out with drill and file, after which a cutting punch, a little smaller than the finished size, is pushed through *D* in a screw press, and then two or three other punches or burnishers are worked through.

316   PUNCHES, DIES, AND TOOLS.

These burnishers or punches are of exactly the same shape as the cutting punch, but the ends have the edge well rounded and smoothed, instead of being sharp, since their office is to smooth and not to cut. They are made in two or three sizes, there being perhaps about .002 inch difference. They are smoothly finished, the direction of the finish being around the punch, or at a right angle to the direction of motion. They are, of course, hardened and tempered to a dark straw.

In use No. 1, which is the smallest and just a trifle larger than the cutting punch, is pushed through first. This is followed by No. 2, which is a little larger, and by No. 3, still larger, and

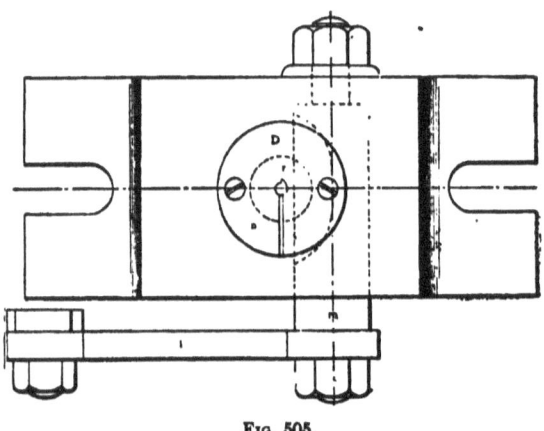

FIG. 505.

the finish size is worked through last. These are best used in a screw press, the punch and die being fastened so that neither can move. Both are greased and the punch is worked up and down, entering but a little at first stroke and being withdrawn, another slightly longer stroke being taken the second time. The process is thus continued, each time the punch being entered a little deeper and withdrawn until it has been entirely worked through. All three burnishers are used in the same way, and the result is a very nicely finished hole in the die. This operation will throw up the metal a little around the hole, but this is easily disposed of. After easing the hole at the top to allow the temple to be easily removed after flattening the die may be annealed and No. 3 punch again worked through.

JEWELRY-CUTTERS AND CUTTER-PLATES.

The above title may not be recognized by many die-makers as having any connection with punch and die-making; but it is the universal name by which that class of tools is known by those engaged in making or using them. The number of dies used and the amount of tool-making required in the jewelry business are very large, and many ways have been devised to reduce the cost of the tools and still secure the desired results, some of which we will try to present for the benefit of readers who "kick" over the cost of their die work.

The cutters, Fig. 506, as punches are called in the jewelry business, are, in nearly every case, made with round shanks, $\frac{1}{4}$ to $\frac{3}{8}$ inch diameter, and used soft, tapered punches being rare

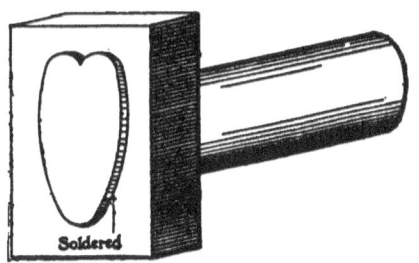

FIG. 506.

tools to find in a jewelry shop. The cutter plates or dies are made in the usual manner, except that they are generally made of very thin steel, $\frac{1}{4}$ to $\frac{1}{2}$ inch thick, and given excessive clearance. This is allowable as the stock, silver or gold plate, is very easy to cut and gives but little wear on the tools. We have often seen dies that had been in use for a year or more that still had the color from the tempering process left on the face. The dies are milled out on an inverted milling machine (every jewelry shop of any size has one of these machines, and, in fact, every die shop should have one), using a very small cutter, often under $\frac{1}{16}$ inch diameter at the point, and leaving very little to file.

The filing is all done vertically, and if brother die-makers would only give this method a good trial and get used to it, they would always prefer to finish a die in that manner. Grasp the file, letting the thumb come well up on the body of the file, and have the left forefinger touch the lower part of the vise to serve as a guide to insure the file moving in a vertical line. Work on the side of the die farthest from you, and after a short trial you will find it a vast improvement over filing in the vise in the usual way.

The punches are often built up as shown in Fig. 506, the working part being got out of $\frac{1}{16}$- or $\frac{1}{8}$-inch steel, and then hard-soldered on to the shank. Tools made in this manner are all right, provided they are to be used for cutting silver or gold plate, or even brass stock. The cost is about one-half that of a solid punch, and when the thin steel is worn down a new piece can be quickly got out and soldered on to the same shank. As mentioned before, although many of the tools are made in this way, yet some of the finest and most expensive tools we have ever seen, as well as some of the best tool-makers, were in the jeweler's shop.

A BIT OF JEWELRY DIE WORK.

The drawings, Figs. 507, 508, and 509, show a punch and die for forming a short piece of German silver wire into the shape shown at A, Fig. 510. This shows at the left the pin as it comes

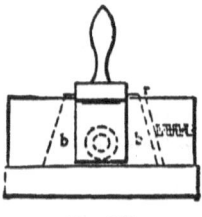

FIG. 507.

from the automatic screw machine with a small tit turned on the end. The pin is inserted between the jaws, $a\ a$, in the die in this shape, and the three other views show the finished article.

PUNCHES, DIES, AND TOOLS. 319

The die bends the pin, flattens a portion of it, and raises a tit on the flattened part. This job was formerly done in two operations—the first bending the pin and the second forming it—but this proved unsatisfactoy as well as slow.

The die is used in the following manner: A pin is inserted between the hardened tool-steel jaws *a a*, which are dovetailed

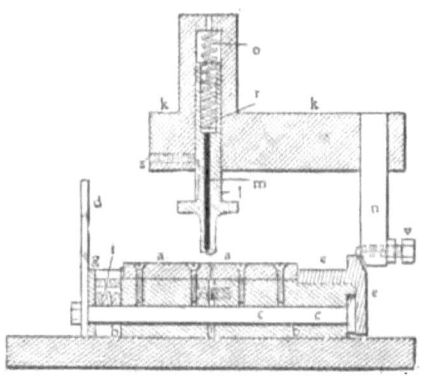

FIG. 508.

into the soft tool-steel slides *b b*, and these in turn are dovetailed into the cast-iron die-bed. The handle *d* is then brought down toward the operator, and, being connected to the cam *c*, this locks the jaws, *a a*, slightly. As the ram of the press comes down, the forming punch *l* strikes the pin with its bevelled

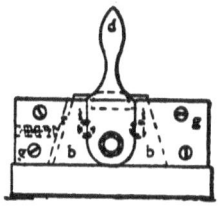

FIG. 509.

front. At the same time the shifter *n* strikes the shoe *e*, and pushes the slides *b b*, as well as the jaws *a a*, forward under the punch *l*, thus bending the wire. During this operation the shifter compresses two stiff springs, *i i*, and increases the pressure between the jaws *a a*, and this pressure continues increasing until the shifter *n* has passed the highest point of the shoe *e*. A trifle later in the downward stroke the shoulder of the forming punch comes to rest against the cast-iron punch-holder *k*, securing a positive pressure, and the pin is formed—that is, the flattening takes place and the tit is raised.

While the punch descends, the spring *o*, between the punch-holder and punch, exerts a stiff pressure against the long steel pin *m*, which is filed down at the lower end to the size of the hole in the punch *l*, and which allows the tit to rise in the

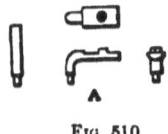

FIG. 510.

punch. Between spring *o* and the steel pin *m* a washer *r* is placed. On the up-stroke this steel pin acts as a stripper to the tit and releases it before the shifter leaves the highest point of the shoe. As soon as this takes place, the slides *b b* return to their normal position and the handle *d* rests against the plate *g*. The cam *c* is then released by turning the handle upward, and the finished pin is taken from the jaws.

It will be noticed that the slides *b b* are kept apart by two springs. These springs separate the jaws only far enough, however, to allow the pin to be inserted or removed, as they would be liable to drop into the open space between the slides if this exceeded the thickness of the pins. The cam portion is shown by the dotted line at its largest diameter. The screw *s* in the cast-iron punch-holder engages in a slot milled into the punch *l*, keeping this from turning and at the same time acting as a stop against the downward pressure of the spring *o*.

The handle *d* was made from soft steel and case-hardened. The soft-steel plate *g* is secured to the die-bed by four screws. The shoe *e* was made from tool steel and hardened without drawing the temper. It is secured to the slide *b* with two screws, not shown. The shifter *n* also is made of tool steel and hardened.

METHODS FOR JEWELRY DIE-MAKING.

There are various methods for jewelry die-making, all very much similar to each other, but each possessing some little advantage or kink, that makes the user of it prefer that method for sinking small dies.

It is much easier, usually, to work on the outside of anything than the inside, with everything easy of access—much easier than the inside, especially of small dies. How much easier it is to make a small tap, and with it tap out a hole, than it would be to thread it by other means. It is much easier to make a master punch just right and then strike up the die blank, than it would be to work out the desired shape with end mills, "cherries," gravers, rifflers, emery, etc. It is very rarely that these means have even to be resorted to in correcting defects in the impression, and only when it is some slight defect that would require less work in the die than on the punch.

A die struck up by this or any other method can be brought to a finish as far as hardening; but before doing this it can be set in the press and a few soft lead blanks may be embossed for trial, measurement, etc., and the usual corrections are made *sometimes* in the die, but usually on the punch, because it is easier and the die can be struck a second time or oftener, as the case may require, to bring about a satisfactory result. There are times when it is a greater convenience to make a slight change in the die than in the punch, because of less work being required to make this correction in but one spot of the die, whereas to make the same change in the punch would require the entire punch, all but this one spot, to be worked.

When making necessary changes in dies, and sometimes in punches, round sticks of India oil-stone will be found to give excellent results. To shape the stone to the work, run slowly in the speed lathe or drill press and file to shape, then for the grinding operation speed up, and use kerosene to prevent clogging.

There are many parts of spectacle mountings that must be pretty close to size, shape, etc., but usually embossed parts require symmetry and beauty of finish rather than exactness of dimensions. In working solid gold, however, when for economic reasons it is not desirable to allow parts to exceed a given weight, this point can usually be more easily taken care of by regulating the thickness of the punchings, the depth of the embossing dies, etc.

Dies for this kind of work often wear through the grit upon the surfaces of annealed punchings. Records of from 7,000 to 12,000 embossed punchings have been made, depending upon the material worked and the cleanliness, etc., of the punchings, gold-filled stock giving the best records. The working surfaces of the dies soon get dulled, and then the embossings will come out no better than the finish of the die, and for this reason, especially for gold-filled work, the die needs to be repolished. Seldom is a die used after its size has been enlarged beyond a certain point, which should be but little larger than the original size, because the embossings from such a die would give trouble

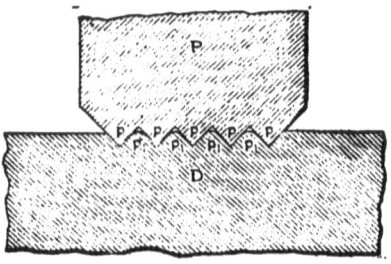

FIG. 511.

in subsequent operations, by coming too large for the chucks, etc.

Grinding off the face of a worn die does not help matters much, because it only has the effect of giving a wider and thinner embossing.

To prove that only a limited number of shapes can be struck up cold it has been shown that a clip with V-shaped grooves running the entire length of it requires dies made in a different manner. The grooves in the piece were required to be brought to a sharp angle. Some die-makers claim that this cannot be done cold, but at the same time admit that it could not be done with a hot die-blank. It would be interesting to demonstrate the producing of a first-class job on a piece of this kind, striking it up cold with all the fine detail and with much less labor than would be put in by sinking much of the impression by hand. It may be said that almost all embossing dies should be struck up cold, and all of the fine details produced in no other way.

It would be foolish to assert that all dies, large and small, could as well be sunk cold as hot, though the impossibility of it is not admitted here, and any one wishing a vivid demonstration of its possibilities would do well to examine the impression that a projectile will leave in an armor plate, and note how faithfully

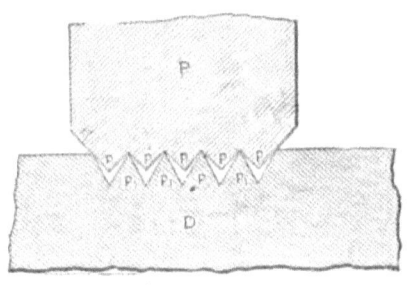

Fig. 512.

the shape of the impression corresponds with the shape of the projectile. Sinking large dies by this method, however, may never become commercially practicable, but this does not affect the case here presented, which deals only with small dies.

It will be profitable to consider for a moment the action of a punch in producing work of the character shown in Figs. 511 and 512. Let us suppose P in Fig. 511 to represent a section of a punch to produce the grooved piece with very sharp angles: D is the die. In action, when D is struck by P the points $p\,p\,p\,p$ cut into the surface of the tool-steel blank, acting not unlike a wedge penetrating the metal, and as P continues downward the material, already pierced between $p\,p$, etc., must slide up the sides of these V-shaped grooves, and as the space between it is continually lessening, the steel must either rise into the apex of these grooves or be compressed, or both—which actually does take place—and finally form up sharply into the sharp angle of these grooves, which is seldom the case. In practice we observe that the ideal is but partly attained, and the reason for it is quite simple enough, too. The metal must flow into the V's, being compressed because of the more confining space it is compelled to fill, until the pressure required to bring these V's sharply becomes greater than the resistance of the metal supporting that portion of the die blank, so that instead of the impression being formed up sharply as desired, it is simply carried deeper and deeper as successive blows are struck, and usually they improve in shape very little by repeated blows. In Fig. 511 the rounded points $p_1 p_1 p_1 p_1$ are intended to represent these partly formed V's.

These are the conditions met with in forming up die blanks cold; and we do not escape them by heating the blank, for in the same measure that we soften the blank by heating do we also lessen the resistance of the metal supporting, and the result is the same in both cases, *i.e.*, an imperfect impression. There may be an advantage with a hot blank in that the texture of the metal is probably less disturbed; but, as shown in the life of dies struck cold, this seems a very slight advantage.

If we use a drop press for striking up these dies better work usually results, for it seems that the quicker the blow is struck the more effective it is in producing detail, and for this reason we usually prefer to advise the use of the drop press.

We have often had occasion to produce the kind of gripping surface mentioned above, and while usually perfectly sharp angles were not secured in the die—trusting to the drop press alone—yet they were all that were required. The die as simply struck up had those points $p_1 p_1 p_1 p_1$, very slightly rounded, and when a punching was embossed in this die it (was of course opposite) had sharp V edges and round bottoms which seemed to meet all requirements, since it is quite doubtful if the human skin would form up sharply therein if they were sharp and still more doubtful if the human mind would favor it thus.

When it is desirable that these details should be brought up sharply, the effect is produced not by trying to overcome this natural habit of metal to resist being formed up thus, but by

PUNCHES, DIES AND TOOLS.

providing conditions which will enable it to be formed easily. This we can easily do as follows: Suppose that we have a master punch of the desired form and detail, and a die blank, both with the working surfaces highly polished. We mount both in a drop press and strike a blow sufficient to make a fairly deep impression, enough to give a fairly good result. We can now take out the die and scrape the grooves until we have brought up the points $p_1 p_1 p_1 p_1$ to a sharp edge and the angle a little more acute than desired in the finished piece, so that it would appear as in Fig. 512, where the effect is shown exaggerated. Polish out, avoiding scratches or rounding the points; return the blank to the press, and strike another blow, or more if necessary, to reach the required depth, etc. The entire impression will now be sunk deeper and the V's will form up to the same angle as those in the punch and exactly just as sharp.

The advantage of this cold-striking method is that much less careful hand-work is required, because we are obliged only to sharpen the V's and pay no attention to the rest of the impression, and when finished it is exactly like the punch. The whole trick lies in making the angle of the grooves sharp and slightly more acute than the finished piece is required to be.

CUTTING, BENDING, AND PIERCING GOLD STOCK.

Fig. 513 represents a piece of punch work that was made from gold stock and which required great accuracy in shaping as well as being bent at a very sharp corner as shown at B. Three operations were performed upon the piece as follows: First, cutting the blank as shown at A; second, bending as at B; and finally, punching the small hole in the bottom as at C. The metal was .020 inch thick and was lapped after being soldered on to a piece of jewelry. Several methods of bending were at first tried, but they proved unsatisfactory, as too much trimming was required to obtain the necessary finish. The problem of bending was solved satisfactorily by the use of the method described here.

The work was produced by the operations in the order above

326   PUNCHES, DIES, AND TOOLS.

mentioned, but in order to obtain the proper size of the blank, the bending was done first. This die is shown in Fig. 514 and, as will be seen, is very simple in construction. The die proper, D, was planed on its face to receive the gauge plates E E. A slot, F, was also planed to facilitate sliding in of the blanks, and another slot, running at right angles with F, served in conjunction with the tongues of E E to locate the blank in position for bending.

In order to obtain the square hole, which was of the exact outside dimensions of the blank, a hob punch was made in the milling machine and the opening broached through the die,

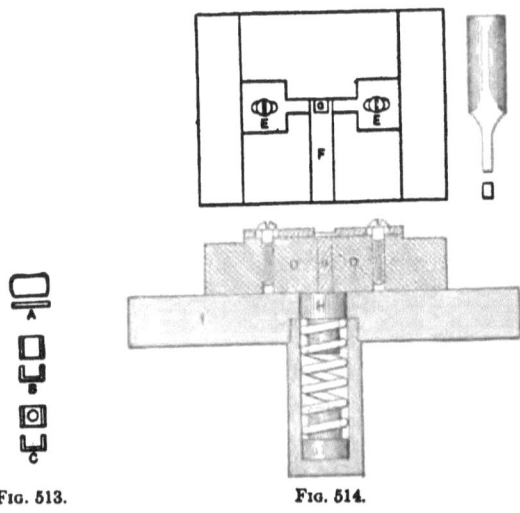

FIG. 513.   FIG. 514.

starting from the back. In broaching a hole of this nature care must be taken not to remove too much metal, otherwise the surface will be rough and the corners will not be well defined. As the broach approaches the front of the die the amount of metal removed should decrease until the corners only are touched. After broaching, the die should be highly polished with a palling stick and hardened, drawing the temper but slightly. It is then again polished brightly with a brass stick and flour of emery. In the opening was fitted the plunger G which served as a seat for the work while being formed, as well as acting as a

PUNCHES, DIES, AND TOOLS. 327

plunger to release the piece when it was completed. This plunger rested on the knockout *H* which was actuated by the coiled spring, upon which it rested. *I* represents the cast-iron bed plate into which the brass casing for the knockout was screwed. The forming punch was twice the thickness of the stock, narrower than the opening one way and of the same size the other way. With this punch and die we were able to bend the metal to a very sharp angle.

By the cut-and-try method, in connection with the bending die, it was a simple matter to determine the proper length of the

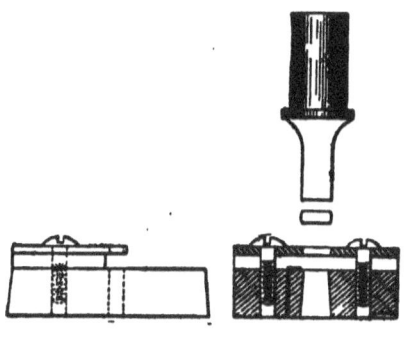

FIG. 515.

blank, and the blanking punch and die were then made. The punch was simply turned down and milled to the proper width, and with this the die was broached. It has been found in all cases to be good practice to make the punch first in all cases where it can be machined and then with it broach out the die; this method of procedure invariably resulting in a saving of much time. The blanking punch and die are shown in Fig. 515, and will be understood without further description.

The perforating die by which the last operation was performed is illustrated in Fig. 516. A stripper plate *K* was fastened on to the base of the die *J*, by screws and dowel pins, and had a projecting arm *L* through which a hole, the size of the perforating punch, was drilled in line with a corresponding hole in the centre of the hardened steel die. The indexing dial was made of steel about $\frac{1}{16}$ inch in thickness. The top and bottom

328   *PUNCHES, DIES, AND TOOLS.*

were accurately faced and the edge knurled to facilitate turning. The stud M was inserted in place with a thickness of paper under the head so that it would bind the disk tightly to the surface of the die. The die was then placed on the milling machine and the slot for the lever N was milled in the disk, using the slot in the hardened die as a guide. The die was then turned one-quarter of a revolution and the next slot milled in the disk; and so on until all of the four had been cut, using the same cutter that was employed for making the original slot in the die. The index lever N is forged from a piece of tool steel and fits snugly into the slot in the die, while the part engaging the disk projects somewhat over the surface of the disk and tapers slightly to take away any possible wear. The lever is pivoted upon the pin O which ends about half-way through the hole P,

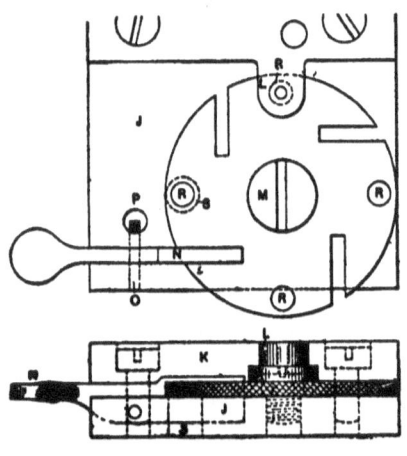

FIG. 516.

thus facilitating its removal should it be necessary. The most particular feature of this die was the location of the four holes, R, R, R, R, which served to hold the pieces while they were being perforated. Great care was observed to have the edges of the slots free from burrs, and the index lever was hardened before the holes were drilled. After the lever had been pressed into one of the slots, a hole corresponding in size to the perforation in the die was drilled through the disk from the back side. The disk was then turned to the next slot and so on, the four holes each being drilled while held in correct position relative to its respective slot. After drilling they were enlarged, by counter-boring or by the use of several drills, to such a size that the punching might be slipped in without effort, yet without any play. The upper side should be slightly countersunk so that the pieces will locate themselves easily. The punch used with this die was of the usual form and is, therefore, not shown.

In operation, a punching was placed in one of the holes R, and the disk moved around by hand until it was opposite and below the hole in the stripper plate. The left hand then allowed the index lever N to engage the respective slot in the disk, and the foot of the operator brought down the ram of the press, thus perforating the punching. Another punching was then placed in the hole next to that in use, the index lever disengaged and the disk moved around a quarter of a turn so as to bring the new punching under the perforating punch. In doing this the punching that had been perforated remained in the disk and was moved around until it came over hole S in the die; then, as the punch descended, a pin fastened in the holder pushed it through the die and it fell into a box below. It will be seen that this arrangement saves a great deal of time, as it is unnecessary to take the punching from the die and it permits the operator to employ any lost time in placing punchings in the front part of the disk while those under the punch are being perforated.

THE MAKING OF DIES FOR EYEGLASS-LENS TRIAL-RIMS.

The term "trial-rims," in the optical vocabulary, means a small ring with a handle, for holding an eyeglass lens selected by the optician where glasses are being fitted to the eyes. A large number of them are used in a set, each holding a lens of different curvature, and adapted to slide into a special frame used in trying the sight. The following described and illustrated method of making these rings with dies is given, as much to illustrate the special dies and tools and methods of working

330 PUNCHES, DIES, AND TOOLS.

as to say anything about the rings themselves. While those made in the dies are cheaper, as they are more readily made, there is a superior style and finish about those made in the turret lathe that makes them more fetching and secures for them a better price.

Presenting the die process, then, in which three sets of dies are used (the metal being fifteen-thousandths thick), the illus-

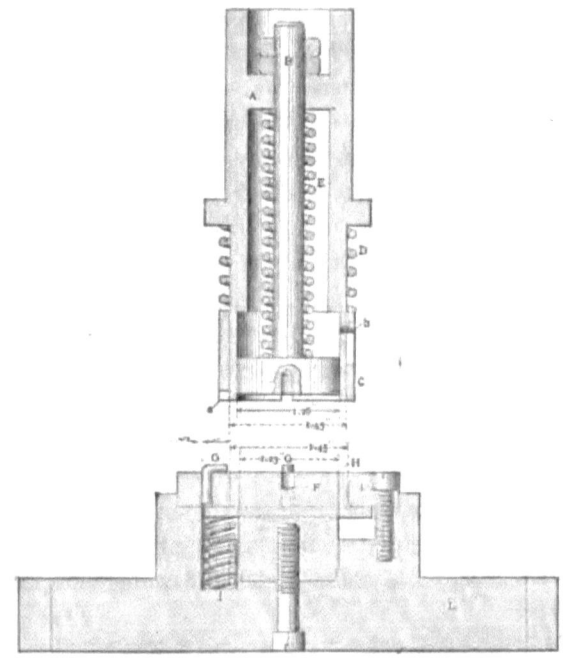

FIG. 517.

trations shown in Figs. 517, 518, 519, and 520 will give a good idea of a common style of spring and compound dies used for such work in a common press.

The first pair of dies, Figs. 517 and 518, combine a cutting-out and drawing die. In Fig. 517 A is a steel punch with shank on the upper end; and the outside diameter at the lower end, 1.45 inches, enters the lower die, I, Fig. 517, blanking out the stock. Then the inside diameter of the lower part of A, 1.26

PUNCHES, DIES, AND TOOLS. 331

inches, turns the rim of the blank down over the plunger F. Figs. 517 and 518, .23 inches in diameter.

As the punch A passes down over F, the shedder H resting on three springs like J goes down, too, and lifts the blank from

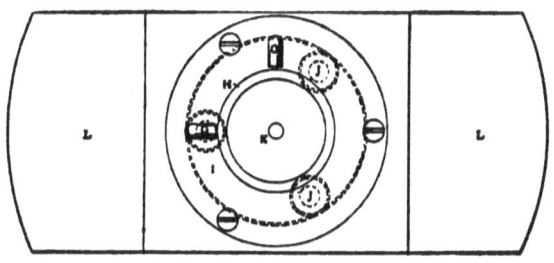

Fig. 518.

F as it comes up. These springs are made of $\frac{1}{16}$-inch wire. The plunger B, Fig. 517, with spring E, makes the upper shedder out of A; and the outside shedder C strips the stock in the strip from the outside of A. The outside shedder C has three slots

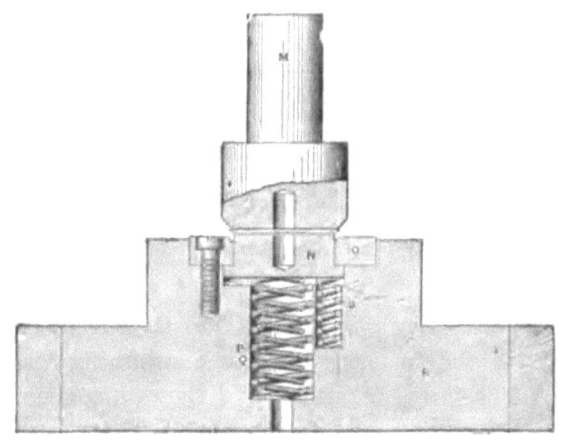

Fig. 519.

with stop-screws like b, and two notches or openings like a to pass over the guide-pins $G\ G$, Figs. 517 and 518. The spring D is made of $\frac{1}{8}$-inch wire, and E of $\frac{1}{16}$-inch wire. The die I is solid, of course, in the base L. The centre holes K are for locating the dies with a pin. Fig. 519 shows the forming die in which the rim of the trial-rim is shown as bulged out by the former. The punch M has a shank which is held by a larger socket used for similar small punches instead of making the latter the regular size of larger ones.

The cap made in the first die is placed on N, which goes down with the work and bottoms on the base S. The surplus

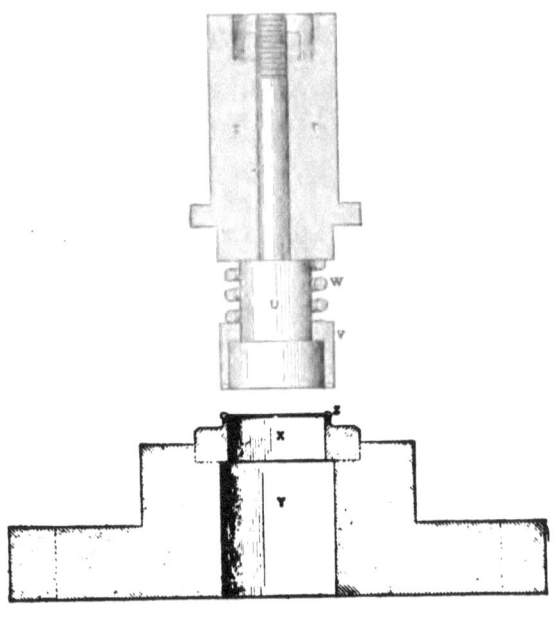

FIG. 520.

stock in the punching spreads out into the curves in the edges of M and O; the springs P, Q, and S returning N as the dies separate. The spring Q instead of P shows a common practice of reinforcing the shedder in die-work. Three springs like S supplement the two in the centre.

Coming to Fig. 520, T is a cast-iron holder with a steel punch U for cutting out the centre of the blank. The die X has a very thin edge, the upper shedder V coming down and preserving the form at Z, while the punch U continues and cuts out the centre Z. The spring W is made of $\frac{1}{8}$-inch steel wire.

PUNCHES, DIES, AND TOOLS. 333

ANOTHER SET OF PUNCHES AND DIES USED IN MAKING RIMS FOR LENSES.

The drawings in Figs. 521, 522, and 523 show three dies and punches that are used to produce the ring shown in Figs. 524 to 527 in three successive operations. This ring is used as a rim for holding lenses, and the construction of the dies will be of interest to those who have occasion to make a set of tools for any similar purpose. The material from which these rims are made is "half-hard brass," as it is called. In Fig. 524, *A* shows

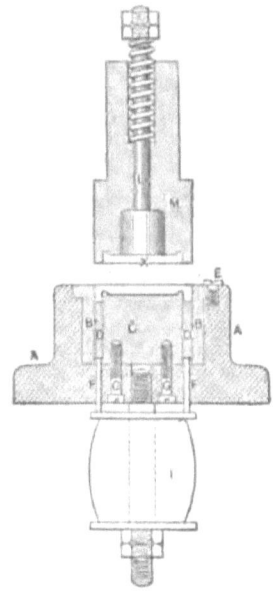

FIG. 521.

the punching after the first operation is performed. The forming punch *C*, Fig. 521, was first turned to the outside diameter of the lens and a templet of soft wire representing the cross-section of the rim, assisted in obtaining the proper form of the face. The hardening of *C* was postponed until the die had been assembled. The punch *M* was then turned, leaving it a little larger than finished size, and while in the lathe the recess and the hole

334   PUNCHES, DIES, AND TOOLS.

for the knockout mechanism were drilled. The templet was then soldered across the face of *C* and with the help of a little red lead, to show up the high spots, this recess was readily finished correctly. To obtain the diameter of the cutting die and punch, *B* and *M*, the wire templet was unsoldered and straightened out. After the outside of *M* had been turned to this diameter it was taken to the gas furnace and hardened. As no grinding facilities were available, great care was taken to heat

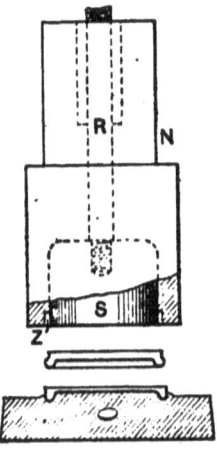

FIG. 522.

the punch slowly and evenly, with the result that it came out of the bath as true as when it was taken from the lathe.

In making the blanking die *B*, it was bored from the back so that the hole was about .004 inch smaller at the cutting edge than at the back. The die-bed *A*, of cast iron, was bored out to receive this die, and after it had been inserted three holes were drilled and tapped, half in the die and half in the bolster, as shown at *E*. In these were fitted three screws to keep the die in place. This die was hardened with the same care exercised in hardening the punch. The next step was to make the tool-steel bushing *D*, which slides between the forming and the blanking dies. The parts of the die were then assembled and two holes for the screws *G G* were drilled and tapped from the back of the die-bed into the forming punch C, which was then ready to be hardened. FF represent two of the steel pins which press against the bushing D, and I is a soft-rubber bolster by which the pins are held up in place.

The die operates in the following manner: The blanking punch M, entering the die B, cuts out the blank to proper diameter. As the punch travels downward it bends the blank over the forming die C and, coming to a stop, it forms the ring shown at A, Fig. 524. When the punch ascends, the bushing D, having compressed the rubber bolster, strips the blank from the forming die C, and the punch carries it upward until the knockout strikes the knockout arm in the press at the end of the up-

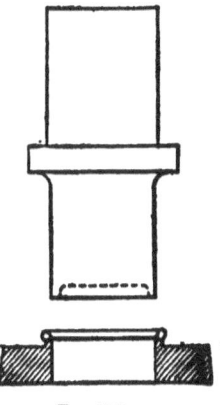

FIG. 523.

ward stroke. As the press is tilted, the blank then drops clear of the die, into a box.

Fig. 522 shows the punch and die used for reducing the diameter of the punching, as shown at B, Fig. 525. The die O is turned out on its face to receive the blanks, one of which is shown in position in which it is placed ready for the punching. In the punch, the recess Z is bored to the required diameter and the hole and recess for the knockout and knockout stem, S and R, are also bored at the same setting. The recess Z is bevelled off slightly to locate the punching. The action of this punch is similar to that for performing the first operation. The punching is carried upward until the knockout strips it from the punch and the tilted press allows it to drop off into a box. Between the punch and die is shown one of the pieces as it is ejected from the punch. Fig. 523 shows the tools used for blanking out the bottom, leaving the work in the shape of a ring shown at C, Fig.

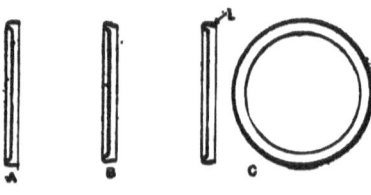

FIGS. 524 to 527.

527. Before the bottom is blanked out, however, the ring is reduced considerably at its rim, as at L. This is done on a monitor lathe, the punchings being mounted on a spring collet and an ordinary cut-off tool being used to accomplish the reduction. The reduced part is afterward bevelled over a lens.

EYEGLASS-STRAP PUNCHES AND DIES AND THEIR MAKING.

The drawings, Figs. 528 to 533, show in detail the dies used for the manufacture of the straps familiar to all who wear eyeglasses. There are, of course, many styles of straps, varying with the different frames; in general appearance, however, they all resemble each other more or less, as shown in Fig. 535. The strap shown is considered to be one of the best, owing to the fact that the threaded portion is reinforced, as will be seen by the cross-section in Fig. 535. To produce the strap four operations are necessary, these being performed in the blanking die, Fig. 530; the embossing die, Fig. 528; the perforating die, Figs. 531 and 532, and the bending die, Figs. 533 and 534. The main condition was that the strap, being made of gold, should not exceed a given weight. Furthermore, every portion of the strap must be touched in the embossing die, thus giving it a smooth surface, and all dimensions of the different parts must be within 0.001 inch.

PUNCHES, DIES, AND TOOLS.

As customary with this class of work, the embossing die was made first. Two samples were filed from a strip of soft German silver. One of these was finished completely, as shown at A,

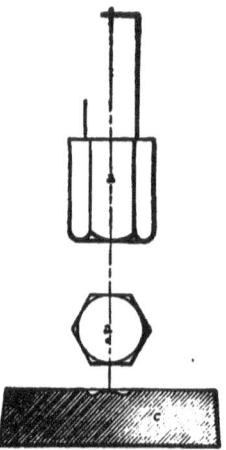

FIG. 528.

Fig. 535, and screwed to a lens, while the other one, at B, was soldered to the face of the hob. This hob, shown at b, Fig. 529, was used to strike the impression into the embossing die C. The

FIG. 529.

hob was roughed out in the milling machine and afterward filed to the exact shape of the sample soldered to its face. In making a hob of this kind care should be taken to have it as short as

. possible so that it will not bend while striking up the embossing die. All the punches used with the various dies were turned

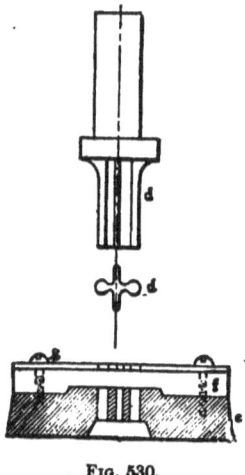

Fig. 530.

from octagon tool steel, 1 inch in diameter, and two punches were always turned together. The embossing die was then

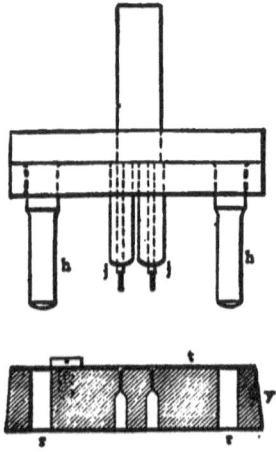

Fig. 531.

planed down, care being taken to have quite a loose fit in the die-bed. To strike up the impression, we proceed as follows:

PUNCHES, DIES, AND TOOLS. 339

The die is put in the bed cold and the hob inserted in the press.
The die-bed is then clamped down and the ram of the press
brought down by hand until the hob is a little above the spot in
the die where the impression is to be made. Now the die is
slipped out and the ram lowered until the ram shows a sufficient
depth in the die, which is pushed back against it sideways. The
ram is then raised again to its full height with jam nuts loose
and the adjusting bar left in the adjusting nut. The die is then
taken out and heated. It has been found good practice to clamp
another die alongside of the one to be struck up to form a gauge,

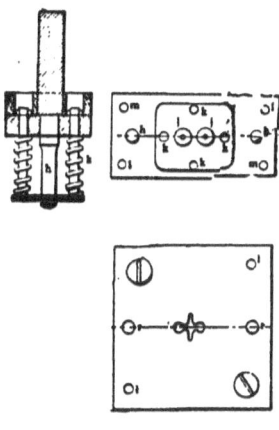

Fig. 532.

as it were, for the heated die. It is unnecessary to state that no
time can be lost in gauging the die while hot. A chalk mark or
any other kind of mark will hardly do, as it is sometimes necessary to strike twice, and, if not properly done, or secured, the
die is liable to shift. It is not advisable to clamp the hot die
with set-screws on the die-bed, as this method invariably leaves
a bad mark on the die, and in most cases the screws are not in
good turning condition. When the die is hot enough, it is
picked up with the tongs and carried to the press at a little live-
lier gait than that of messenger boys and slipped into the die-
bed. The fly-wheel of the press is then turned by hand and the
ram brought down. If the impression should not be deep

enough, the adjusting bar left in the ram can be operated rapidly and a second stroke will bring about the desired result.

As soon as the die is cooled off sufficiently, it is taken to the shaper and several light cuts are taken off. This is done to bring the corners of the impression, which have been drawn inward by the striking-up process, level with the rest of the die. We now proceed to finish the impression. In order to obtain greater symmetry of the spherical portions, we make two end mills of Stubs steel, filing the ends to the desired shape, and putting teeth into them by means of a three-cornered file. These end mills also assist in giving the spherical portions the proper diameter. The impression is then scraped and polished

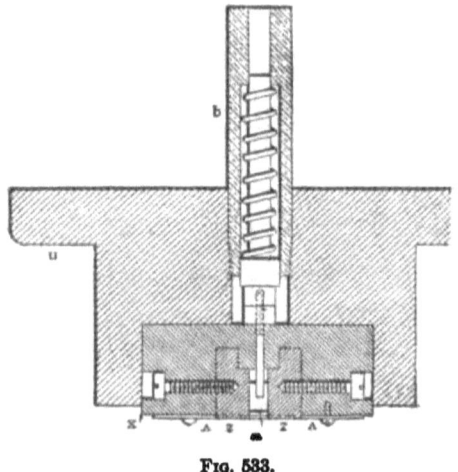

FIG. 533.

to a high finish. The embossing punch is made next and hardened glass-hard. In ordinary practice we would now harden the die, but in this case a different plan was followed.

It has been stated before that one of the spherical portions is heavier than the other, although both are of the same diameter. This would make it essential that one of the round sides in the blanks be larger in diameter than the other, and to get the various dimensions a little experimenting is necessary. To try the embossing die a thin strip of lead is used until a perfect strap is produced. The die is then hardened as hard as fire and

PUNCHES, DIES, AND TOOLS. 341

water will leave it. A little yellow soap rubbed into the impression before heating will cause it to come out of the bath very hard and bright around that spot.

The blanking die is next in order. Two samples are filed to size approximately. One of these is tried in the embossing die and struck up. Any defect can be corrected in the other one, and a third sample, or master blank, like the corrected one, is made. No. 2 sample is then tried in the embossing die and any corrections necessary are made in No. 3. This method is con-

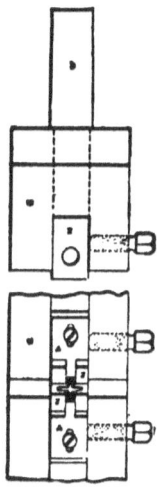

FIG. 534.

tinued until a perfect sample is obtained. C, in Fig. 535, shows the correct article, which is soldered to the face of the punch after the die e, Fig. 530, has been made accordingly. Both gold and German silver are used in the production of the straps, and as these metals are very hard on delicate-edged dies they are filed almost straight all the way through. The face of the die is relieved, as shown in the drawing, so that it is not necessary to grind the entire surface of the die. The gauge is shown at f, while g is the sheet-steel stripper.

Figs. 531 and 532 show the punch and die used for perforating the strap as at B, Fig. 535. A correct sample is drilled and

342   PUNCHES, DIES, AND TOOLS.

the holes are spotted off on the die *y*. These holes are also marked on the back of the die and drilled, with a larger drill, a little more than half-way through, as shown in the section in Fig. 531. They are then drilled with a drill several thousandths smaller than the desired size and finished from the back with a taper reamer. Small reamers of this kind can be made from Stubs steel, filed taper and then square. The gauge-plate *t*, of sheet steel, is made as follows: After it has been filed square it is clamped to the die and the holes for the screws are drilled from the back of the die. The clamp is now taken off and the holes for the dowel-pins *i i* are drilled in the same manner. The dowel-pins are riveted in place, the plate is screwed to the die, and the perforating holes are drilled through the back of the die

FIG. 535.

into the gauge-plate. This is then taken off and the holes are enlarged to the proper size. Before doing this, however, the outline of the strap is scribed off on the plate after inserting two drills or pins through the sample into gauge-plate and die. The shape of the strap is now filed out, and if these operations are gone through carefully as described, the perforations will be in the centre of the strap. When it becomes necessary to grind the die or to remove any dirt or small chips from the gauge-plate, the latter may be taken off and put back again without fear of disturbing the alignment. Figs. 531 and 532 illustrate two different views of the punch. At the left hand it is without the stripper device. The section shows the stripper at rest, while at the right hand punch and stripper are shown in plan. Two stout guide-pins *h h* slide in the die at *r r* when in operation.

PUNCHES, DIES, AND TOOLS. 343

The perforating punches fit in the Stubs-steel punch-holders and are riveted over and inserted from the back. The two parts of the punch-holder body are held together by screws ll and dowel-pins $m\,m$. The holes for guide pins $h\,h$, as well as for punch-holders $j\,j$, are spotted off through the back of the die, the lower plate of the body being clamped to the face of the die y. This perforating die operates in the following manner: An embossed blanking is placed in the opening of the gauge-plate, and as the punch descends, the guide-pins $h\,h$ enter the die until the stripper plate strikes the gauge. The four springs $k\,k\,k\,k$ now compress and their respective guide rods slide up through the holes in the upper part of the punch-holder body. During the upward movement of the punch the springs open again, thus stripping the blank from the punches.

Figs. 533 and 534 show the tools for the last and bending operation. The cast-iron die-bed n is slotted lengthwise to receive the snug-fitting die-body x of mild steel. Into this die-body are fitted the hardened jaws $z\,z$, as well as the hardened former w with its pin and jam nuts resting on the plunger g. The jaws $z\,z$ are slotted to the width of the spherical portions of the embossed strap and thus gauge the blank sideways, while the tongued gauge-plates $v\,v$, being adjustable, gauge it lengthwise. The construction of this die should be plain from the drawing. To bend the strap, it is placed in the groove in the jaws $z\,z$, the spherical parts only being held. With the down-stroke the forming punch o bends the strap inward until it rests on the female former w. This moves down until it strikes home and the strap is bent in the shape A, Fig. 535. At the same time the pin in w, or, rather, the jam nuts at the end of it, compresses the spring around the plunger in the brass-holder g. On the up-stroke of the punch o the female former, being forced upward by the compressed plunger spring, keeps the strap on the punch, where it is stripped off with the fingers. The brass-holder g is screwed into the die-bed n until the female former almost touches the strap lying at rest in the gauge. It is well to anneal the punchings after the embossing operation, as the sharp corners of the punch o are liable to break them. This reinforced style of eyeglass straps is patented.

THE STAMPING OF SMALL MEDALLIONS—THE COINING OF MEDALS.

The small medallions used throughout the world by the Catholic Church, are made of aluminum, brass, copper, gold, and silver, as the occasion requires, and may come under the same category as coins, as similar machinery and tools are used in their production.

We will take the most common or "scalloped" style, as shown in the blanked section of stock, Fig. 537, mostly made of aluminum, but sometimes of silver and gold. The aluminum

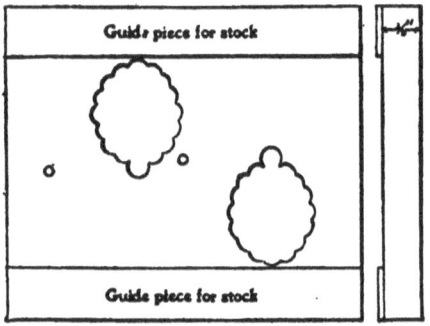

Fig. 536.

comes just wide enough to blank two rows and perforate at the same time, and usually the roll is about 200 feet long.

The blanking die is made as in Fig. 536, is usually about $\frac{3}{4}$ inch thick to insure uniform hardness throughout, is worked out with barely any taper at all, and is fastened to the die-block with screws and dowel-pins. The punches (except the small perforating punches) are left soft and fastened in the usual way in a steel block or plate and fastened to the punch-holder with screws and dowel-pins. The common spring stripper is used, attached to the punch, leaving the die perfectly clear with only the guides for the stock and scrap, Fig. 537.

The press used was an ordinary Bliss No. 19, equipped with a ratchet feed roll (single plan), operated from the crank shaft

PUNCHES, DIES, AND TOOLS. 345

"home-made," easy to adjust for the various sizes and usually blanked about 175,000 punchings a day. The blanks were then dipped to clean them of all dirt and oil and sent to the embossing department, which consisted of several presses.

Considerable trouble was experienced at the start to overcome the wear, and especially the sinking, by the continual compression of the die-block at the base of each die, and after considerable experimenting the die bolster was made of steel bored

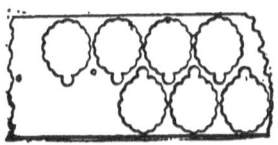

FIG. 537.

to take a hardened and ground-steel centre with five steps, as Fig. 538, which proved satisfactory in every respect.

The dies (two for each medallion) were made of the very finest hammered tool steel, were then finished 1⅜ inches diameter by 1¾ inches long over all, and were as follows: The bar was cut into pieces 1⅞ inches long, faced off perfectly true on the bottom

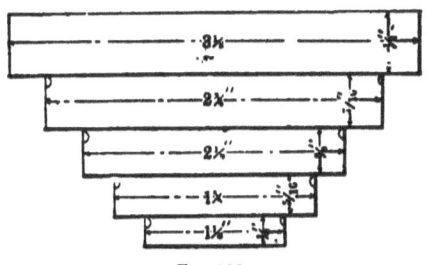

FIG. 538.

end, and the other end slightly raised toward the centre and highly polished, after which it was ready for the impression.

The "hob" is made very shallow, engraved by hand and carefully hardened, after which it is sunk into the die the full depth. This was formerly done in the drop hammer, but it never proved entirely right, because of the inability, when striking several blows, of keeping the bottom of the impression perfectly flat. It was finally done in the same press as the embossing, and done cold, usually in three blows, and always turned out successfully.

After the impression was made the die was carefully chucked in a special device made for all oval shapes, and turned back on both sides of the face to a shoulder ¼ inch deep, as Fig. 539, locating the impression exactly in the centre; after which the die was rechucked in an independent jaw chuck and the face or bottom of the impression set perfectly square and central, regardless of the wall of the die. The die was then faced off to leave the impression the proper depth, and the wall was turned to size very carefully, after which two high points of the oval

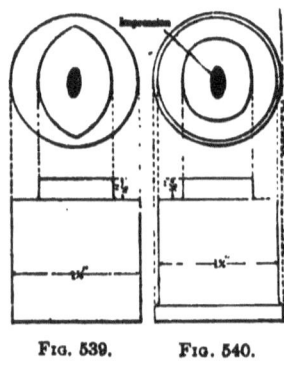

Fig. 539.   Fig. 540.

on the face were turned back to the shoulder, which left a raised "tit" similar in shape to Fig. 540, when the die was practically ready to be scalloped, which scalloping was worked back to the shoulder perfectly straight, leaving a small fillet at the shoulder sized with a templet, Fig. 541, and the edges of the face slightly rounded so that the medallion when finished had what is called a "raised" edge.

The end of the die previously held in the chuck was then turned to size and length in a special "boss" chuck.

A groove was then milled in the wall, central and straight with the figure, and a piece fitted to same for embossing the ear or lug, and fastened with a screw and dowel-pin (see Fig. 541). This was not a particular job.

PUNCHES, DIES, AND TOOLS. 347

It may appear peculiar to a great many tool-makers that the die was not made with an ear all in one piece. As a matter of fact, the very first dies were made in one piece, but they proved anything but satisfactory, as the ear in hardening had a tendency to split away from the body, which necessitated the annealing of the die to patch it.

The screw and dowel-pin holes for the lug were carefully plugged with fire-clay and the die was hardened. The water was burnt off and it was left to cool slowly. The die was then

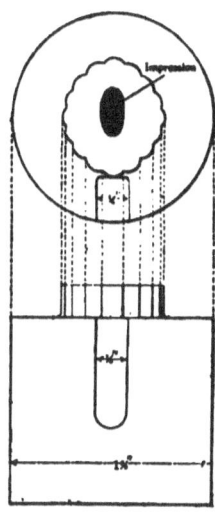

FIG. 541.

highly polished on all the raised parts with what is called "diamond dust" and a cork, and the base ground flat and true with the face, after which it was ready to put into the press.

The dies were held in the bolsters by special clamps, with two screws through the lugs to insure the base of each die being perfectly flat against the hardened steel centre.

A hardened-steel sizing ring, A, Fig. 542, which prevents the metal from expanding beyond a certain limit, was fastened in another steel ring or plate B, also hardened and ground, with a screw collar C, not hardened, with two holes for a pin wrench. Plate B had two grooves D on opposite sides or ends into which the stripping lugs (not shown) were placed, which, after the blank had been embossed and the press is on the return stroke, push the sizing ring down to the shoulder, and a little below the face of the lower die, as in Fig. 543, and thus enable the discharge of the embossed blank. In this case this was rigged up in connection with a blower, so that the valve opened at the

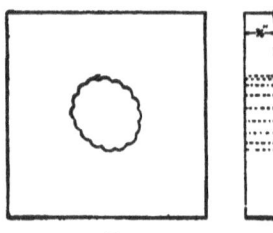

Fig. 542.

right time and a sharp puff of air blew the blank away down a chute and into a box.

The feeding of the blanks to the die (described and illustrated in Section XVII.) was done with a slide actuated by the movement of the press, and all the operator had to do was to keep the hopper filled with blanks, to see that the ear was in the right position, as the slide, on the completion of

Fig. 543.

the stroke backward, picked up only one blank at a time and carried it forward into the sizing ring, which, except in stripping, was always level with the bracket on which the slide moved, and the hole in the slide, when the press was at rest, would be right over the sizing ring or die, and on the commencement of the press movement would travel backward out of the way and pick up a blank, as detailed above.

MANUFACTURING METHODS OF SPOON-MAKING.

The following described methods and illustrations that go with them pertain to the manufacture of the common or "garden variety" spoon. Most graded and stamped spoons are made of German silver, which is a stiff, strong metal for these articles. First, we order our sheet stock from the rolling mills to the correct sizes needed for each size of spoon to be made. The width, thickness, and length, being determined by previous experience, gives us stock ready for the blank chopping-press, this being the first press operation. The cutting of these first blanks is one of the most interesting operations, as they are cut without

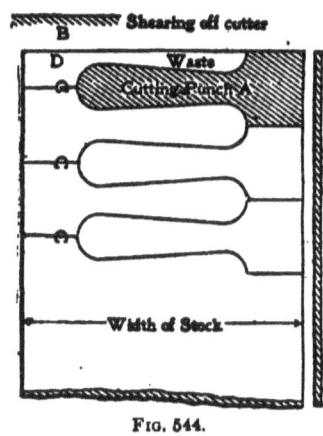

FIG. 544.

any waste, cutting two blanks at each stroke of the press, while only one cutting die and punch are used. A shearing-off cutter is attached to this die, which makes a parting cut at the proper distance from the die to make a second blank. Fig. 544 will show this blank-chopping process. *A* gives the shape of the parting shear at the lower end of this die. The small strip of metal *D* is all the waste there is at each end of the long strips. *C* shows where the parting cut comes as the strip is moved forward for the succeeding cuts. The dies are of course fitted with proper gauges, stripper, and necessary means for quickly handling this work.

Having finished the first operation of the blank-chopping, we leave the press work for a while and go to the cross-grade rolling mills, which prepare the stock for what is termed "grading." By this is meant that the blanks will be stretched to the proper length, as they are only about one-half long enough for the fin-

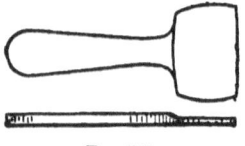

Fig. 545.

ished spoon and of an even thickness when first cut. We cross-roll both the bowl and handle ends of these blanks, this usually being done on one end at a time, but sometimes spoon blanks are of such shape that both ends can be rolled at the same pass.

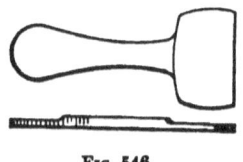

Fig. 546.

Our process will be the one end at a time. We first cross-roll the part making the bowl. Fig. 545 shows the shape the blank takes in this rolling. The third operation will be that of cross-rolling the end which makes the handle, giving the blanks the

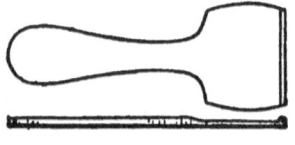

Fig. 547.

shape, as in Fig. 546. We now have both ends of the blank cross-rolled, giving them nearly the proper thickness, for the first grading or lengthening process.

The next handling is the first pass of grade-rolling, which gives the metal the first stretching operation (see Fig. 547).

PUNCHES, DIES, AND TOOLS. 351

After this operation the blanks receive their first annealing. Then they are pickled and cleaned and made ready for their final rolling, which stretches them to their required length, giving the various parts their proper thickness. This last rolling leaves the stock as in Fig. 548.

Now we come again to the power press, the blanks having been again annealed and ready for clipping or trimming to shape. This is done by dies and punches. Fig. 549 shows the spoon and the waste around the clipped shape.

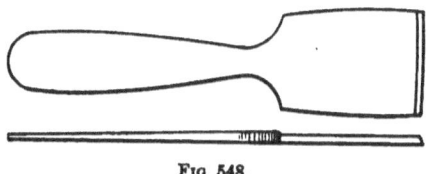

FIG. 548.

This trimmed, spoon-shaped blank is next taken to the buffing wheel, where it is made smooth and free from the trimming-cutter burrs, or other roughness caused by handling, as a very clean surface is required for the next operation, which is called "stemming," that is, the stamping of the handle part only. From the graded blanks, Fig. 548, many different shapes or styles are cut or clipped, and various patterns are stamped on the handle, that being one of the reasons why the entire spoon is

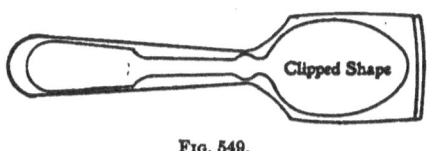

FIG. 549.

not stamped, bowl and all, in one die. Some designs are made in one blow of stamping hammer, but they are usually very plain, in fact, they need be. When the handle stamping is done in a separate die a single bowl-stamping die may be used on a great many different stamped handle designs. In this manner many expensive dies are saved and still a great variety of spoons may be made. Our spoon now looks like Fig. 550.

352 PUNCHES, DIES, AND TOOLS.

The spoon, as we will now call it, is again taken to the polisher, who, with a rather tough wheel, "fins" the edges, or cuts off the toughness caused by the stamping dies. These fins are very small on German-silver work, as the blanks are cut so close that they just fit into the dies and fill out the design, leaving

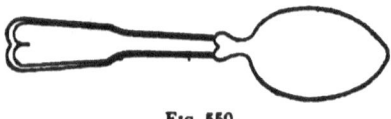

FIG. 550.

scarcely anything of flash or fin, at least not enough to cause them to go to a trimming or second clipping die. The expert polisher can cut and polish this finely, leaving the spoon or work very smooth.

After this polishing we take these stamped spoon handles to

FIG. 551.

a small bench foot-press which has suitable set of handle-end shaping dies in it. With these the proper end form is given this handle, more for the convenience of the bowling operation than for any other reason. By this means the stamped design on the handle can always be placed into the bowling die the

F a. 552.

right way. Were it not for this end shaping, which acts as a guide for the stamper, many spoons would be stamped wrong way about, spoiling them (see Fig. 551).

Now we have the spoon ready for bowling; this is done in

PUNCHES, DIES, AND TOOLS. 353

a drop hammer. This tool stamping also gives the stem a partial shape which is shown in Fig. 552.

The spoons are now ready for their final polishing and buffing, which is a trade in itself, requiring a great deal of skilled labor, as no hand filing or finishing is done to German-silver goods as usually made. After this last polishing the spoons are placed in a shaping die, which is often made of hard wood, and sometimes copper-faced, having the exact shape of the finished spoon. Hard wood is used for these dies, to prevent the fancy designs being marred or destroyed by this operation, as would

FIG. 553.

be the case were iron or steel dies used. These dies are usually used in a foot-press or a light bench hammer. Sometimes special bench hand-hammers are made, each hammer having its own form or shape to it. This makes a very convenient tool for the purpose, as no die or shape changing is required. Fig. 553 shows the finished shape of a teaspoon. The next operation is that of stamping the trade mark or maker's name or branding them with such marks as may be customary for the many different patterns or styles. After this the spoons are ready for silver plating, burnishing, inspecting, and packing.

THE MAKING OF GERMAN-SILVER FORKS.

German-silver forks are made practically in the same manner as the spoons, only instead of bowl-stamping we have what is called the tining operation, that is, cutting the tines into the end of the clipped blanks, as shown in Fig. 554. This fork-tining is usually done in a special press for this purpose, which cuts the tines one at a time, cutting the centre one first, automatically working the three cuts and then stopping. This is a very ingenious device for the purpose, doing the work with great accuracy and rapidity. The ends of the tines are left tied, to prevent their spreading while being stamped. It also prevents the points of the tines, which would be sharp were it not for the front piece, from piercing the operator's hands or tearing the polishing belts. This end tie piece is cut off just before the last tine polishing is done. A small foot-press is generally used for this purpose. Sometimes when the forks are of a plain pattern

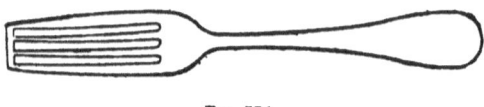

FIG. 554.

they are stamped in a heavy drop, both handle and tines in one die, with one drop blow, resulting as shown in Fig. 555.

The shapes shown are as they actually appear when taken from the dies in which they are cut or stamped. Sterling-silver spoons and forks are made in much the same way, only that more hand work is applied to them; there are several more

FIG. 555.

stamping and trimming operations as well as a great deal of hand filing. Graded-steel spoons are made more along the lines of German-silver ones. There are, of course, steel spoons made which are not graded, but are made from stock of even thickness and stamped with heavy ribs to strengthen the handles. The power press and the drop hammer are great factors in spoon-making. The grading rolls also perform important parts.

THE MAKING OF SOUVENIR SPOONS.

The majority of souvenir spoons are made of sterling silver, which is purchased from an assayer in an ingot at a certain price per pennyweight. The silver is melted in crucibles, and

PUNCHES, DIES, AND TOOLS. 355

the furnaces are usually gas-burning, as the heat has to be great. As the silver melts in the crucibles a small handful of charcoal is sprinkled over the silver in the crucible; this clears all sediment and draws it all to the surface. When the silver is melted, it is poured into a mould which varies in size, the most common being 3 inches wide, ¾ inch thick, and 12 inches long. This mould is polished and free from all blow holes and small pits that usually appear in cast iron, of which this mould is made. The mould is always made on a slight taper with a false piece of steel fitted at the bottom; this is used in loosening the ingot by driving on the wedge or false piece. The mould usually is heated and swabbed with a piece of oily waste to prevent the metal from adhering. After the silver has cooled in the mould to a certain degree it is removed, and if found to be free from pits and air holes, it is scrubbed with hot water, soap, and a stiff brush to remove all substance adhering to the ingot.

The next operation consists of breaking down, which is taking the ingot of silver after it has been scrubbed, and rolling it through rolls made of tool steel, hardened, ground, and polished. They are usually 6 inches in diameter with a 10-inch face, the size being generally determined by the size of the ingot to be rolled. The silver, passing and re-passing through these rolls, becomes thinner and greater in width at each consecutive rolling, and the rollings are continued until the silver begins to crack at the edges. The operator then knows that the metal has become so brittle that it must be annealed before any more rolling can be done. The ingot of silver is now about ¾ inch thick, 5 inches wide, and 20 inches long. The operation of annealing consists of grasping one end of the silver with tongs and passing it back and forth through a gas furnace until it is red-hot; it is then allowed to cool. After cooling, it is placed in an acid bath, which removes all the oxide and discoloring caused by heating. After the acid bath it is plunged in another bath of boiling water and scrubbed to remove all traces of the acid, in order to protect the rolls from being disfigured by same. The silver is dried by covering with hot sawdust, and is then put through the same operation of rolling and re-rolling again, the annealing process taking place as often as the silver requires it, which is

356 PUNCHES, DIES, AND TOOLS.

determined by the silver cracking on the edges as mentioned before.

Great care must be taken during the operation of rolling to have the rolls exactly parallel with each other; if the rolls are not parallel the stock will curl and not roll straight, thereby stretching the silver more at one side than at the opposite, which causes it to break. Fig. 556 shows the shape of the rolls used. These rolls have to be ground and polished from time to time as they wear irregularly, and therefore do not roll the silver perfectly. Again, at times an imperfection appears on the surface of the rolls which must be removed as it will always show on the silver.

After the silver has been rolled to its desired thickness, it is then annealed for the last time, pickled, and dried, and then taken to the finishing rolls, which are made the same as the breaking-down rolls, but smaller in diameter and ground more accurately and also polished to a better finish. These rolls do no heavy rolling, but are used simply as sizers to bring the silver uniform in thickness. After passing through the finishing rolls, the silver, which is now in a rolled sheet varying in thickness

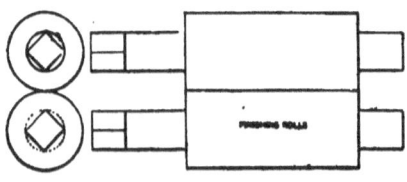

FIG. 556.

about from 0.025 to 0.035 inch, is taken to a pair of rotary shears where the rough edges are trimmed and the silver sheet stripped into the desired widths. The rolls of silver, together with the scrap that comes from the shears and that which remains in the crucible after melting (called the "button") are taken to the office and weighed to determine the shrinkage and waste. The silver is now ready to be made into spoons.

The silver being in rolls it is taken to the shears and cut into pieces the desired length which vary according to the size of the spoons. The next operation consists in what is termed

PUNCHES, DIES, AND TOOLS. 357

"grading," which is tapering the ends of the pieces of the silver for the handle and bowl. The pieces of silver are always cut shorter than the spoon itself, as the "grading" operation lengthens the piece. The operation of grading is accomplished by the rolls shown in Fig. 557. These rolls are cut away in the centre to about ⅛ inch deep at one side and gradually lessening until the cut rises to the face of the roll. This section of the roll is usually about 2 or 3 inches wide, the length being about one quarter the circumference of the roll. The rolls always run the direction of the taper, or, in other words, the rolls revolve, so

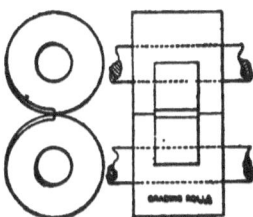

Fig. 557.

that at one given point the deepest portions of the recesses meet, having at this point a combined depth in the two rolls of ¼ inch. It is at this point that the spoon is inserted for grading. As the handle and bowl are differently graded, there are two recesses in each roll. The pieces of silver are taken between pliers, the jaws of these being of a length to determine the length to which the silver is to be placed in the rolls. When the rolls have come to the point where the recess space is greatest, the operator is ready with the blank adjusted in the pliers. He pushes the projecting portion of the blank in the opening until the pliers come in contact with the rolls. In this position he holds the blank until the rolls grasp it and reduce the taper. The rolls are mounted in housings and are adjustable to any thickness.

Having graded both ends of the blank, it is then blanked out in a die made as shown in Fig. 558, which cuts the former outline of the handle, leaving the bowl end in the state it left the grading rolls. We will explain the object of performing this operation in this manner. Generally a shop manfacturing spoons of this nature (which, by the way, must not be confounded with the manufacture of common commercial spoons) has various designs of handles, also it has a line of bowl dies of various sizes and designs. A mixed order is received in which so many spoons of the order are to have a certain handle and bowl, and a certain number to have another design of handle and bowl. By making the dies as shown, any style of handle can be made on any bowl or fancy spoon. Having blanked the handle, another die, Fig. 559, is used to blank the bowl. The blanks are now ready to have the design stamped on the handles.

This work is done in a drop-hammer between dies shown in Fig. 560. In the dies is cut the design that is required on the

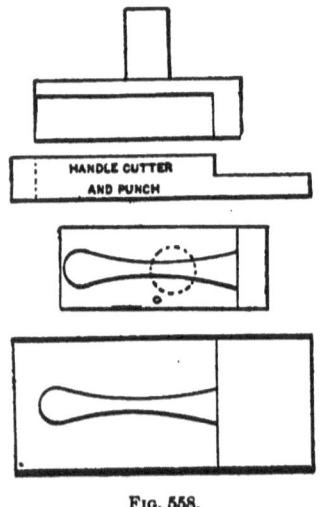

FIG. 558.

top and bottom of the handle. These dies are called matched dies. The depth of the design always determines the thickness of the silver to be used—the greater the depth the thicker the stock. The blanks then pass through the operation of trimming which consists of being forced through dies similar to Figs. 558 and 559, but instead of the punch being flat, the centre is cut away, leaving a knife-edge on the outside. This is done to relieve the punch, so it will not disfigure the design, and at the same time allow it to cut a clean edge. Having trimmed the handle and bowl, the next operation consists of bowling, or, in other words, to bend the bowl from the flat to the desired shape.

The work is also done in a drop-hammer with the dies, Fig. 561, and the force Fig. 562. The die is placed between the poppet on the drop-hammer bed and the force is secured in the jack-bed, which is fastened to the hammer. The operator now places a blank on the face of the die, releases the hammer, which descends and the force presses the spoon bowl into the die. An automatic drop-hammer is the best for this operation, as it

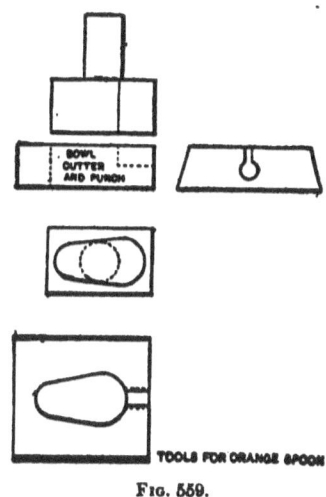

Fig. 559.

strikes every blow with the same force. If the name of any city or town or the outline of any particular object is desired in the bowl of the spoon, it is cut in the force as shown in Fig. 562. Doing this brings the two operations into one, for, as the bowl is formed, the design appears.

The spoons are now passed on to the bench, where the filing is done. This operation consists of filing off the burrs and ragged edges on the bowl. The spoons then go back to the press department, where the handles are curved by being pressed between two hard-wood blocks formed to the shape desired on the handle. The spoons then pass into the polishing-room to be "boled"; this work is done on an ordinary polishing head, using a cloth wheel with a coating of rotten-stone, which has to be renewed at intervals. The boling removes all sharp edges and scratches on the silver, after which the spoons are polished by using another cloth wheel coated with crocus. The spoons are then sent to the coloring-room for the next operation, which consists of plating bowls with gold in an electro-plating bath.

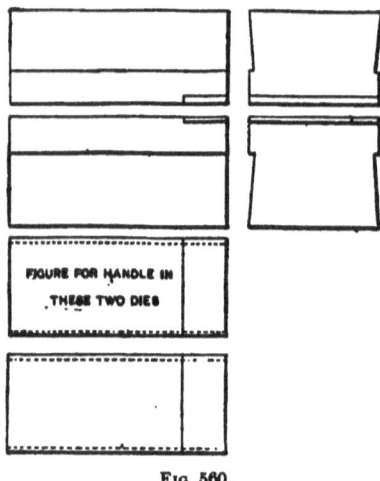

FIG. 560.

Some bowls are oxidized; others pass through the operation known as sand-blasting, which gives the spoon a dull finish. The spoons that are gold-plated are sent back to the polishing-room and again polished, after which they are washed in a solution of hot water and ammonia to remove all stains, and finally they are

FIG. 561.   FIG. 562.

dried in hot sawdust, when they are ready for the shipping-room.

The souvenir-spoon industry is very large, one firm alone having over 1,000 designs for bowls, which include the names of every city in the United States and Canada of any size, and a large number of cities in other countries.

SECTION XIV.

Design, Construction, and Use of Sub-Press Dies for Watch, Clock Work, and Accurately Perforated Blanks of Irregular Shape.

THE MAKING OF A SUB-DIE AND PUNCH.

IN the author's work, "American Tool-Making and Interchangeable Manufacturing," Chapter XXVIII. treats only of the construction and operation of the sub-press and its dies, and as the methods explained would not do for watch and clock work, it is thus fitting that there should be embodied in this work a chapter which would treat exhaustively on these and other methods of procedure in the making of sub-dies and punches for watch and clock work, and also more improved processes for the production of pieces of irregular outline.

The methods outlined in the above-mentioned work do not describe the making of the dies and punches, but assume that the punch would of necessity have to be sheared, which on wheelwork for watches and clocks (and also accurately perforated irregular shaped pieces) would be very objectionable, owing to the fact that all punches are milled to size, and the spoke-wheel punches are ground and lapped inside the rim, which fact does not permit any shearing to fit the punch to the die. The methods outlined in the other work are just the thing for irregular shapes where the dies and punches are filed to fit, but where they are machined entirely the methods are not applicable, therefore we submit the following methods of making sub-dies which have proven very successful for the work mentioned:

Fig. 563 is a sectional elevation of the sub-press and its die and punches complete. The frame A is faced on the bottom and bored to fit a special face-plate having a boss on it of the desired size, the frame is then fastened to this face-plate and bored ½-inch taper to the foot, and with the same setting (after

362   PUNCHES, DIES, AND TOOLS.

locking the back gears) four grooves are splined lengthwise within the barrel to hold the babbitt from turning.

The frame is now put on an arbor which fits the taper just splined, and the barrel is turned and a thread cut at R to which the ring C is afterward fitted. The bottom of the frame is again faced and bored while on the arbor to insure perfect alignment with the inside of the barrel.

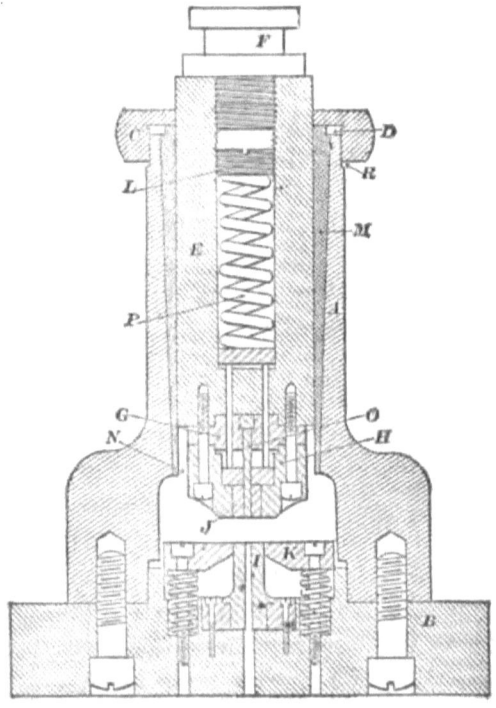

FIG. 563.

The ring C is then bored through about .005 inch smaller than the desired finished size of the plunger, recessed, and a thread cut to fit the barrel. A ring (used for construction only) is now made about $\frac{1}{8}$ inch thick to fit within the ring C, the inside diameter being $\frac{1}{8}$ inch smaller than the bore barrel at D. The ring C is now screwed on the barrel against the small shoulder within and bored true to the required size of the plunger.

The base B is planed and screwed to the face-plate, faced

PUNCHES, DIES, AND TOOLS.

true, the boss turned to fit the bottom of frame, and the centre recessed for the base of punch I. The plunger E is then centred, held in a steady rest, and bored to accommodate the shedder spring P and a thread cut for the tension nut L and button F. The button (of machinery steel) is then screwed into the plunger against the shoulder, and the plunger is roughed down in a lathe and then ground to size on centres, leaving it with a highly polished surface which should not be touched with an emery stick.

Punch plate G is now made by holding a piece of stock in a chuck, turning and facing a boss on the end of the right size, and (before cutting off) roughing down another boss on the opposite side from the finished one. The plunger is again put in a lathe, supported by a steady rest and recessed to receive the boss on punch plate just made, which is rapped in against the shoulder, and, driven by the plunger, the unfinished boss is turned to fit the back of the die, which in place runs absolutely true. Three grooves are now milled lengthwise on the plunger to prevent it from turning in the babbitt.

We now take base B and drill, tap, and counterbore it for the stripper springs and screws, also drilled for dowels and counterbored for the screws to secure the base and frame. The base, frame, and plunger are now completed.

Die H is made by holding the stock in a chuck and with steady rest is turned and recessed to accommodate the shedder base J, and boss on punch-plate G; then drilled and bored the size of the root of the teeth on the sample wheel. The die is now cut off and held in spring chuck and the front faced true with the back.

The broaches are made with either three or four teeth or steps (according to the size of the die), increasing from .0015 to .0025 inch to each step, and the teeth milled (great care being taken that the cutter is milling exactly central and straight). The broaches are hardened and drawn, and the blank die is broached (much care again being required). After the die has been drilled for screws and dowels it is packed in bone-dust and carefully heated and immersed in oil, and drawn to light straw temper. The bottom of the die is next ground and lapped.

Punch I and shedder J are then milled (the miller not having been disturbed) to fit the die, then drilled on centres (using steady rest) to the size of the centre hole in sample wheel; the hole in the punch is reamed from the back $\frac{1}{2}$-inch taper for clearance, and the hole in the shedder is left straight.

A base is made for the punch I which fits tightly into the base proper where it is held by screws and dowels, also one for the shedder J which works freely within the recess in the back of the die. The object of putting the shedder into a base is to allow the milling cutter to traverse its length entire when being milled, and forms a square shoulder for it to strike against when it sheds the blank. The punch is next set in position in the base and screwed down tightly, and slightly ground on a surface grinder to give a perfectly square top.

The die, shedder, and punch-plate are assembled, being screwed and dowelled to the plunger, the tension nut is so adjusted as to allow the shedder to come within $\frac{1}{64}$ inch of the face of the die. The frame is put into position on the base, the screws are tightened, and after the plunger has been lightly wiped with an oily cloth and sprinkled with flake graphite, it is carefully let down through the frame and the punch I allowed to enter the die until it is checked by the shedder. The ring C is now screwed down (the small ring being within) which brings the punch, die, and plunger exactly in line, and central within the frame.

The frame is inverted upon a level and rigid bench (free from vibration), slightly warmed with a Bunsen burner and poured with babbitt M from two sides simultaneously, and allowed to cool. The plunger is afterward withdrawn, the base and frame separated, and the babbitt turned out at N (where it overflowed in pouring), allowing a further downward motion of the plunger.

Frame and base are again assembled (the small ring within ring C taken out) and the plunger driven down and a blank cut from brass stock about .010 inch thick. After the plunger is withdrawn the die and punch are carefully examined to ascertain if there is even the slightest shearing, for, if any occurs, the frame must be re-babbitted, but if care and exactness have been

PUNCHES, DIES, AND TOOLS.

exercised in the making of the frame and base, also in the proper temperature of the babbitt, the die will invariably "line up." The punch-plate G is now drilled and counterbored for the piercing punch O, which is turned to fit within the shedder J, and the punch I, the former supporting and strengthening it when the die is in operation as well as shedding the blank.

The lower stripper K is made, being broached similar to the die, drilled for screws and pins for guiding the stock, then hardened and ground. For a general line of watch work eight springs under the stripper are required to strip the stock. The stripper is properly adjusted in the base, allowing it to stand .005 inch above the face of the punch, the frame screwed down into position on the base, proper tension brought down upon the shedder spring P, and the sub-die is ready for its work.

MAKING A COMPOUND SUB-DIE FOR PUNCHING AN IRREGULAR PIECE.

The best way to make a punch and die for sub-press work for an irregular piece similar to that which is shown in the lower punch, Figs. 564 and 565, is to proceed as follows:

First step to secure accurate results is to make a master-plate, Figs. 565 and 566, to drill the holes by. Having secured a round, flat piece of tool steel of a diameter sufficiently large for

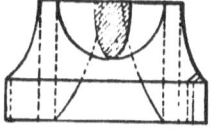

Fig. 564.

the work in hand (in this case 1½ x 2 inches thick will do), it should be turned and then ground perfectly straight and parallel, for a good deal depends upon the accuracy of the master-plate; one side of which should be marked "Top" and the other side "Bottom," in order to avoid confusion in the following operations. The sample piece should be marked in a similar manner. If working without a sample, the master-plate should be laid out to suit a drawing or sketch, centre punched, clamped to face-plate of lathe, trued with indicator and all holes drilled,

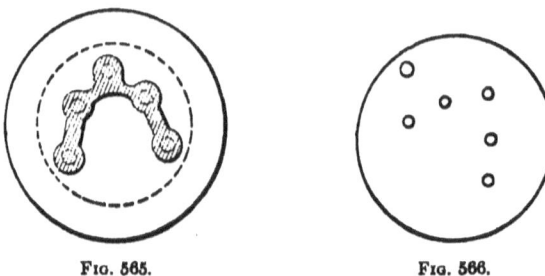

Fig. 565. Fig. 566.

tooled, and reamed to one size. If working from sample, the top side should be soldered to the bottom side of the master-

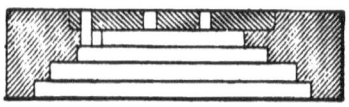

Fig. 567.

plate, inserting a pair of parallel strips in order to be able to drill through the master-plate. A bench lathe with draw-in

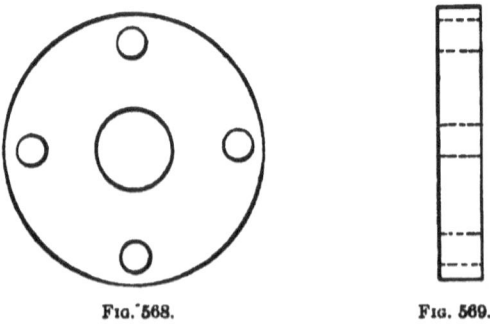

Fig. 568. Fig. 569.

spindle is best suited for this class of work. Having a suitable chuck with small taper in end, a guide-pin for same should be turned, hardened, and ground to fit the largest hole in sample piece. Clamp the master-plate with sample attached to the face-plate of the lathe, swinging it on the guide-pin in the largest

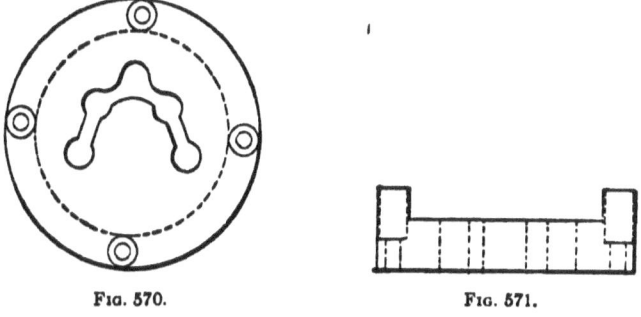

Fig. 570. Fig. 571.

sole, taking care not to cramp the plate. The first hole should now be drilled, tooled, and reamed about 1 inch diameter.

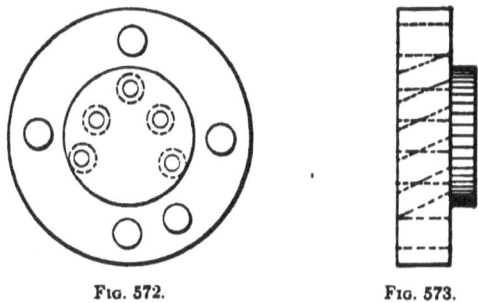

Fig. 572. Fig. 573.

Having finished the first hole, grind the guide-pin to the next smaller size and proceed as before, until all are completed.

Fig. 574. Fig. 575

Have all the holes one size in the master-plate. Having finished the master-plate the next step is to get out a concentric ring,

PUNCHES, DIES, AND TOOLS.

Fig. 567, of machinery steel, recessed on one side to suit the diameter of master-plate; after which it should be reversed and have four recesses turned into it for each of the following pieces: lower punch, Figs. 564 and 565; lower shedder, Figs. 568 and 569; upper die and punch-plate, Figs. 570 and 571; upper shedder, Figs. 574 and 575. The five pieces composing the die should be rough-turned and then turned and ground to size on the bench lathe, fitting them a snug fit to the recess in the concentric ring.

MAKING THE UPPER DIE.

Take the piece for upper die, Figs. 570 and 571, recess to proper depth, and allow for grinding on inside (about .006 inch). Drill a dowel-pin hole through the upper die blank, Figs. 570 and 571, and punch-plate, Figs. 572 and 573 (these two should be dowelled with the same pin) and concentric ring; also the lower punch, Figs. 564 and 565, and master-plate, Fig. 566,

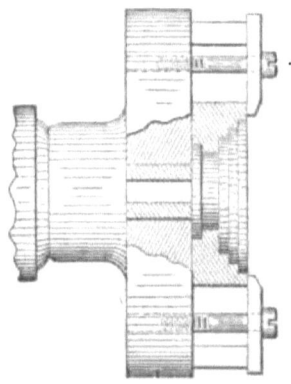

FIG. 576.

each in its proper recess. The other two pieces do not require dowelling, as they are not hardened. Place the master-plate with the side marked "Top" in the ring, so that the top side will be next the face-plate; now place the upper die, Fig. 567, into the recess in the ring with its dowel-pin in place, swing it on the guide-pin and clamp securely to the face-plate, as shown in Fig. 576. Drill, tool, and ream the five holes in die, as shown

PUNCHES, DIES, AND TOOLS. 369

in cut, allowing .002 inch to grind after hardening, when the four screw-holes are drilled and countersunk. The balance of the die may be worked out with a small end mill in the slide spindle grinder, reducing the speed of the latter. Finish with fine file and harden. The upper shedder-plate should be drilled the same as the upper die—that is, with the top side of the master-plate next the face-plate—only that the holes should have 1° taper; the guide bushings being made separately, and after

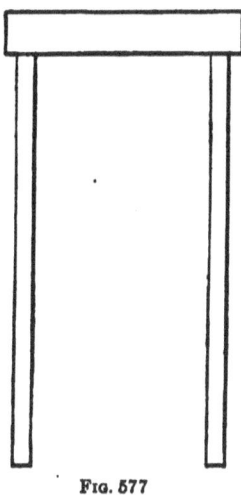

FIG. 577

hardening and grinding on the inside to the size of punches, and on the outside a snug fit to the upper die, should be forced into a shoulder in the shedder-plate. Filling-in pieces may also be inserted in the plate, although it is not exactly necessary, as the guide bushings will force the punching back into the strip. The punch-plate should be placed with the hub side next to the face-plate, as the holes have a back taper, and the side of the master-plate marked "Bottom" next to the face-plate. This plate should have the same dowel-pin that has been drilled in the upper die, in order to have the holes in line, when it may be swung in the lathe on the guide-pin and have the holes drilled and tooled to fit a plug gauge with a 2° taper. Do not use a taper reamer, unless with extreme caution.

MAKING THE SMALL PUNCHES.

The small punches are rough-turned on a rod and partly cut off with a parting tool, then hardened and drawn to suit; after which the punches should be ground straight to size and the taper fitted to the plate, all at the same grinding. The cut

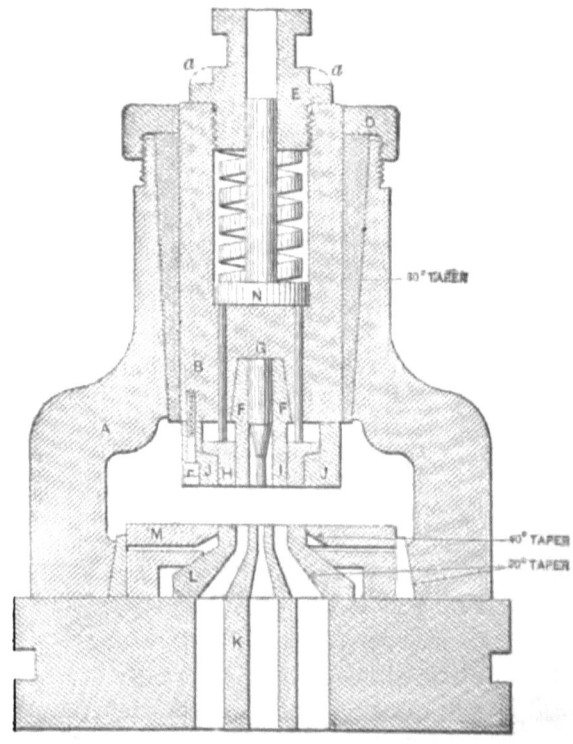

Fig. 578.

shows two tapers on the punches, which is done to strengthen the punches as much as possible. The guide bushings in the shedder should be cupped to suit the short tapers on the punches, to allow the shedder to come up as close as possible. Having the punches all fitted to the plate and driven down solid, the plate should be secured in a brass chuck that has been re-

PUNCHES, DIES, AND TOOLS. 371

cessed to clear the punches, and the back or broken sides of punches ground smooth and even with the plate. The small ones of punches, also the guide bushings, are ground after they are all assembled together. The screw-holes are drilled after the upper die has been ground and fitted to the plate, lining them with dowel-pin.

MAKING THE LOWER PUNCH.

The lower punch, Figs. 564 and 565, and concentric ring must be dowelled together, and with the side marked "Bottom" of master-plate next to the face-plate, should be swung on the guide-pin and the holes drilled, tooled, and reamed about $\frac{1}{4}$ inch deep and about .003 inch smaller than the size of the punches, to allow for grinding, when it may be removed from the lathe. A small guide-pin to fit the holes just drilled must now be provided, upon which the lower punch must be swung and the clearance holes for punchings drilled in the back. The punch, being fitted in with a taper, will require no screw-holes.

MAKING THE LOWER SHEDDER.

The lower shedder is placed in the recess in the concentric ring with the side of the master-plate marked "Bottom" next the face-plate, swung on the guide-pin and drilled, tooled, and reamed to suit the respective diameter of holes of sample, and now removed from the ring to have screw-holes drilled and tapped. The holes forming the hole should be well flared, and the rest drilled and fitted free over the lower punch. Two holes should also be drilled in the plate for guide-pins to guide the strip to be punched, and so placed that they clear the upper punch, and then hardened and drawn.

GRINDING THE UPPER DIE.

The upper die having been hardened and drawn, should be secured in a brass chuck by the outside diameter, and the face ground square across, and recess ground true and a snug working fit for the upper shedder; the taper should be ground to fit the taper on the small punch-plate. It should now be reversed and held by the recess on a brass chuck and have the outside diameter ground true.

The dowel-pin hole may require a little lapping, and if the outside diameter of the die be too small to suit the recess in the concentric ring, use prick punch and hammer and close in the recess enough to allow for truing out again. Place the die in the recess with the dial-pin in position and remember which side of the master-plate goes next to the face-plate. Swing up again on the guide-pin in the lathe and grind the holes to size with a diamond lap.

LOCATING AND FITTING THE PARTS.

The die is now ready to be fitted to the plunger of the sub-press, the construction of which is fully described in preceding pages of this chapter, with the exception that the straight sides may require lapping with a diamond lap in order to correct any light bulge caused by hardening. Three small guide-pins must now be turned, the small ends of which are fitted to the lower punch and the large end to the holes in the upper die, in order to position the lower punch in the base of the sub-press. A thin washer about $\frac{1}{16}$ inch thick should be fitted below the lower punch and the punch left larger, so that it will force down about .04 inch into the base of the sub-press positioned by the three small guide-pins and the upper die; drill dowel-pin in the base, remove the lower punch, and give it a thin surface of solder on the face, upon which to give an impression of the upper die, scraping away as much of the surplus solder as possible and getting a sharp, clear impression. Mill away the sur-

PUNCHES, DIES, AND TOOLS. 373

plus stock as close to the impression as possible, then remove the solder from the face, place back again into the sub-press and get another impression on to the steel itself. Mill and file the punch, leaving about .001 inch to shave (the less the better), shave and finish smooth with scraper and file. Harden and draw; remove the upper die from the plunger and see how the punch fits the die, and if it requires any fitting use a flat diamond lap carefully so as to not round the edges.

ALIGNING THE PUNCHES AND DIES.

Now make sure that the upper die bottoms squarely and place the lower punch into the die, using a parallel ring to fill in the space between the face of the die and the punch; grind off the bottom of the punch, also true up the taper and fit to the base, and you have the punch in line with the die; all that remains to be done now is to place the punch back again into the recess in the concentric ring, with the master-plate in proper position, and grind out the small holes in to size, giving them $\frac{1}{4}°$ back taper. Remove the thin washer and place the die and punch back again in the sub-press, and see if the punch lines up fair with the die; if you have made any mistakes now is the time to find them out. Fig. 576 shows a bench-lathe head, and how to swing the master-plate on the guide-pin. We might add that this kind of work requires constant watching.

MAKING A SET OF SUB-DIES FOR CLOCK WHEELS.

The dies for a compound wheel punch, Fig. 578, consists of seventeen pieces including the top and bottom shedder. There are five segment punches *F*, one centre punch *G*, five filling-in pieces *O*, one spider *H* to go in between the five segment punches, one ring *I* to fill in the space between the outside of the wheel, and also three round rods forming the connection between the upper shedder and the bottom of the spring piston. These are all to be fitted to the plunger. The five segmental punches are planed

374 PUNCHES, DIES, AND TOOLS.

in one piece to the proper angle (in this case 75°, in order to have the necessary taper to the arms of the wheel); a groove is planed in one side to suit the key filling-in pieces. They are then cut off long enough to allow turning on the ends, after which they are turned on a special chuck in a bench lathe.

MAKING THE CHUCK FOR THE FIVE SEGMENTAL PUNCHES.

The best way to make the chuck, Figs. 579 and 580, is to make a blank chuck of a diameter somewhat larger than the inside diameter of the wheel that you wish to punch, bore a hole in the chuck the same diameter as the hub of the wheel, and fit

FIG. 579.

in a plug, allowing it to project about ¼ inch, which end should be turned in diameter to suit the thickness of the arms. It is necessary now to cut the clearance in the back of the chuck before you can plane it. A fixture must be provided for the shaper on which the head of a bench lathe can be fastened, so

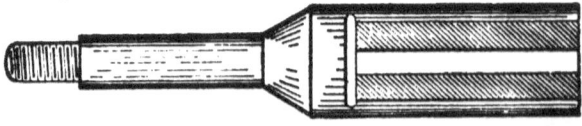

FIG. 580.

that the chuck may be shaped to the proper angle when in place in the lathe head. It may be planed by throwing the shaper 37½° to each side, using the guide-pin to plane to, or leaving the shaper-head in vertical position and planing across, using the guide-pin to plane to, and then swinging the lathe head through an angle of 105°, when the other side may be planed in a similar manner.

FINISHING THE SEGMENTS.

Having finished the chuck, we now proceed to turn the outside of the segments, Figs. 581 and 582, by clamping each fast with a ring in which we have previously tapped a set-screw. Turn the segments half-way, allowing .01 inch for grinding, and finish the other half. Bore the segment of hole, allowing .003 inch to grind, and turn the ends all to one length. Harden at as low a heat as the piece of steel will harden, and draw enough

FIG. 581.

FIG. 582.

to take out the shrink. Now carefully lap the sides down on a good cast-iron bench block, so as to get good smooth sides, being careful to maintain the proper angle. The segments are set into the plunger on a 1° taper. After lapping, take the segments one at a time, clamp in V-chuck, and proceed to grind each segment square on end, also grinding it lengthwise on a 1° taper, taking care to have them all one diameter, using a slide-spindle grinder on a bench lathe.

MAKING THE FILLING-IN PIECES.

The filling-in pieces, Fig. 583, are planed in one piece, and of the proper taper, with a key on one side, and cut off long enough to come flush with the recess. A chuck blank is made, 1 inch larger than the inside diameter of the rim of the wheel, and long enough to form the spindle H, Fig. 584, on the end. With the bench-lathe head fastened to the fixture that was made for the shaper, proceed to mill out the arms with a suitable roughing cutter, and finish milling with a special cutter, which must be made for each different size of wheel. It should be milled and fitted a snug working fit, between the five segment punches. Return it to the lathe, and drill and bore the large hole in the centre to suit the centre punch G, and cut off to length, allowing for facing in brass chuck, when the small hole should also be drilled. The small hole in the spider serves as a guide for the centre punch, and should be ground after hardening. Harden and draw to a purple, and should it require any fitting after hardening, use a copper hand-lap charged with dia-

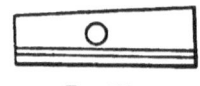

FIG. 583.

mond powder. The ring I, Figs. 584 and 585, into which the spider fits should be turned with a recess on the inside, as shown, and secured in a brass chuck fitting the bench-lathe head, and have grooves planed on the inside, deep enough for the spider

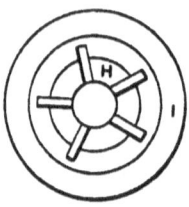

FIG. 584.

arms to go in a snug fit, after which it is hardened and drawn, and then ground inside and outside, after which the spider may be soldered at the joints. The outside ring die J should be made as shown, hardened and drawn to take out the shrinkage, and

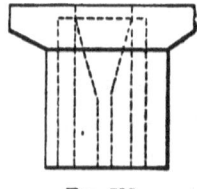

FIG. 585.

ground all over, fastened with from four to six steel screws, which should be hardened and drawn to spring temper. The die should have a fine saw-cut passed through the screw-holes on the outside, to prevent it from cracking in hardening. Now back-

PUNCHES, DIES, AND TOOLS. 377

rest the plunger, in the bench lathe, letting it revolve in a ring; bore out the recess for the ring die and make it a tight drive fit; also bore out the taper hole 1° taper for the punches and make them a good driving fit.

The segmental punches may now, while still in the back rest, be ground in place on the outside to suit the inside diameter of the rim of the wheel, and the hole ground out to suit the diameter of the hub of the wheel, and also that the hub of the spider fits. The hub of the spider must be the same diameter as the hub of the wheel. The small hole in the spider may be ground at the same time by securing it in an extra brass chuck. Drill the three holes in the plunger for the round rods, reassemble the dies, and grind them off flush.

MAKING AND FINISHING THE BOTTOM PUNCH.

The base having been turned and bored, the next step is to get out the bottom punch K, Figs. 586 and 587. The blank should be well secured in a brass chuck milled out with a roughing cutter, and finished with the same mill that finished the spider. After milling, it should be fitted to the segmental punches, and with the punches as guides, it should be forced

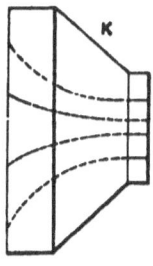
Fig. 586.

Fig. 587.

into the proper position in the base, with the base and stand fastened together, after drilling through the die K into the base. Harden the top end of the punch, draw and grind the arms to finished size, giving the small centre hole $\frac{1}{4}$° back taper. The round blanking die L is plain sailing, also the bottom shedder M. The base should have the clearance holes cut through to allow the punchings to fall. The wheel punchings strip back into the brass strip. There are eight springs under the bottom shedder, and it is secured by four screws. Guide-pin holes should be drilled in the shedder before hardening, and which should be drawn to a blue.

STEEL FOR SUB-PRESS PLUNGERS.

There is a difference of opinion among users of sub-presses as to the best material to use for the plunger, some preferring cast iron and others machine steels.

Experience covering many years has demonstrated that it is always best to specify steel whether the piston be $1\frac{1}{4}$ inch or 5 inches in diameter.

First, the steel will cost less than one-half as much as the iron, making a slight financial gain, but this should not be considered at all, if made at the expense of efficiency or durability of the tool. In this case the gain is a double one.

It is not necessary to figure on occasional loss on account of bad casting, which may not show until the plunger is nearly completed. After finish grinding, the fine grain of the steel will soon wear down to a very smooth surface which, it seems reasonable to suppose, will not wear the babbitt lining of the press as fast as would the porous cast iron.

The argument has been used that as the outer shell is of cast iron, so should the plunger be of the same material, in order that the expansion of each may be equal when the press warms up under steady use. This may be all right, but we do not think that this argument carries much weight.

Now for what we consider conclusive facts that should convince. Probably more often than otherwise a sub-press piston contains a die which is built up of quite a number of pieces, which must be fitted and secured in place with a nicety which approaches as near perfection as anything in practical mechanics.

Compare a piece of steel of uniform texture throughout with an iron casting, the centre of which is reasonably sure to be full of flocks of from very open pores to actual blow holes.

PUNCHES, DIES, AND TOOLS.

Fitting small screws and dowels in the centre of a casting on a job which requires great accuracy will not tend to improve the quality of the job or the tool-maker's temper, thus steel should be used in all cases.

A DEPARTURE FROM ESTABLISHED SUB-PRESS DESIGN.

The Authomometer Company, of St. Louis, Mo., manufacture a machine known as the "Burroughs Adding Machine," the construction of which involves a large amount of accurate punch-and-die work. For all of this intricate work they have adopted the sub-press as a regular thing. The adding machine contains

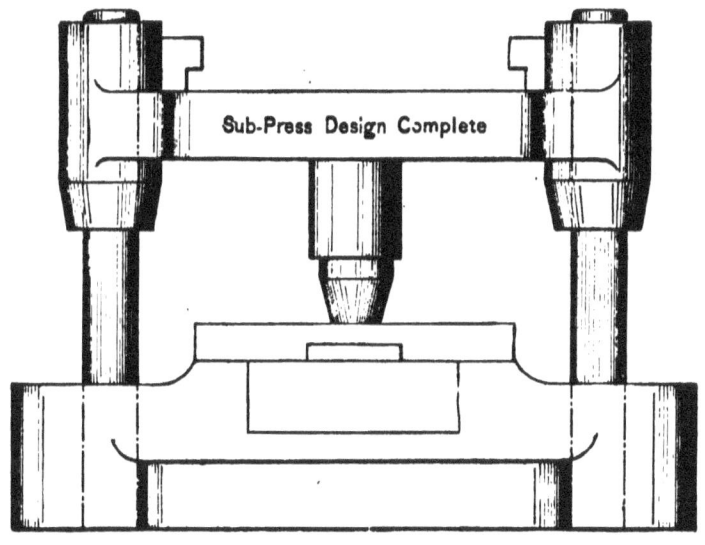

FIG. 588.

a wilderness of wheels, levers, and palls, as well as gears, etc., which are all made in a press from cold-rolled, brightly finished steel, and the collection of sub-presses for these various pieces makes quite an imposing display. They are of uniform pattern and are stored in cases to the number of three hundred.

Figs. 588 and 588 *a* show an engraving of the regular pattern of sub-press used, fitted with a set of dies for making a washer.

As will be at once noticed, the design is a distinct departure from the standard type as constructed by Pratt & Whitney and the E. W. Bliss Company. Duplicates of this sub-press are used for nearly all of the pieces to be made.

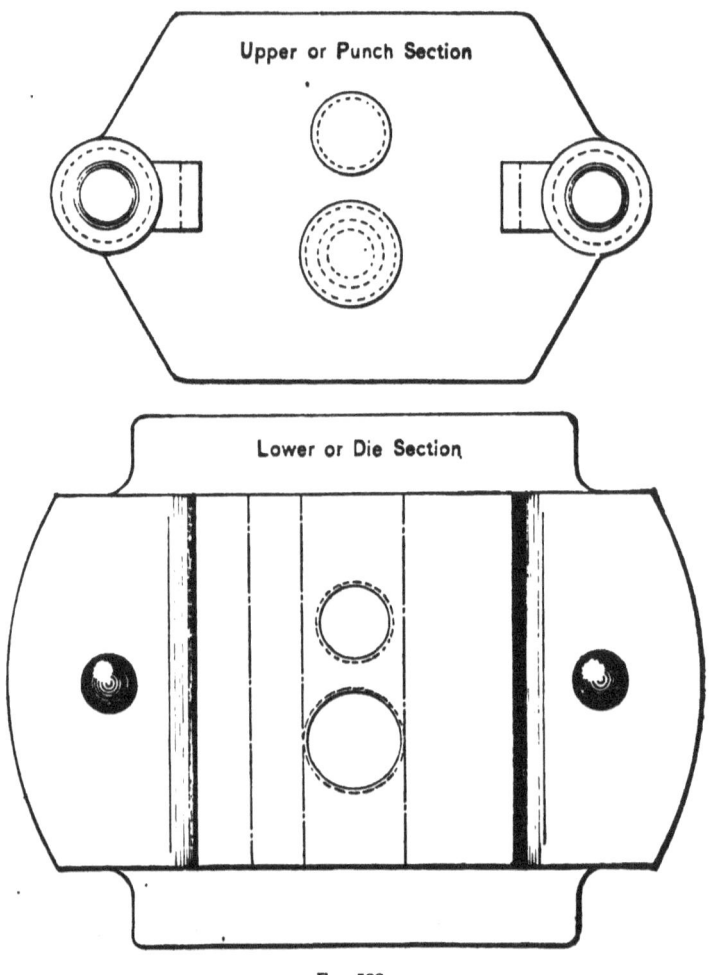

FIG. 588 a.

The great advantage of the sub-press system as installed and in operation in the Authomometer Company's plant is, of course, that it insures positively accurate setting of the punches and dies. The wear of the guides is distributed through all the sub-

PUNCHES, DIES, AND TOOLS. 381

presses in use, and this wear is very slight, as these guides have to endure a side thrust of the connecting rod. Moreover, there is no possibility of improper setting of the dies in the machine or doing any damage to them while being set, and while the first cost is increased, this is largely compensated by the fact that the dies are much more quickly changed, leading to direct economy of time, as well as increased output through the reduction in the time the machine is standing idle. With the sub-press system the adjustment of the dies to each other is made once and for all on the bench, where everything is accessible and in sight, and proper dowel-pins and screws enable them to be exactly and quickly re-located after removal for grinding.

These sub-presses are very carefully made, especially in the parallelism of the guide-rods and in the fit of the rods in their sleeves. The punches here shown are of a form suitable for direct attachment to the moving part of the sub-press, but for punches of most other shapes a punch-plate or pad is used. It is the usual practice in the works referred to in the foregoing to leave the punches unhardened, and, if the blank is to be used, the face of the punch is left perfectly flat. The rule followed for clearance in the die is to make on all sides the clearance equal to 5 per cent of the thickness of the stock to be worked.

It is unnecessary to give figures regarding the performance of dies mounted in sub-presses. It is sufficient to state that it exceeds that of dies mounted directly in the press many times. The sub-presses themselves are almost everlasting. The punching made in the type of sub-press illustrated in Figs. 588 and 588 a are of the finest order of workmanship.

PROPORTIONS OF SUB-PRESSES, ETC.

It is really surprising when going through shops engaged in the manufacture of small light goods in quantities to observe the limited use of the sub-press for making many of the smaller parts of thin sheet metal. Further reflection leads to the conclusion that outside of the watch and clock industries the merits of these extremely useful tools have not been fully appreciated. They are stiff, accurate, reliable, and convenient, and are quite

382 PUNCHES, DIES, AND TOOLS.

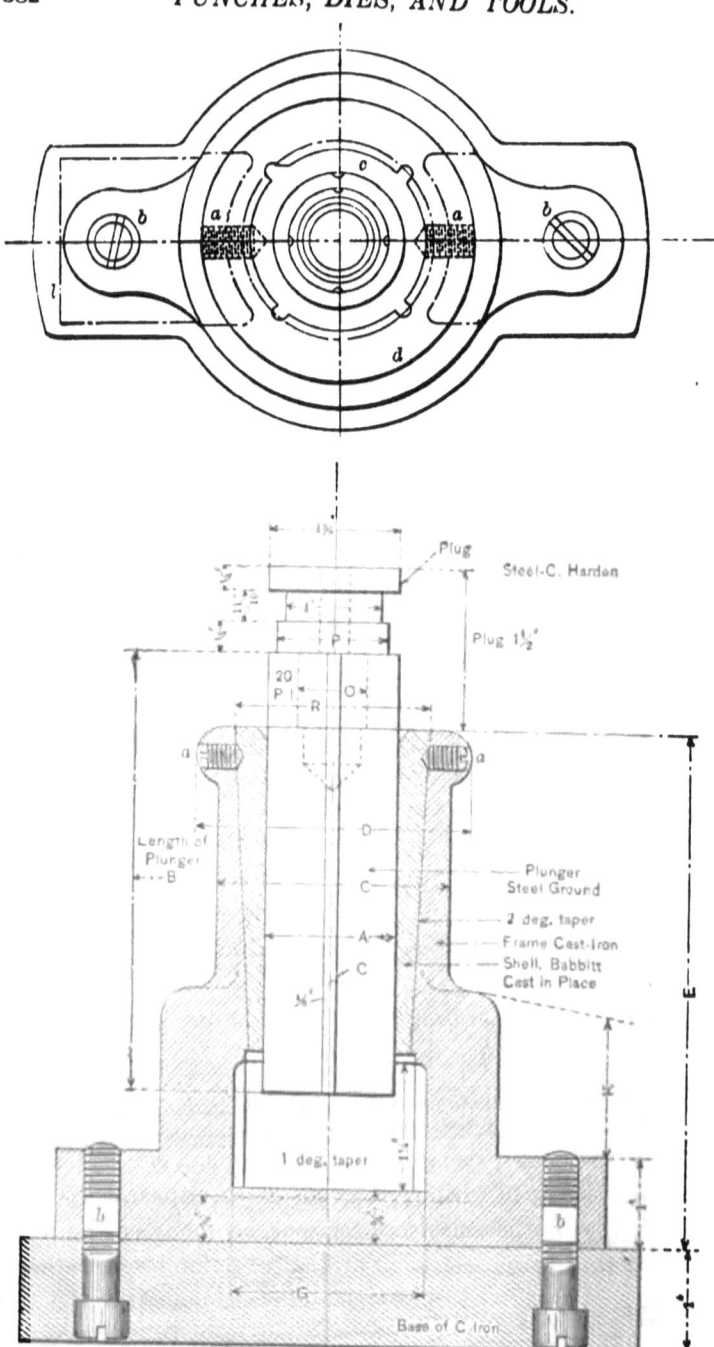

Fig. 589.

PUNCHES, DIES, AND TOOLS. 383

independent of the wear of the gibs and slide, with the attendant accuracy of the parts of larger presses.

The accompanying engravings, Figs. 589 and 590, and the tables of dimensions, Figs. 591 and 592, are the outgrowth of the designing of a line of sub-presses fulfilling the following conditions, named in the order of their assumed importance:

(1) Accuracy.
(2) Strength and stiffness.
(3) Lightness—within the limits imposed by 1 and 2.
(4) Convenience.
(5) Cheapness and maintenance.

The frame is cast with a cored hole in the centre, tapering 2°. The shape of the core provides four semicircular grooves

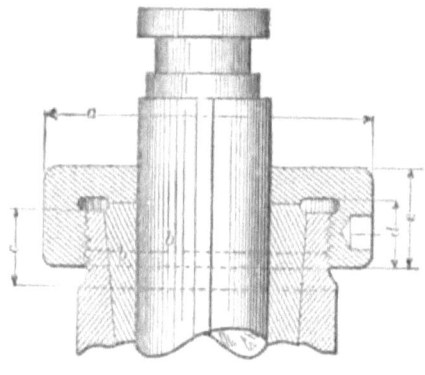

FIG. 590.

down the whole length of the hole, for the purpose of anchoring the babbitt shell. The hole is made tapering to facilitate the removal of the babbitt when removal becomes necessary. The bottom is faced and hole G bored with an inward taper of 1°. To this hole the stem of the base, having a height of ¾ inch, is carefully fitted and the base fastened to the frame by two screws $b\,b$. The plunger is turned and tapped for plug at O, and ground to gauge. Four semicircular grooves are cut in the plunger, inequally spaced to insure its proper subsequent entrance in the shell. The plunger is then held central in the cored hole in the frame by suitable means, and the babbitt shell poured around it. From two to four pointed, headless screws are rapped into the frame as shown, to preclude all possibility of disturbance of shell from accidenal causes. The bottom of the slide of a larger press is arranged to receive the plug, which is screwed into the plunger, the frame strapped to the bed of the press and then run until the plunger is a proper working fit. An oil pocket at the top of the shell is for convenience.

Plugs for the different sizes of sub-presses are of the same dimensions, so far as length and fit in the bottom of the slide of

Table of Dimensions of Sub-Presses.

Press No.	A	B	C	D	E	G	H	J	K	L	M	O	P	R
1	1¼	5	2⅜	2⅞	7	2	5½	3¼	1¾	1½	2¼	¾	1⅛	1¾
2	1¾	5	2⅞	3⅜	7	2½	6	3¾	1¾	1⅞	2½	1	1½	2¼
3	2¼	5	3⅜	3⅞	7	3	6½	4¼	1¾	2¼	2¾	1¼	1¾	2¾
4	2¾	5	3⅞	4⅜	7	3½	6½	4¾	1⅞	2¼	3	1½	2	3¼
5	3¼	5	4⅜	4⅞	7	4	6½	5¼	1⅞	2¾	3½	1¾	2¼	3¾
6	3¾	5	4⅞	5⅜	7	4½	6½	5¾	1⅞	3	3¾	2	2½	4¼

FIG. 591.

Table of Dimensions of Sub-Presses.

Press No.	A	B	C	D	E
1	2¾	2⅛	⅝	9/16	¾
2	3¼	2⅜	⅝	9/16	¾
3	3¾	3⅛	11/16	⅝	1 3/16
4	4¼	3⅝	11/16	⅝	1 3/16
5	4¾	4⅛	¾	11/16	⅞
6	5¼	4⅝	¾	11/16	⅞

FIG. 592.

the larger press are concerned, varying only in diameter of collar *P* and thread *O*. The pitch of the thread is the same also—20 to the inch.

In use the tools are preferably left in the sub-press, as a permanent fixture; but they may, of course, be removed readily and quickly interchanged with the certainty of alignment independent of any changes that may have taken place in the mean time in the larger press.

PUNCHES, DIES, AND TOOLS.

In a form of construction preferred by several firms making these tools, the rounded top of the frame is dispensed with and the sides are made straight to the top instead. A thread is cut in the cast iron to which is fitted a case-hardened spanner nut which screws hard against the top of the babbitt sleeve, presumably squeezing it into the taper in the frame. So far as our experience goes, the only point this design possesses that the form shown in Fig. 589 does not is the expense. But in case this form is preferred proportions for it are given in Fig. 590 and the table, Fig. 592.

INDEXING SUB-PRESSES.

In Figs. 593 and 594 are reproductions of two index cards showing the system used by the American Authomometer Company,

	Symbol No. 212. 212 N. 212 FR. Stock, $8\frac{1}{16}$ x $8\frac{3}{4}$ x $\frac{1}{16}$ Galley Brass. Number of Operations 4.
Number of pieces to machine..	1	Duplicate Dies....	Sub-press.
Number of pieces to one foot...	1		
Number of pieces to one pound.	1.47		
Operations.	Sub-Press.		
212–1 Round Holes..........	4–12		
212–2 Slots.................	3–4	Old or Repair Dies.	
212–3 Perf. for 227..........	P–5		
214–4 Perf. for 215..........	P–5		
212–N–1...................	L–11	212–2 Old slots...	V–C–I
212–N–2...................	S–4	212–N–2.........	V–C–I
212–N–3...................	P–5		
212–N–4...................	P–5		
	FIG. 593.

of St. Louis, and developed by Mr. E. Dean, foreman of the press and die shop, for keeping track of dies used in manufacturing the above-named company's plant. There are about seven hundred operations covering the adding machines as now made, and also all repair work on all machines that have been made. All pieces have a symbol number, all sub-presses have a location number. By referring to a symbol card No. 212, Fig.

Sub-Press No. 3–4. Style Special.			212-2 slot. Remarks.
Presses.........	P–4		Change stops for 212–213
Bolster.........	1		Change studs for 212 N, 213 N
Eccentrics......	2		Oil punches lightly for 2–4
Plunger Pin.....	On Die		Keep paper cover on work
Drawings........	H–2	1160	Use Sub-Press V–C for fractional
Patterns	A–97	A–98	
	A–99	A–100	
Other pieces on Sub-Press.........			
212–2			
212–N–2			
213–N–2			

Fig. 594.

593, it will be seen that there are three pieces of that number—212–212 narrow and 212 fractional. The operations on the piece are given in their order, and opposite each operation is the location number of the sub-press the work will be done on, and this number is also the key number to the sub-press card shown in Fig. 594, which will explain itself. Sub-press S-E, for example, will be found in section "S," locker H of the die rack. Both sub-press and rack have brass plates with the location number on them, and when the sub-press is in place, the numbers will correspond. All dies used in this shop are sub-pressed.

FINE POWER-FEED PUNCHING WITHOUT SUB-PRESS.

In Figs. 595 to 597 we describe and illustrate a method for using a very fine power feed for piercing, which it will be well to adopt when the use of a sub-press will not pay.

Fig. 595 is the die blank; B, Figs. 596, 597, the stripper; C, the trimming die; D, the piercing die; E, the blanking die; F

FIG. 595.

are the screws which hold the stripper down on A, and Fig. 595 is the piece produced; the latter shows the actual size, while the die is reduced. The holes in the finished piece are .054-inch diameter, and the radii of the circumscribing circles are all $\frac{1}{16}$ inch.

The operation consists of the stock entering at H, and, pass-

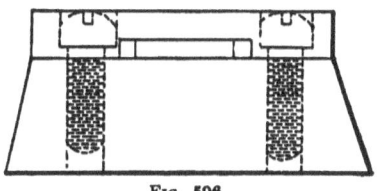

FIG. 596.

ing to the reduced part of the stripper at the inner end of C, it is trimmed on the edges to pass the reduced part mentioned. It is perfectly plain that when the rear part of the trimmed stock advanced to the shoulder in the stripper it can advance no farther, and if the roll feed has been adjusted to carry it, say, $\frac{1}{16}$ inch longer than the trim, the stock will be held rigidly in position both on the sides on account of the trimming and at the end for the same reason, thus advancing the piercing to its exact location in regard to the blanking cutter.

The die is simple to construct, and similar ones have been used a great deal for light embossing where it was essential that the die should trim very closely to the engraved work and leave no "fin," using the embossing dies, it being understood that the

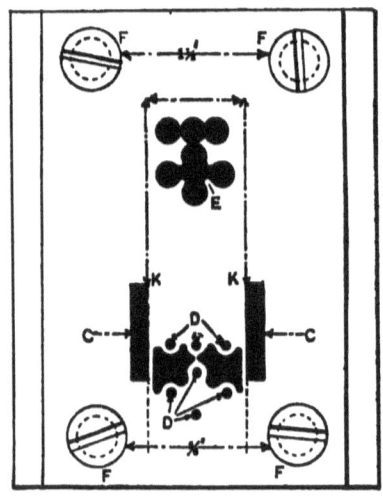

Fig. 597.

piercing punches must have "entered" before the embossing force shall draw the metal out of shape.

These tools are far superior for the production of work of the class indicated than the common "combination" die, both in accuracy and cheapness.

SECTION XV.

Drop Forging and Die-Sinking, Together with Making of Drop Dies, Steam-Hammer Dies, Number-Plate Dies, and Dies for Bolt Machine.

DROP FORGINGS.

THE successful manufacture of drop forgings necessitates the meeting and deciding of many difficult questions. They may be enumerated as follows: The proper heats to employ, the number and force of the blows to be struck, the effects upon metal of rapid blows as contrasted with the slower squeezing action of the hydraulic press, etc. The proper deciding of this question entails on the part of him who must do the deciding, a mind well equipped with mechanical knowledge and methods of procedure as well as much practical experience in the art of drop forging.

It is claimed by some that drop forgings are not as strong as hand-forged ones. This objection will not stand except where positive tests have been made with identical metal, under exactly similar conditions. To be sure, there is much more work done on the metal when hand forging than in drop forging; and there is less bending, too, in the latter process, fewer blows to be struck and less finishing, all of which improves the forging, provided the work is done with the metal at the correct temperature.

It is now well and generally known that the compression which drop forgings are subjected to—if done at a high temperature, which is often a welding heat during the first stages—should consolidate the metal and improve its structure just as the lighter blows of the hammer in the blacksmith's hands or his sledge do. Therefore we find that the fundamental point to be considered is to have the metal at the proper temperature during the entire action of dies so that it shall flow into shape as desired. If the metal is given sufficient heat, and the hammer is of sufficient power, it matters little what shape a piece of steel is before being forged, as the hot metal will flow under pressure like any plastic substance, following the lines of the least resistance.

With this point of view before us, we find that there are three methods of attaining the desired results. The temperature of the metal may be increased, to increase its flowing qualities and the mobility, bringing it nearer to the condition of foundry metal—a device to which the temptation is very strong, though the risk of overheating and burning is ever present; or the pressure may be increased.

In the use of the drop-hammer we have examples of the light blows applied with extreme rapidity, in strong contrast to the slow squeeze of the hydraulic press. Light, rapid hammer blows are better for shallow work, as the press is better for deep work. Light blows in deep work do not consolidate the interior; the press does. Light blows are transmitted to light forgings confined in a die, and this resembles the character of the work done by the smith on the anvil.

Experience alone is the sure guide to the proper intensity and number of the blows struck on a given forging. The temperature varies from a welding heat, at the commencement, to a low red heat or in some cases even a black heat at the termination. During the first stages the metal is caused to flow like a soft substance. At the latter the blows are of a consolidating and surface-finishing character. A forging in which considerable differences in mass exist in adjacent parts will not be so uniform in quality as one in which all parts bear a more equal proportion to each other.

Even if there were any force in the objection to die forgings on the ground of their not being as strong as those made on the anvil, that would not weigh against the demands of modern manufacture. It is necessary, commercially and economically, to produce hundreds of similar parts at a merely nominal cost for labor. If it could be proven that such forgings were inferior to hand-made ones, the remedy would clearly be better metal and closer attention to the matter of the proper temperature at which the best results are obtained.

CHARACTERISTIC OF DROP FORGINGS.

A characterstic of all drop-forged parts is a little irregular fin, surrounding the work all around it. This comes about because of the surplus metal creeping out between the dies—the only opening whereby it may escape. It is true this fin might not occur, but it usually does in all cases. Its absence may be attained only by the blank being placed exactly in the right position, remaining there during the blow, and containing the right amount of metal. These fins are always present in some degree, but are trimmed off afterward in a trimming-press, in which dies are used that are ordinary cutting dies with the punch hollowed out, as far as practicable, to conform to the upper surface of the work. Obviously, by this process such articles can only be made as will deliver freely from the dies, by reason of their having considerable taper and no high vertical walls.

PRACTICAL APPLICATIONS OF THE ART.

The practical applications of the art of drop forging are so numerous as to make their enumeration impracticable here. By the process many thousands of small tools, parts of machines, hardware, cutlery, etc., are rapidly made, with the uniformity of punched-out work, but of far better quality as regards smoothness and density of structure. Most of the parts are of round-up form which could not be made at all from flat sheet-metal with power-press tools. Such forgings are usually better than the best hand-made forgings, as well as cheaper and more uniform. They are often cheaper than castings of like form, as they require much less finishing and are more interchangeable, and therefore for most purposes much better suited than castings.

DIES FOR MAKING DROP FORGINGS.

The dies used for making drop forgings are made in two parts. One part (the upper) is fastened in the ram or hammer of the drop, which moves vertically between two uprights or guides and is raised by means of friction rolls controlled by the operator. The other part of the die (the lower) is fixed in the anvil on the base of the hammer. The ram rises until released, when it falls instantly, striking with the upper die the heated bar of metal placed on the bottom die and then forcing it into the impression in both dies. By a series of such blows the complete article is formed.

Necessary sets of dies to produce drop forgings of special shaping and sizes are usually made from a drawing or model, usually the latter as it facilitates designing the dies and allows of figuring the cost of the tools much more easily than could be done from a drawing.

In making drop-forging dies the die-sinker must know whether the drawing and model show finished or forging size; he needs also to know the allowance desired in machining. It is usual to add $\frac{1}{32}$ inch on each surface to be machined unless the piece is to be finished by polishing or grinding only, in which case $\frac{1}{100}$ is allowed; surfaces not to be machined or ground are made as close as possible to size. Forgings vary slightly in thickness—say from $\frac{1}{100}$ to $\frac{1}{32}$ inch—depending on their shape and the material used. They can, however, be made to gauge, by a re-striking operation; this operation requiring separate dies and entailing additional expense.

In addition to forging dies, the cost and endurance of which depend upon the work required of them, trimming dies are necessary to remove the surplus metal thrown out between the forging dies in working.

Before using the finished set of dies for forging a lead proof is struck up which is compared with the model. The proof often varies from the model or drawing by what is called draught. This is the taper necessary in the forgings to allow of drawing them from the die while working, and it averages about 7°. It can be obtained by adding or taking off from the forging; usually the draught metal is added.

METHODS OF DROP FORGING.

The drop forging of iron, steel, and copper pieces is a far more rapid method than hand forging, and will not only produce large quantities of work, but make possible shapes which are next to impossible by hand forging.

There are two methods of dropping: to blank out, drop, and trim, either with combination drop dies having flashed edges, or with separate trimming dies, as the nature of the piece will permit. In blanking pieces which are to be dropped afterward make a careful study of the piece, and if you have time cut the drop dies first, hardening and finishing ready for use. Make a wooden pattern just as you think the blank should be, and from it make a lead casting in sand, which you may try as a test piece in the drop hammer. The result will show you if you have made any errors, which you can correct and try again. For nearly all drop dies it will be found necessary to "flash" the edges, that is, to provide a narrow edge round the mould of the die. The purpose of this is to cut off the surplus metal which flashes or squeezes out, as well as to reduce the pressure which would occur should this surplus, in the form of a thin flat or rim, come between two flat surfaces. It depends much on the shape of the pieces as to the form of flash edge.

To keep the dies, when in operation, free from both excessive heat and scales, conduct a pipe from the forge-blower or compressed-air supply pipe, and arrange it to blow against the face of the lower die. A very powerful blast will sometimes be needed to keep the dies free from small scraps of iron or steel. Much hard work can be avoided in cutting drop dies by getting out steel pattern pieces, and after fitting the dovetails, heating the dies, and dropping the cold steel pattern between them, thus forming a perfect mould of the piece. In doing this great care must be taken to see that the pattern rests properly between the two dies to prevent bending of the piece. It is generally necessary to partly rough out the dies, and give them a final shaping

394   PUNCHES, DIES, AND TOOLS.

by the hot method. Some allowance for shrinkage must be made in dies intended for long or very accurate pieces—the two shrinks, that of the die and the piece to be dropped, combined together amounting to $\frac{1}{16}$ inch per foot.

Difficult pieces with several projecting parts are sometimes roughed out under a hammer and dropped to a finish afterward, but may, where large numbers of one piece are made, be trimmed or cut from a rolled bar, leaving the projections of the forging represented by continuous flanges along the bar. By cutting away part of these flanges the outline of the forging will appear, which in one or more dies can be shaped up to the proper form. The crank hanger or bearing of a bicycle is an extreme example of a forging of this kind.

THE MAKING OF SETS OF DROP-FORGING DIES.

The dies are made of .45 to 60 carbon steel, and are, usually, from 5 to 8 inches thick. Figs. 601 and 602 give a general idea of their appearance when finished. They are marked T and B (top and bottom) to prevent their getting mixed up in the laying out. The front and left-hand sides are squared up, and from these sides the centre lines of the impressions are laid out

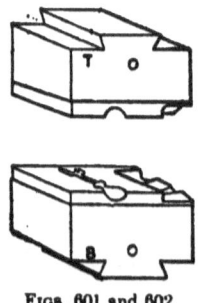

FIGS. 601 and 602.

and the dies set up when ready for use. The edger, or breaking-down impression, is on the right-hand side of the die. It is for breaking down the rough heated stock into something like the shape required before it goes to the finishing die. Sometimes a separate breaking-down die is used. The heaviest part of the

PUNCHES, DIES, AND TOOLS. 395

forging is always nearest the front. In deep dies shapes which show parallel sides on the drawing are given from 5° to 7° taper on each side, to prevent the forging from sticking in the die. For machining the forging $\frac{1}{32}$ inch is usually allowed and for shrinkage .012 to .015 per inch. When the dies are finished a specimen casting of lead is made in them for ascertaining whether or not they will give the desired result.

The round portion of the impression is sunk first. Swinging the die blank in a lathe, when there is much stock to remove, is

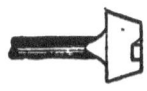

FIG. 603.

a convenient method. In the large drop-forge shop of the J. H. Williams Company, Brooklyn, N. Y., a cast-iron bolster for the lathe face-plate is used. This bolster has a web on its back, which fits the slot in the face-plate. The face of the bolster has a dovetail slot, which is identical with those in the hammer, and is at right angles to the web in the back. By this means a circle is quickly trued up. When the round portion is under $1\frac{1}{2}$

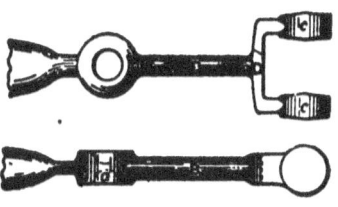

FIG. 604.

inches in diameter, a profiling machine is better adapted for the work, using the half-round cutter shown in Fig. 605 to finish with after having used the two-lipped cutter, Fig. 603, to rough out the stock. The half-round cutter is very useful, being stronger, easily made, and is easily ground by hand. The one illustrated in Fig. 603 leaves a point in the centre of the impression for spotting the centre of the boss on the forging. A die for forging a ball is sunk with a two-lipped spherical cutter. If there is to be a large hole drilled in the forging a plug is left, when sinking, or is afterward inserted in the die to lighten the forging at that point, as shown at a, Fig. 604. This plug should have a taper of 15° on each side, and the top well rounded. In making the dies for forging the piece illustrated in Fig. 604 the round portion B can be machined out after part D is sunk. This may be done with a spherical cutter. In some drop-forge die shops a patented machine for die-sinking is used, by means of which a cutter can be sunk into its centre in the work. The cutter is held in fixture on a short arbor between half-round centres, around which the cutter is rotated by means of a rawhide gear. The teeth of the gear engage the back of the teeth of the cutter. This is only used for finishing. Parts c c, Fig. 604, would have to be typed out, i.e., sunk by hand. Some circular impressions are sunk in the malling machine, using one long half-centre, and a forming cutter with a small shank, a groove being first cut to clear the shank. The parts c c, Fig. 604, would have to be typed in most shops. A type is a hardened steel template, of the size and form that the impression is to be, with the top left soft to prevent the steel from flying when struck with a hammer. Portions of the die that cannot be machined, owing to their irregular shape or the lack of shop facilities,

FIG. 605.

must be sunk by hand. This requires special skill with the hammer and chisel, scrapers and gravers, as well as a good eye for judging form.

The chisel used in hard die-sinking must be ground to the proper angle on the cutting edge, for chipping the curved surfaces and awkward corners. Scrapers must be made of various lengths and shapes to suit the requirements of the work. One of the most useful is the three-cornered scraper, with two edges rounded; the third, being the cutting edge, as shown in Fig. 605, is left sharp and is curved toward the point. Another handy one is shown in Fig. 606. It is a leaf or heart-shaped, and is convenient for getting at small curves and corners, especially those at the bottom of the impression. The type is covered with a thin coating of Prussian blue, or red lead, and driven into the die from time to time, as the work progresses, and the high places worked down until the correct form is produced. For the fillets and small corners a graver or scraper is made of Stubbs steel drill rod, of the desired radius. To save room in the tool-box a $\frac{1}{16}$- or $\frac{1}{4}$-inch handle may be threaded on one end, and these scrapers fitted to it. A hole is drilled in the cutting end to save time in grinding. The scrapers leave small ridges in the work, which are filed out with riffles or bent files. Some

FIG. 606.

impressions are polished with a soft pine block and powdered emery, but this is not the usual practice. When the impressions are worked out to the lines a lead casting is taken to see where they need matching or evening up. The lead is tested for size, and if all right a half-lead is taken from the top die to be used as a template in laying out the trimming dies, that is, in shops where sheet-metal templates are not used. If the lead is overheated, or is heated too often, it will not flow freely and will chill before the impression is filled. Powdering the impression with chalk causes the lead to flow more freely.

The edger or breaking-down form, on the right of the die, is made from $\frac{1}{16}$ to $\frac{1}{32}$ inch smaller than the horizontal cross-section of the forging, and has no abrupt shoulders or curves. The idea is to get the heated stock smaller in width than the finished impression, so that the bottom of the impression strikes the stock first, and spreads it to the sides, filling the die. Cast-iron dies are also used for breaking down heavy work.

The flash, which is a recess .015 to .025 inch in depth, and about $\frac{1}{8}$ inch wide, milled around the outline of each impression, allows the surplus stock to escape from the dies. This surplus is afterward trimmed off in the trimming dies. The top die also has a groove about $\frac{1}{16}$ inch in depth milled around the impression, ¼ inch from the edge. The gate for clearing the stock tapers gradually toward the front from the impression so as not to weaken the die at that point.

In dies for making small forgings in large quantities, there are several impressions sunk, one of which is used for a rougher, and should be about $\frac{1}{32}$ inch narrower and deeper than the finishing impression. Some dies have to be interlocked when difficult shapes are to be forged, that is, the faces have to be shaped to suit the offset in the forging. Care must be taken to have the interlocking parts high enough so that the dies will not glance off when striking the stock, and making an imperfect impression and therefore a useless forging. When the face of the

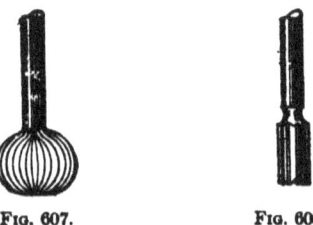

FIG. 607.   FIG. 608.

dies is curved special cutters are made, similar to those in Figs. 607 and 608, for surfacing and flashing. As a guide for machining curved surfaces or impressions some mechanics transfer the lines on the side of the die blank and lay out the curve there, then clamp a surface gauge to the profiling machine, and with the needle set on the face of the cutter, work out the stock, by following the lines with the needle point. Dies for forging gears, or similar work, are finished with a branch having the teeth machined in it, which is then driven into the die.

Drop-forging die-sinking, on the average, pays better than tool-making and machinist work. It requires more manual skill and labor, but not such a general mechanical knowledge as is necessary in tool-making. If one becomes expert in sinking dies for drop forging, and has a taste for the artistic, he can easily work himself into ornamental die-sinking and steel engraving. To follow this up, and become expert at it, requires time and patience, but it holds out the prospect of better wages and better conditions than ordinary tool work.

PUNCHES, DIES, AND TOOLS.

THE HARDENING OF DROP-FORGE DIES.

In hardening all kinds of drop dies, especially those of large size, on no account plunge the whole die into water, but as soon as properly heated set in a convenient place, and direct a stream of cold water from a hose on to the face of the die. The stream of water should be ample to cover nearly the whole face of the die, the water being kept flowing till about the temperature of boiling water is reached, and then the die should be allowed to cool gradually. In arranging the die for cooling, place it on a slight elevation in a manner that will prevent the surface water from cooling off the lower part of the die at the same time with the top. This method is intended to shell-harden only the top of the die, thereby preventing fire cracks and undue strain. The block can be drawn, if desired, while chilling by back drawing. When the proper color is reached again turn the hose on the die.

METHOD OF REPRODUCING DROP DIES.

To reproduce drop dies which are to be used on one piece of work right along, and must be constantly renewed, master-dies can be made from which to coin or drop new dies at any time. Should the die be very deep proceed as follows: First heat the die block and drop the master-die into it, giving a sufficient number of strokes to insure a perfect mould. Carefully anneal the die and remove scale by immersing in muriatic acid.

Wash the die in warm water, dry off, and oil all over. Set master-die in mould and examine for tight places resulting from shrinking. Lower down all flat surfaces outside of the die, trimming up all the edges and surfaces as when finished. Place under master-die, and drop and rub lightly with oil between the surfaces, producing a finish by a number of light blows rather than a few heavy ones. Take die from drop and give the final finishing touches and harden as already described. By this method dies which would cost, if cut by hand, several hundred dollars, can be easily produced in a day at a few dollars' expense a piece. Coupled to this is the certainty of every die being a duplicate, a feature very difficult to reach by hand work, if the details of the die are to be very intricate.

SET OF FORGING DIES FOR PNEUMATIC AND STEAM-HAMMER WORK.

The first set of dies described and illustrated in the following are for use on a pneumatic hammer, and the second pair on a steam hammer while performing the operation shown, and they afford a good example of methods and means used to obtain results in an up-to-date forge shop. The forging illustrated in

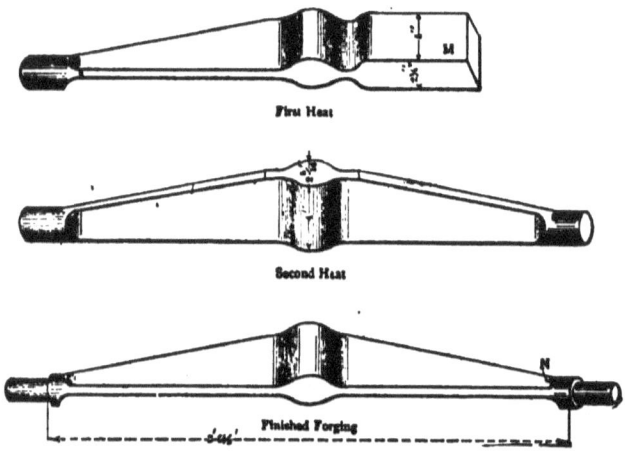

FIGS. 609 to 611.

Figs. 609 to 611, in its successive stages of first heat, second heat, and finished forging, is a brake beam as supplied for the bogie and four-wheeled stock of the New Zealand railways.

Fig. 612 shows the dies for performing the operations under the first and second heats, the *modus operandi* being as follows: A piece of soft steel, $2\frac{1}{4}$ x 4 inches and $17\frac{1}{4}$ inches long, is brought to almost a white heat and placed under the dies at A, where the eye is just stamped out to be formed, when by means of the flat faces B and the inclined surface at C the material is drawn out until it assumes the form shown in the first heat, Fig. 609. Of

PUNCHES, DIES, AND TOOLS. 401

course, it requires the repeated application of all three parts of the die to produce this result. During the finishing blows the faces of the dies are sprayed with water, which produces a very smooth surface on the forging.

The part A of the die for forming the eye and the inclined surface C were such that when the faces B came together the eye was brought to size and the rake or taper of the beam determined at the smallest section. The second operation under the

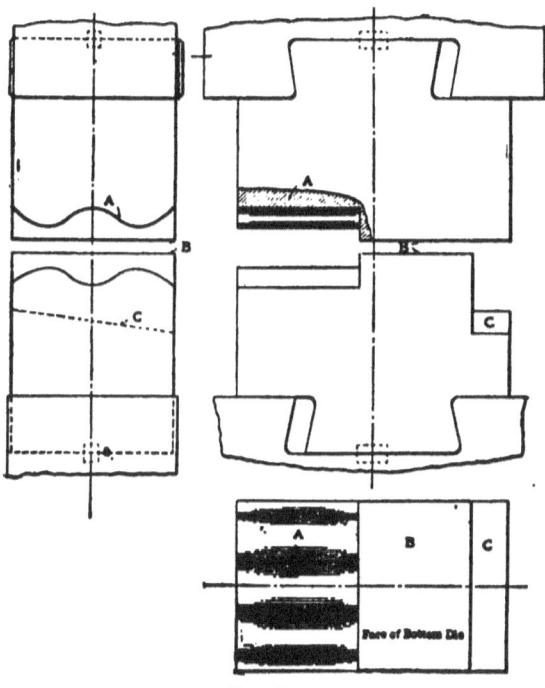

Fig. 612.

same dies gave the beam the form shown in second heat, Fig. 610, but entailed less work, as the eye was already formed, leaving the body of the metal (as shown at M in the first heat) only to be drawn out and shaped as shown.

The ends of the beam, as shown in the illustration of the finished forging, where formed in the dies shown in Fig. 613, which were fixed in a 300-pound hammer (steam), one heat being required for each end. The fillet marked N on the finished forg-

402 PUNCHES, DIES, AND TOOLS.

ing, Fig. 611, is formed by that part of the dies indicated at *D*, the correct length of the beam being determined by the location of this fillet, a gauge being used to fix the same. The swages at *E* are to reduce the end to size, both shoulder and journal being formed at one operation. The inclined surface at *F*, on the bottom half of the die only, adjusts any irregularity on the edge of the beam caused by the forming of the end.

The finished forging weighs 40 pounds and can easily be produced for 1 cent per pound in the forge. The dies were of tool

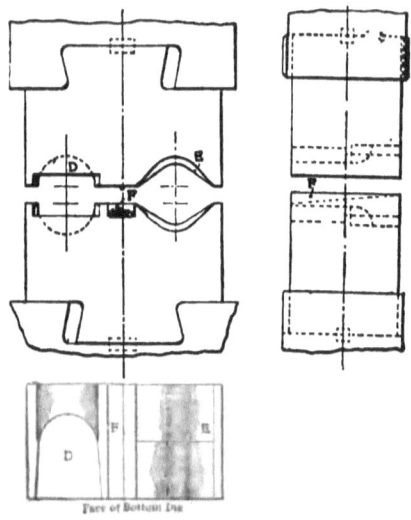

Fig. 613.

steel, not hardened, and they produce over 4,000 forgings before being re-faced. A set of dies, shown in Fig. 612, were made from steel castings and gave very good results.

Formerly these brake beams were made without any of these fixtures, ordinary spring swages being used. A great deal depended upon the skill of the smith who made them and also the time taken by him in producing them. Some smiths could form the eye only in the first heat, while one smith utilizing an 800-pound steam hammer with two helpers to heat the blanks and handle the swage, produced a state of finish similar to that shown under first heat, Fig. 609, in one operation. A very ordi-

PUNCHES, DIES, AND TOOLS. 403

nary smith, after a little practice, could produce the forgings with the dies shown, both the forging and the time taken being satisfactory.

The hammer used was a Massey 800-pound pneumatic manufactured in England, the air mechanism being driven by a 20-horse-power induction motor. Two smiths worked alternately, that is, when one had completed an operation, the other began his, so that the hammer worked continuously, or nearly so. The operation occupied about 4½ minutes. The peak load, as shown by a wattmeter, was caused by the stamping out of the eye, which was always the first part of the operation to be performed. It took the full capacity of the machine to forge this piece, considerable overload being carried during the first blows, and for the type and capacity of this hammer, the forging operation illustrated was about the limit.

STRIKING UP FORMING DIES FOR GLOVE FASTENERS, ETC.

The casual observer seldom thinks of the trouble it is to get the fasteners on gloves and suspenders, etc., finished so that they will act just right when in use. To look at them it seems as

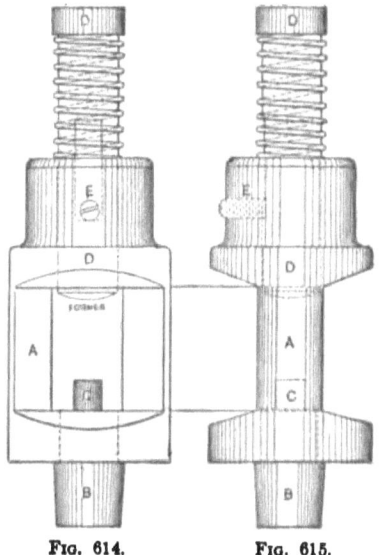

FIG. 614.   FIG. 615.

though all that is required is to set them any old way, so long as they will hold, but this is far from the truth. In applying these fasteners to light or thin goods, if the dies set them so that they snap hard, the goods soon tear in trying to open them.

A mechanic was given charge of making the setting dies for this work. Now these dies were all made with forming tools

Fig. 616. Fig. 617. Fig. 618.

and had to fit to gauges; and they were polished with emery cloth afterward by apprentices. It frequently happened that the boys held the emery cloth too long on some of the dies, and when they were put in use this made so much variation in the work turned out as to cause trouble.

A scheme was adopted that stopped all trouble from the source mentioned above. Figs. 614 and 615 show what might

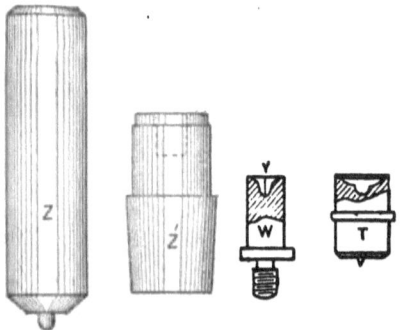

Fig. 619. Fig. 620. Fig. 621. Fig. 622.

be called a small forming press or sub-press. This was used under a drop, which was controlled by the foot. *A* is the body of the forming press. *B* is a taper plug which fits tightly in *A*, the taper end fitting in the anvil of the drop. *C* engages the die when being struck up by the hob *D*. A die similar to Fig. 616, or any of that class, Figs. 617 and 618, being placed on the stud *C*, the drop is then raised to the height required and dropped on *D*, which descends on the die and produces an exact

PUNCHES, DIES, AND TOOLS.

shape. The spring raises the hob D so that the finished die can be removed and another inserted. The screw E stops D at the proper height. In striking up a die like W the parts Z, Z' are used. The blow required to form W is so slight that it was not necessary to use the drop. A plain hole was drilled at Y and a light hammer was all that was necessary for striking the piston of the press. With a slight experience the right blow can be easily gauged, so that the dies will come out alike. Of course, different hobs were made to suit the various shapes, a few of which are shown in Figs. 619 to 622.

SET OF DIES FOR STRIKING NUMBER PLATES.

The punch and die shown in Figs. 623 to 628 are for making number plates which are used to number stoves and to designate the different styles of steel range made in a certain shop. The tools illustrate a rather novel method of making the number plates, and the description and illustrations will benefit others who are engaged in this line of work.

Fig. 624 shows the forming die, and the number holder, as it is called, which is of hardened steel. The mortise R holds the

FIG. 623.

number blocks shown in Fig. 623, which are interchangeable, and is made to hold three blocks at a time. The number blocks are $\tfrac{11}{16} \times \tfrac{11}{16}$ inch, and are made of brass; by using two $\tfrac{11}{16}$-inch pieces to make up for the third block, two numbers can be made on the number plate instead of three. The piece J is set on the cast-iron piece I which is made to fit a 4-inch shoe, and held on the block I by the $\tfrac{3}{8}$-inch fillister-head screws H, to keep the metal from buckling while being formed. Fig. 625 shows a sectional view of Fig. 624 with the cushion rubbers in position. Fig. 627 shows the cushion ring, which is mortised out to the

406   PUNCHES, DIES, AND TOOLS.

shape of the forming die *J*, and sets on top of the cushion ring, which projects above the top of the die about 1/16 inch, and, as the ram descends, it forces the cushion plate *H* down, and the tension of the cushion rubbers keeps the metal from buckling.

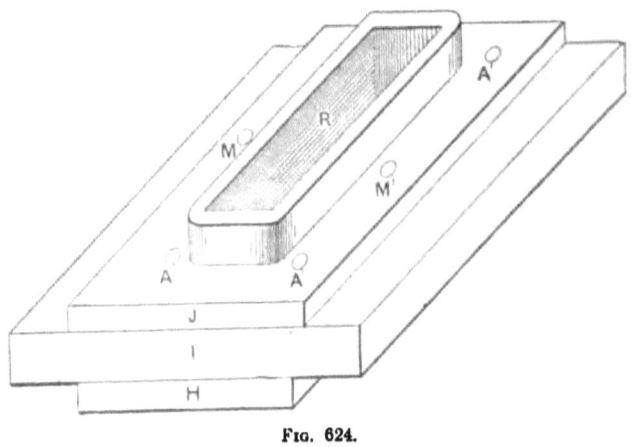

Fig. 624.

Fig. 626 shows a sectional view of the female part of the die. The upper half of the die *E* is made of cast iron and a ¾-inch hole is drilled through the shank for the spring *N*. Fig. 628 shows a sectional view of the lower half of the die *G*, Fig. 626;

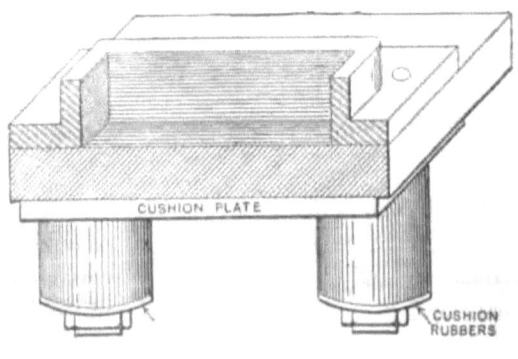

Fig. 625.

it is made to conform to the shape of the forming die *J*, Fig. 624, and is made twice the thickness of the metal larger. Fig. 626 shows the female number blocks *D* in position. These are

PUNCHES, DIES, AND TOOLS. 407

held in place by the rod C, the blocks having a hole $\frac{1}{8}$ inch diameter drilled through them, as shown in Fig. 623, at P. The two screws K are to keep the rod C from coming out. These blocks act as a kicker, which accounts for the hole being drilled

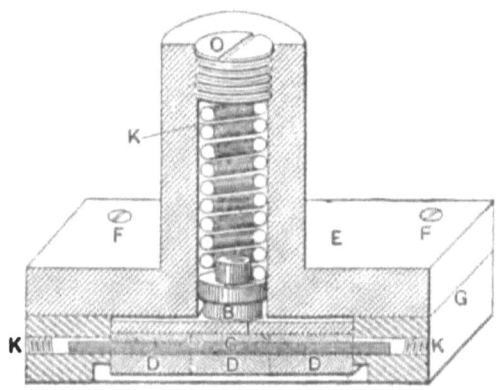

FIG. 626.

$\frac{1}{8}$ inch in diameter, as the large size allows the blocks to work free and bottom on the upper half E, or shank-holder. The piece T is a loose piece laid on top of the blocks D, so as to get even pressure on all of the blocks. The piece B rests on the

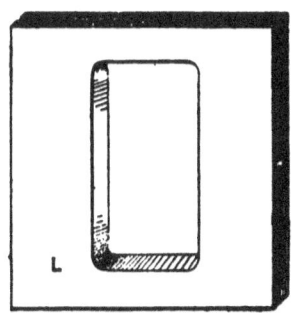

FIG. 627.

top of piece T. As the width of the blocks is only $\frac{11}{16}$ inch, it was necessary to use this means of making an effective ejector, as the spring hole being $\frac{3}{4}$ inch in diameter, it would not allow the spring to come in contact with the piece T. The lower half of the die *G* is fastened to the upper part by means of the four fillister screws *F*. *O* is a tension plug and keeps the spring from coming out.

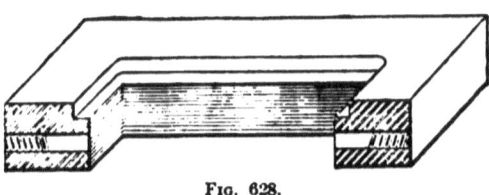

Fig. 628.

One advantage of this die lies in having the numbers interchangeable; it only requires a few blocks, whereas if each number was all made in one it would require a great many number blocks. Constructed this way it makes a very neat tool and produces just as good work.

SPECIAL DIES FOR THE BOLT MACHINE.

The old method of making dies for the bolt machine is quite an item, not only in the first cost, but in the renewing and patch-

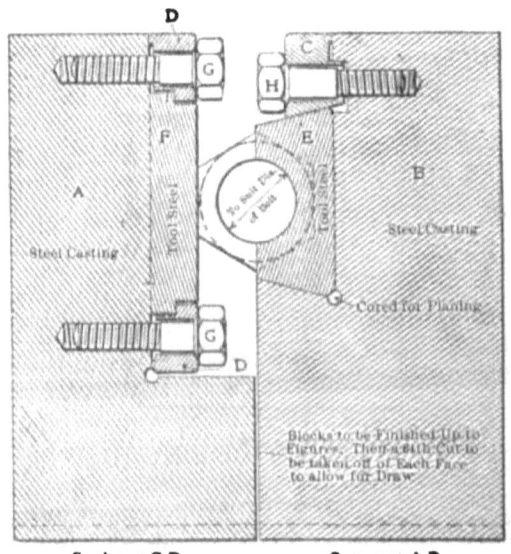

Fig. 629.

PUNCHES, DIES, AND TOOLS.

ing up of the old ones; especially is this so when they take in a range of sizes from half an inch to three inches. Each pair of dies which we have in mind, Figs. 629, 630, and 631, were the size of the blocks marked A and B—$15\frac{3}{4} \times 10$, $\frac{1}{2} \times 8$ inches—in the illustration, and could be used for making two sizes of bolts.

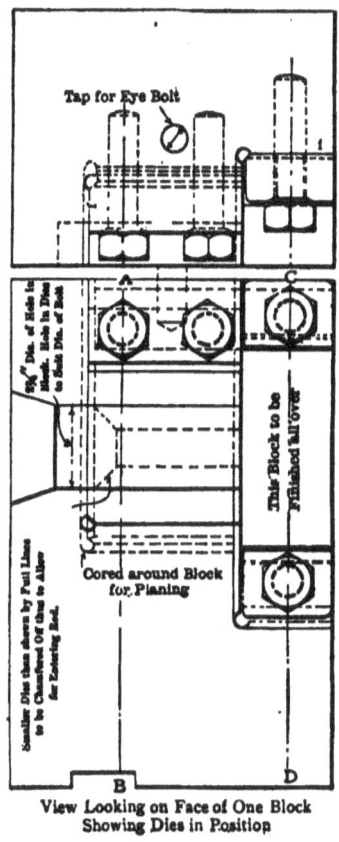

View Looking on Face of One Block
Showing Dies in Position

Fig. 630.

This made it necessary to have a great many heavy pieces to handle every time a different size of bolt was to be forged. To make a set of dies for the different sizes of bolts not only took a great deal of steel, but the machine work and hand work on them in the tool-room was excessive. These blocks being soft steel soon became worn and the corners rounded off and unfit for bolt-making, especially so for those for Government work. The die-holder and dies shown are the outcome of trying to overcome those obstacles and to make something that could be renewed with little expense and also whose faces could be hardened to withstand the wear and tear of the work.

The holder consists of two steel castings A and B, machined to take the tool-steel dies F and E. After they are clamped in place in the machine they do not have to be handled again for inserting or removing the dies.

The dies consist of two tapered pieces of tool steel hardened, marked E in the illustration, which are used to grip the stock while the plunger in the machine is forming the head between

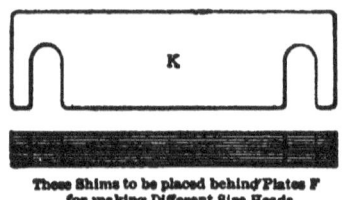

These Shims to be placed behind Plates F for making Different Size Heads.

FIG. 631.

the faces of the pieces marked F. One pair of tapered dies E can be used for making two sizes of bolts; for instance, the dies used for making 1⅛-inch rough bolts can also be used for making 1-inch finished bolts, as the same stock is used for both. The difference in the size of the hexagon head is made up by placing the shims K under the tool-steel piece E and clamping it in position with the bolts $G\,G$. Thus with different thicknesses of shims, Fig. 631, any width across the flat of a hexagon can be made.

To insert or remove the dies, the bolts $G\,G$ and $H\,H$ have merely to be loosened, and they can be slipped out the end. The shims can also be put in from the end of the block, the holes being made horseshoe shape to slip over the bolts.

The method of making the bolts in the machine is the same as with any other die, so it requires no explanation to be understood.

SECTION XVI.

Methods, Designs, Ways, Kinks, Formulas, and Tools for Special Work; Together with Miscellaneous Information of Value to Tool- and Die-Makers and Sheet-Metal Goods Manufacturers.

AN IMPROVED WASHER PUNCH.

WASHER punches of the design shown in Fig. 632 may be made of all sizes from $\frac{1}{8}$-inch hole and $\frac{3}{16}$-inch outside diameter up to 2-inch holes and 4-inch outside diameter, to punch all

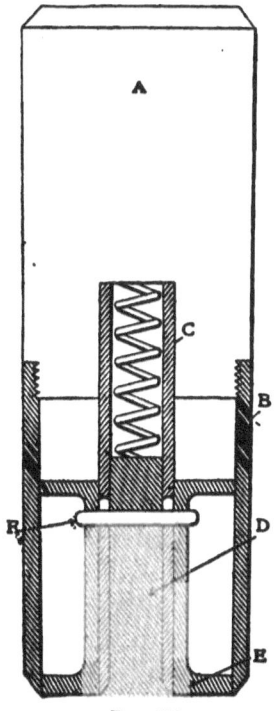

Fig. 632.

412   PUNCHES, DIES, AND TOOLS.

kinds of material, rubber, felt, paper, and leather. Using a mallet and hand-punch of this sort as high as 800 pieces may be punched per hour.

All punches patterned after this design should be made so that they can be used in power presses, if desired; but usually this will not be found necessary, as even the 4 x 2-inch punch will cut clean and easy at one blow of the mallet in $\frac{1}{8}$-inch harness leather.

Referring to Fig. 632, shank A is threaded to a shoulder and bored to receive C, which is the same diameter as the hole in the washer. It is a drive fit in the shank, with an elongated hole to receive pin F, which is fitted into plungers D and E, so that they move together. To assemble the punch, insert C into A, drop spring into bore of C followed by plunger D. Place plunger E in position; locate pin F and screw on B.

BLANKING AND PIERCING A FELT WASHER.

The tool shown in Fig. 633 is designed to be used in a foot press, and is made of tool steel throughout, except the holder which is of machine steel. It was used for punching felt washers, $1\frac{1}{8}$ inch in diameter with a $\frac{1}{8}$-inch hole and accomplished all work in one operation.

A is the outer punch or die, tapered to a knife edge for the outer diameter of the washer to be cut. B is the inner punch, tapered to a knife edge for its inner diameter. The head of B is a drive fit in A. The cutting part of B is but $\frac{1}{64}$ inch thick. A is screwed into holder C, bottoming as shown. D is a circular shedder plate fitting around B and A, and is held in place by three $\frac{1}{8}$-inch fillister-head screws (one only shown) screwed into D, as shown, and is held down by three helical springs d, which, by pressing on D, shed the washer from between A and B. E is a tool-steel plug, turned to fit the bore of B and prevented from dropping out by a head on its upper end, and above which is a spring which fits in a hole drilled in C and also A, and pressing on the head of E, sheds the scrap made by the inner punch B.

In use the punch is placed in the gate of a foot press, and a

PUNCHES, DIES, AND TOOLS. 413

piece of soft maple, grain end up, is fastened to a bolster in the press. A sheet of felt is placed on the block, and a quick shove with the foot on the treadle of the press drives the punch through the felt, until the punch just touches the wood, where

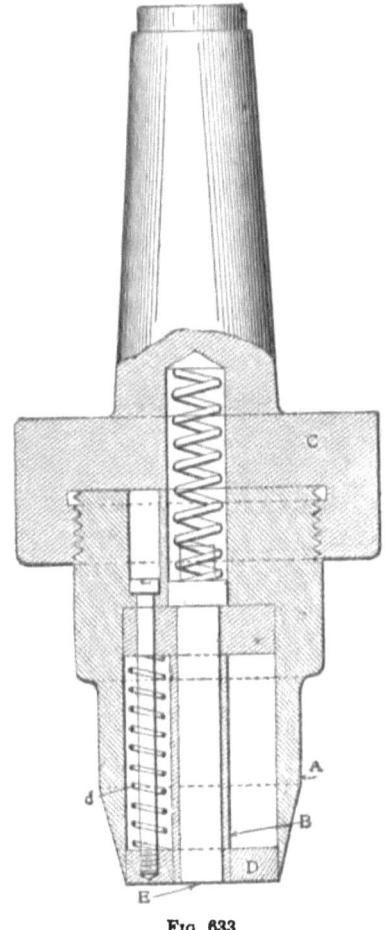

FIG. 633.

it is stopped by the adjustable stop on the back of the press, thus preventing the punch from penetrating the wood. On the up-stroke the strippers shed the washer and scrap from the punch, and the operation is repeated.

Tools of this kind may be used on almost any kind of soft

material, such as leather, thin rubber, paper, cardboard. In fact, one has been known to penetrate brass .020 inch thick.

In tempering the punches care should be taken not to overheat the steel, and a bath of fish oil gives the best results drawing the temper to between a dark brown and blue. After hardening and tempering, the edges should be stoned with a fine stone until keen.

DEVICE FOR CONTROLLING SCREW-BLANKS IN A THREAD-ROLLING MACHINE.

Figs. 634 and 635 illustrate a device for controlling small screw-blanks while entering the dies used for rolling the threads. The screws were about $\frac{1}{2}$ inch long over all, and made

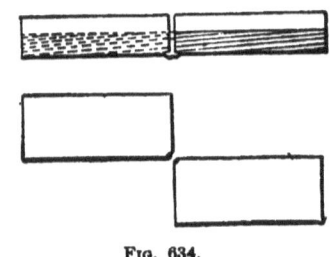

FIG. 634.

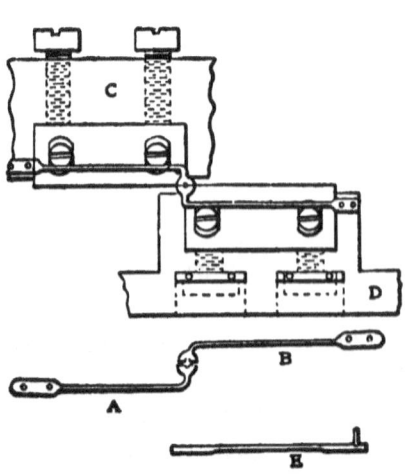

FIGS. 635 to 637.

of gold wire .040 inch in diameter, with 90 threads to the inch. The trouble was that the screws were so short that they would twist about in entering the dies, thus stopping the machine.

Fig. 634 shows top and side views of the dies without the device, and Fig. 635 a side view with the device in place. The springs were made as shown at A and B, Fig. 636 and 637, respectively, and fastened one to the sliding block C, and the other to the bracket D which holds the dies. The screw blank, being pushed in by the plunger, passes between the springs, which close over the outer end of the blank, as shown at E.

The device is perfectly practicable and well worth adopting for any similar work.

EXPERIENCE WITH THREAD ROLLING DIES.

Figs. 638 to 643 show a die for rolling threads. The first dies made were like B, with one cutting edge or face, and it was found on testing after hardening that the cutting face was convex, as shown, exaggerated, in the dotted line, at least $\frac{1}{100}$ inch. This may appear a small thing to complain of, but it was suffi-

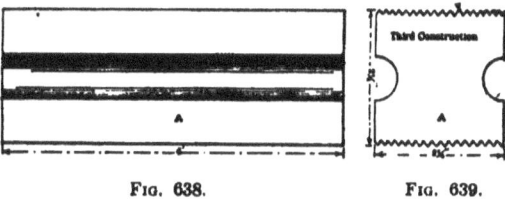

FIG. 638.   FIG. 639.

cient to spoil the thread. The threads are cut 24 per inch and $\frac{1}{100}$ inch convex in the die makes $\frac{1}{100}$ concave, or smaller, in the centre of the threads than at the ends, and a nut fitting tight at the end will shake in the centre. To overcome the trouble in the die, threads were finally cut on the opposite faces, and with grooves, as shown at A, Fig. 638. A perfect die was secured in this way, and at the same time two dies in the one block of steel.

Some mechanics may have doubts as to clamping on the toothed faces, saying that the dies will move and cut grooves in the clamping jaws in the machine. This was not the case, as the clamping surface is much larger than the points of contact on the stock, so that there is no danger of the dies moving while doing the rolling. *D*, Fig. 642, shows another way to make the die for the same purpose with four cutting faces on one steel block *for short stock threaded short*. The reader may ask, "Why do you not cut on both sides of the groove and have

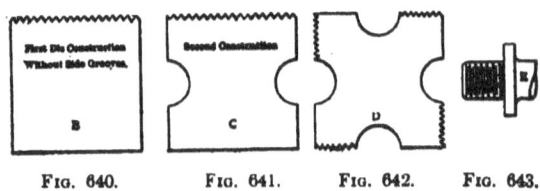

FIG. 640.   FIG. 641.   FIG. 642.   FIG. 643.

eight cutting faces on one block?" The die *D* is for cutting threads in stock with a shoulder at *E*, Fig. 643. A die with eight cutting faces would be useless for this class of work, as the corner threads would not be perfect, and when hardened would break off, but where the thread ends about $\frac{1}{8}$ inch from the shoulder the die can be made with eight cutting faces by rounding the corners. It should be noted that in *D*, Fig. 642, the metal is equally distributed, causing an equal expansion and contraction when hardened.

MAKING THREAD-ROLLING DIES.

The sketches, Figs. 644 and 645, show some tools used in making dies for rolling threads. One of the most important things in making such dies is to cut the thread or teeth properly on the die blanks so that they will form the correct thread on the work. We have been told that a pair of dies made for a $\frac{7}{16}$-inch 26 thread will answer a $\frac{3}{4}$-inch 20 thread. They may do for common rough threads, but do not pay. One cannot expect to have a 60° angle on a $\frac{3}{4}$-2p bolt cut with a $\frac{7}{16}$-20 die. When one die is used for various sizes of stock it requires to be adjustable, and an adjustable die is not easy to make, neither is it easy or practical to set in the machine.

PUNCHES, DIES, AND TOOLS. 417

Experience and much experimenting have taught us that it is much cheaper to make a pair of dies for each size of stock.

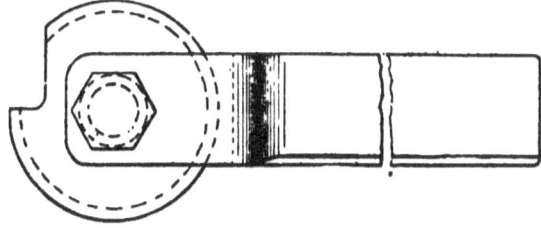

FIG. 644.

The dies are standard, simple to make, and the machine operator can take care of changes and setting without requiring the atten-

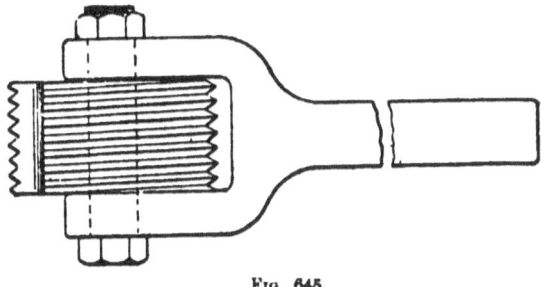

FIG. 645.

tion of a skilled mechanic. Such dies have been made in the shaper, using a circular cutting tool. This cutter is made in the

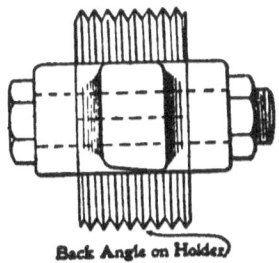

FIG. 646.

lathe in the same way as an ordinary piece of work is threaded, and must be as wide or a little wider than the blank.

Figs. 644, 645, and 646 show the cutter and holder. It is

very important that the bottom of the shank be planed at right angles to the pitch of the thread shown, otherwise the cutter will have no clearance on one side.

Fig. 647 shows a special chuck for holding the blanks while cutting the thread. The chuck is swivelled on a cast-iron base-plate, and the latter is made fast by clamps at the shaper knee.

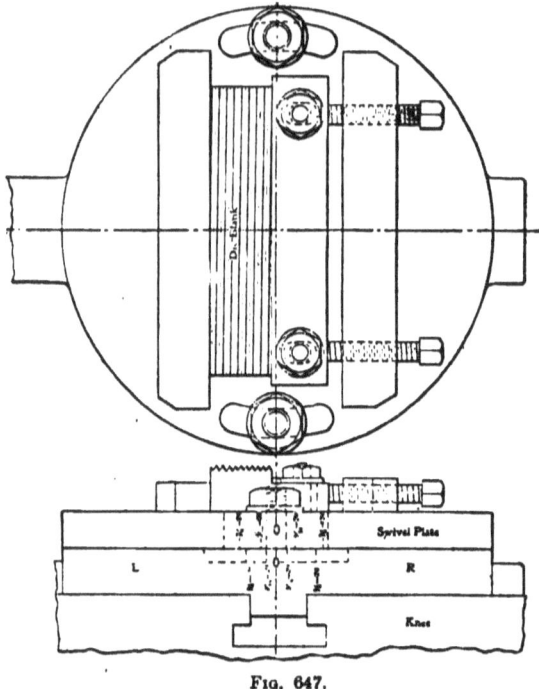

FIG. 647.

After finishing the proper angle for the die to be cut, the chuck is clamped tight; then a mark is put on the outer edge, and also the size and number of threads as shown. It is then but a few minutes' work at any time to set the chuck to duplicate dies. On one side of zero stamp "R" and on the other "L," indicating right- and left-hand threads.

FINDING DIAMETERS OF SHELL BLANKS.

Various formulas and tables for finding the diameters of blanks for given sizes of shells have been published. Most of them are based upon the presumption that the thickness of the sides of the drawn shells is the same as the original thickness of metal from which they are made, or, in other words, that the sides and bottom are of the same thickness.

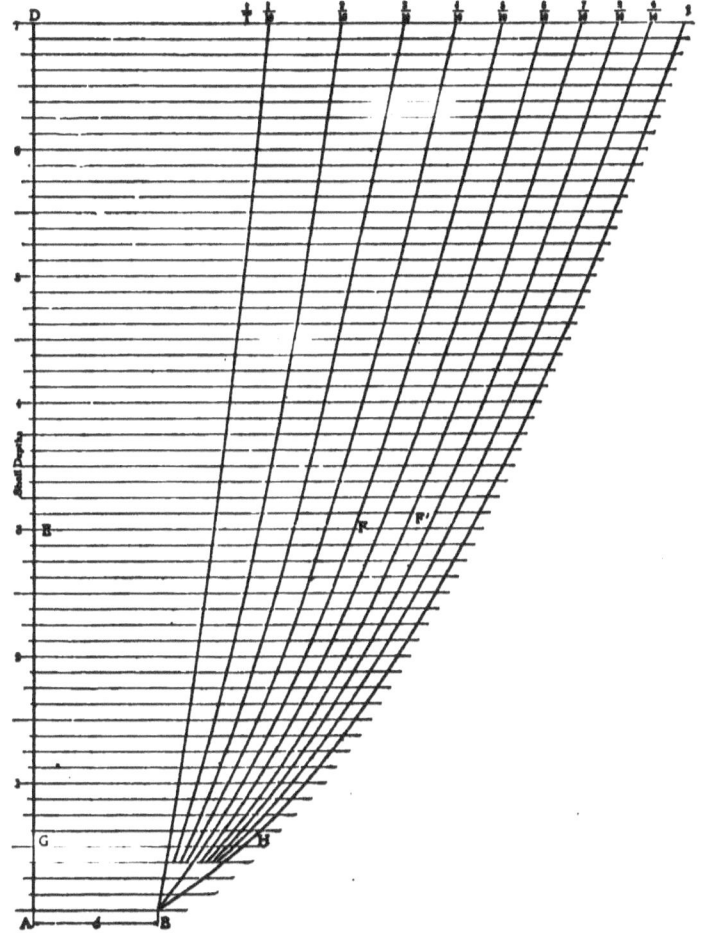

Fig. 648.

Besides, since most of these formulas work from the surface of the shell instead of from the cubic contents, they contain an error which, while being small on light material, is too considerable to neglect when heavy shells are to be drawn. The tendency of the times is toward heavier work of this nature.

The formula 1M is the result of a plain volumetric speculation. It will give a correct answer under small limitations, viz., that the shells are round and the bottom edges sharp drawn, not rounded.

$$1. \quad D = \sqrt{d^2 + (d^2 - d_1^2)\frac{h}{t}}$$

where
 D = diameter of blank,
 d = outer diameter of shell,
 d = inner diameter of shell,
 t = thickness of shell at bottom,
 h = depth of shell.

The formula does not make any allowance for trimming the shell after it has been drawn, but this can readily be covered by adding as much to h as it is desired or necessary to trim off.

Formula 2 has been utilized in constructing the diagram shown in Fig. 648. It is not absolutely correct in a mathematical sense of the word, but the error is very small, especially on light material.

$$2. \quad D = \sqrt{d^2 + 4\,d\,h\,\frac{t}{T}}$$

where
 D = diameter of blank,
 d = diameter of shell,
 t = thickness of shell at side,
 T = thickness of shell at bottom,
 h = depth of shell.

The curves of the diagram are parabolic, and as marked at the top of the diagram, they stand for various proportions of t to T. It is obvious, however, that proportions not directly given by the curves can readily be estimated between the two neighboring ones.

The diagram is drawn to the unit of 1 inch. It gives the blank for 1-inch shells directly, up to a depth of seven times the diameter. Distance AB stands for the diameter of the shell on the line AD, and the distance from the point so found to the corresponding curve gives the desired diameter of the blank.

Example:

To find the size of blank for shell 1 inch diameter, 3 inches deep; thickness of bottom $\frac{1}{8}$ inch, thickness of sides $\frac{1}{16}$ inch. $t \div T = .5$, consequently the curve .5 is to be applied in this case.

Measure 3 inches from the point *A* on the vertical line *AD*; this gives point *E* and the distance from *E* to the curve .5 gives the desired diameter *EF*, or $2\frac{5}{8}$ inches.

To use the diagram for other diameters than 1 inch, it is to be remembered that since the distance *AB* stands for the diameter, we simply determine a new scale for the diagram and measure the height *AE* in this scale and also the horizontal line from *E* to the given curve.

Example:

To find the size of a blank for a shell $\frac{5}{8}$ inch diameter, $1\frac{7}{8}$ inches high; thickness of bottom $\frac{1}{16}$ inch, thickness of side $\frac{3}{64}$ inch. $t \div T = \frac{3}{4}$, or .75. The curve for .75 not being drawn in the diagram, we can locate it very readily half-way between curves 7 and 8.

Since distance *A B* represents $\frac{5}{8}$ inch, the scale to be used in this case is $\frac{8}{5}$. The depth $1\frac{7}{8}$ measured in this scale is $1\frac{7}{8}$ times $\frac{8}{5}$, or 3 inches. Three inches from point *A* we locate *E*. From *E* to *F'* measures $3\frac{3}{16}$ inches. This multiplied by $\frac{5}{8}$ gives 2 inches for the diameter of the blank.

As a third example, we select a diameter of large size. The diameter of the shell is 4 inches, the depth 2 inches; bottom and sides are of the same thickness. Since $t \div T = I$, the corresponding curve is the last one on the diagram.

The scale of the diagram in this case is $\frac{1}{4}$. The depth of the shell in this scale is 2 times $\frac{1}{4}$, or $\frac{1}{2}$ inch; this gives point *G*. Line *GH* measures $1\frac{23}{32}$, in $\frac{1}{4}$ scale; consequently the diameter of the desired blank is $1\frac{23}{32}$ times 4, or $6\frac{7}{8}$ inches.

In looking over the diagram it will be observed that it shows the various proportions of thickness in a very clear manner.

Formula 2 being more convenient for practical use than No. 1, we give below formula 3, which is a modification of 2, with the object in view to reduce the error. As mentioned above, this is of consequence only when heavy material comes in question.

$$3.\ D = \sqrt{d^2 + 4 d_a h \frac{t}{T}}$$

d stands here for the outer diameter of the shell less one thickness of the side.

Formula 3 can also be used to good advantage when the thickness of the side of the shell is not uniform. In this case it is only necessary to substitute the average thickness of the side for t.

FINDING THE SIZE OF BLANK FOR A CUP.

It was necessary to build a double hydraulic press for making cups, as shown in Fig. 649, the cups being the first operation for nickel-jacketed steel bullets. A sample piece was secured from which it was necessary to determine the diameter of the blank. It was found that the thickness of the sheet metal from which the sample had been drawn was 0.99 millimetre at

Fig. 649.

the bottom of the cup. The diameter was found as follows: The physical principle made known by Archimedes was made use of. This principle is that any body immersed in water weighs as much less as the weight of the water it displaces. Therefore, in the metric system, the volume of water displaced expressed in cubic centimetres times 8 is the weight of this volume in grammes. This fact makes the proceedings simple. First is found the absolute weight of the cup in air, and second, when suspended in water, by means of a chemist's scale. Weighed in air it was 2.923 grammes, and in water 2.55 grammes, the difference in weight being 0.375 grammes or cubic centimetres. From these

PUNCHES, DIES, AND TOOLS.

data the diameter of the blank was figured by the following formula:

$$V = \frac{0.09\, d^2 \pi}{4}, \text{ in which}$$

$V =$ volume,
$d =$ diameter of disk,
$\pi = 3.14$.

Transposing d^2, extracting square root and substituting 0.373 for V we have, successively:

$$d = \sqrt{\frac{4\,V}{0.09\,\pi}} = \sqrt{\frac{4 \times 0.373}{0.09 \times 3.14}} = 2.298 \text{ centimetres diameter of disk.}$$

This method of obtaining the diameter of the disk described above is also interesting in that it shows the facility with which computations in volume and weight can be interchangeably effected in the metric system. The fact that the units of weight, capacity, and extension are all based on the metre saves the laborious process necessary with the awkward English units.

SIZING BLANKS FOR DRAWING INTO SHELLS.

In turning a flange around a head or an end, the material is forced from a larger to a smaller diameter, and this surplus stock must be worked in somewhere, either in making the flange thicker or in making it deeper. In making the ends in dies it

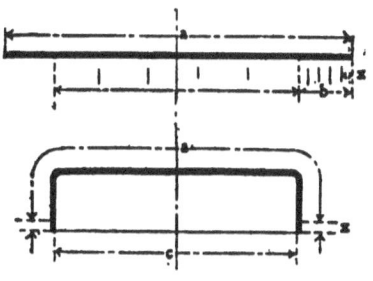

FIGS. 650 and 651.

is common to keep the thickness nearly uniform and to draw the surplus stock to the edge, thus making the flange deeper than the width of the ring from which it was made.

In making dies it is often necessary to know how deep a

PUNCHES, DIES, AND TOOLS.

flange will draw from a certain blank. In Figs. 650 and 651 the upper part represents a round blank and the lower part and end made from this blank. If we represent the diameter of the blank by a and the inside diameter of the end by c, then b will be half the difference between a and c, while x will represent the increase in depth of the end. To find x we may solve the proportion:

$$c : b :: b : x.$$

Or, when we have a drawing, we may use the following rule: Take b in the compasses and step off c, then divide b into the same number of steps, and one of these whole steps will be x. Add x around the edge and we have the depth of the finished shell.

b is contained in c about $4\frac{3}{4}$ times, so x is contained in b $4\frac{3}{4}$ times. This rule assumes that the thickness of the flange will be the same as that of the blank, but when it is from 2 to 5 per cent thinner it will be proportionately deeper.

LUBRICANTS USED IN RE-DRAWING CYLINDRICAL SHELLS.

The following recipe has been used with great success and perfect satisfaction for lubricating cylindrical shells in their successive re-drawings. The recipe is as follows: 12 pounds of green olive soap, 2 pounds of tallow. Each should be cut up into thin slices and put into a sheet-iron tank made for this purpose. This tank ought to be 2 feet high and 2 feet square. The two ingredients must be thoroughly mixed and boiled together in the above-mentioned tank, and should be afterward allowed to stand for at least three or four hours, after which the mixture may be thinned to suit one's judgment.

The best course to pursue is to have a steam pipe and valve connected with this tank, so as to warm up the lubricant when we are about to use it. The temperature of the lubricant when in use should be about 130° F. Let us assume that the shells have been cut and drawn on a double-action power press, and then carefully annealed, pickled, and washed. Now we are ready to proceed to the operation of re-drawing. The pressman opens the valve connecting with the tank and warms the lubricant, after which he puts a quantity of the annealed shells into a wire basket and immerses them in the tank.

It is usually required, to hold the shells, a receptacle attached to the die-bed in front of the press, and the man running the press takes the shells from the wire basket and puts them into this in quantities easy for him to handle. This lubricant has been used very successfully in the re-drawing of shells or tubes used for bicycle foot-pumps, and they are 12 inches long and 1 inch in diameter.

LUBRICANT FOR FIRST CUTTING AND DRAWING OPERATION.

Use good lard oil for a lubricant in the first, or cutting and drawing operations, on double-action cam presses, it being understood that the first shells or cups are made directly from the sheet stock, with dies of the usual compound type used in this style of presses.

Good lard oil is also a first-class lubricant to use in compound dies connected with the sub-press.

LUBRICANT FOR DRAWING COLD-ROLLED STOCK AND SHEET STEEL.

The following mixture has given very good results as a lubricant on drawing dies when drawing cold-rolled stock and sheet steel of mild grade: Boil together until thoroughly mixed, 1 pound of white lead, 1 quart of fish oil, 1 pint of water, and 3 ounces of black lead. Apply with a brush to the sheet metal before entering it into the dies.

COPPER COLORING BRASS FOR LAYING OUT PUNCHINGS.

To apply a copper coloring upon brass for laying out blanks and punchings, put a few drops of the ordinary coppering solution upon the brass and then dip a piece of iron or steel into the solution and touch the brass with it.

HARDENINGS, BLANKINGS, AND CUTTING DIES.

We present here a method of preparing dies for tempering, which are apt to crack when plunged into a bath, that will give the steel the proper degree of hardness. Iron wire is first wrapped about the die for the purpose of causing clay to stick to the smooth surface of the work, this substance being put on to cover all parts of the die not required to be hardened. This

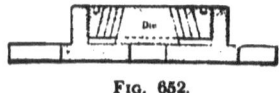

FIG. 652.

method may be used with success on almost any class of die work. Use common stove clay, such as stove dealers use, mixed with water and made just wet enough to stick well to the steel, then place the work in a slow fire until the clay becomes dry, when more heat may be applied. Never allow the clay to remain on the work longer than necessary, for if left on for any length of time rust spots will appear on the finished work. This method has been a great help to many die-makers.

ACID BATH FOR HARDENING DIES AT LOW HEATS.

To make an excellent hardening solution, mix pure rainwater and salt strong enough to float a raw potato, and to twenty gallons of brine add three pints of oil of vitriol. Tool-steel parts may be hardened at a surprisingly low heat in this solution, a very great advantage when hardening difficult shapes. The solution, however, has one slight disadvantage in that it causes the steel to rust quickly unless the steel is thoroughly scrubbed in strong hot soda water immediately after hardening. Tools and dies hardened in this solution should come out of the bath a beautiful silver-gray color, and if there are any black spots they are likely to be soft.

PUNCHES, DIES, AND TOOLS. 427

BOLSTER FOR HOLDING ROUND DIES OF DIFFERENT DIAMETERS.

It is generally the case that in looking for a die-bed, or bolster, in the press-room one always finds everything but what is wanted. The sketches, Figs. 652 and 653, will show how to overcome this difficulty to a certain extent, as one bolster takes the place of many.

The block a, Fig. 653, is made of cast iron; the rings c, d, and e are of machine steel, each 1 inch larger than the one it fits

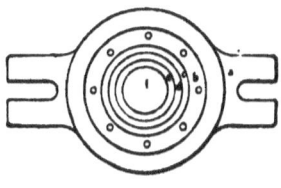

Fig. 653.

over, and turned taper inside and outside, the ring b screwing into a, having spanner holes drilled in its upper face.

The screwing down of ring b clamps all rings securely together. At f is shown the smallest die the block will hold in place. The object in having the rings is to enable one to remove any ring and replace the same with the die. The die blanks can be turned to fit duplicate rings and kept in stock. This arrangement will be found very handy when trimming shells or handling work of a like nature.

TEMPLET TURNING OF PUNCHES AND DIES.

Fig. 654 illustrates a method of turning punches and dies to templets. If a die is to be sunk to fit templet a, strap the work b on the face-plate and face it off true. Scribe a line on the face of the die of the largest diameter of templet. Then clamp the templet on the tall stock of the lathe, being sure that it is parallel with the cross-carriage of the lathe. Set a pointed turning tool to the line on the die. Clamp a surface gauge *c* to the tool-post shoe, or, what is better, use a small tool made for that purpose called a *tracer*. Then all the operator has to do is to watch the contact of the pointer and templet. On large, heavy work time can be saved by first drilling out most of the stock.

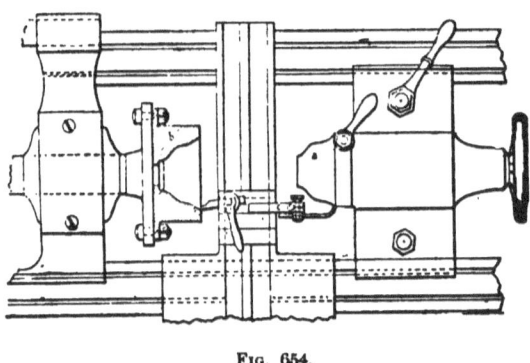

Fig. 654.

The punch may be fitted to the same templet under the clamp-screw, and proceeding as before. In this way a templet may be fitted so closely that a very little hand-tooling will finish the work. A great many dies have been fitted in this way, and it has been found much easier and quicker than the old way of cut and try. The operator must watch carefully that the tracer does not leave the templet as the tool follows the tracer.

FASTENING PUNCHES AND DIES IN HOLDERS.

Make the punches like Figs. 655 and 656, if the tools are to be used for disk-notching, the same shape all the way up, and with no shank nor special holder for each punch. Instead of that have a holder of machinery steel that will serve for all, within certain limits. Take a piece of $2\frac{1}{2}$ x 3 x $3\frac{1}{8}$-inch material and turn the round shank to fit the ram of the press, smooth the outside in the shaper, and drill and ream a $\frac{7}{8}$-inch hole about $\frac{7}{8}$ inch from one side and at an angle of $\frac{1}{4}$ inch in 3 inches with the other side, as shown in Figs. 657 and 658, and parallel with the

PUNCHES, DIES, AND TOOLS.   429

bottom face. Then drill a clearance hole for a slotting tool at B and cut slot C parallel with the sides and the bottom face ⅝ inch deep and almost ᵣ⁷₆ inch wide, as that is the width of the widest punches that should be held in it. Next cut slot D the same depth and ⅜ inch wide. A ⅜-inch drill rod should be cut

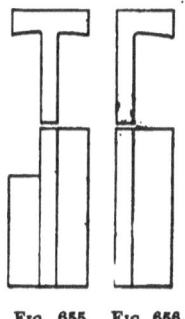

Fig. 655.   Fig. 656.

off 3½ inches long and tapered flat by filing line the cotter-pin of a bicycle crank, and hardened and drawn to a dark blue. All that is necessary to do now in fastening the punches is to set them into the slot in the holder, drive in the pin, strike a couple

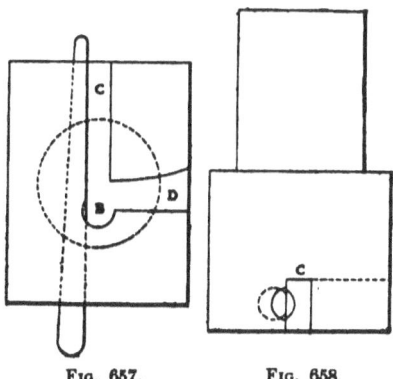

Fig. 657.   Fig. 658.

of blows on the punch, to make sure it rests on the bottom of the slot, and drive the pin home. For the thinner punches put some sheet-iron shims between the pin and the punch. Fig. 659 shows an improved bolster for dies.

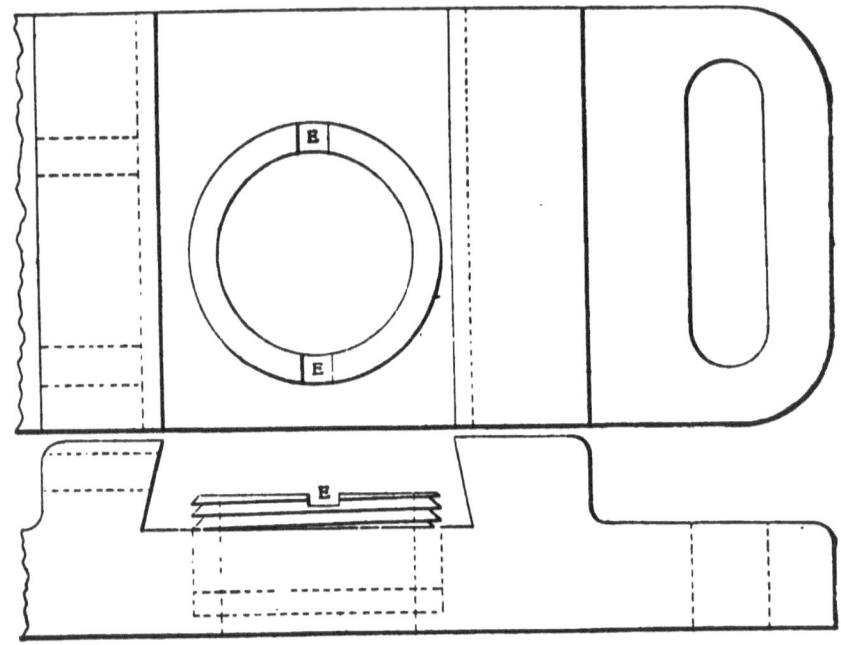

Fig. 659.

CAST-IRON BOLT DIES.

When running bolt-heading machines dies are often a constant source of trouble for the tool department, as well as a rather large item in the expense account. In one shop where such machines were used it was found that after trying various brands and tempers of steel the best output that could be gotten out from a pair of dies was 30,000 half-inch bolts (7,500 per

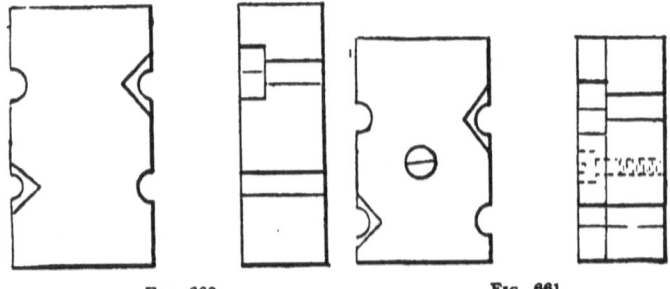

Fig. 660.   Fig. 661.

day) when the dies had to be annealed and worked over again. Trying to overcome this a pattern was made as shown in Fig. 660, all in one piece, of course, while the steel dies were made of two pieces and held together with a screw, as shown in Fig. 661. From this pattern white-iron castings were made. When they came from the foundry all that was required to be done was to grind them a little and they were ready to go into the machine. To the surprise of all concerned they stood up for 75,000 half-inch bolts, or $2\frac{1}{2}$ times as much as the costly steel dies ever did, while all the cost was the casting and a little grinding.

A STRIPPING KINK IN A DRAWING DIE.

Fig. 662 shows a die for making a small cup of sheet steel from No. 12 to No. 22 gauge. It gave much trouble, sometimes in stripping from the punch and at other times sticking in the

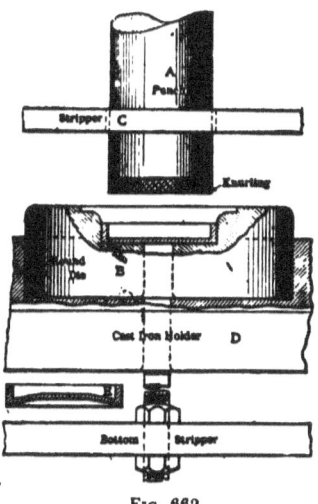

FIG. 662.

die, and it was a difficult piece to handle. Stripping the punch was an easy matter to overcome, but to remove from the die gave trouble. Some of the cups had a small hole in the centre

432   PUNCHES, DIES, AND TOOLS.

and others were made with the bottom solid; in both cases the bottom was required to be as flat as possible. The bottom of the drawing die was made with a small hole for a stripping pin from the bottom, as shown in Fig. 662. The cup with the hole prevented stripping by this means, and the solid cup would strip

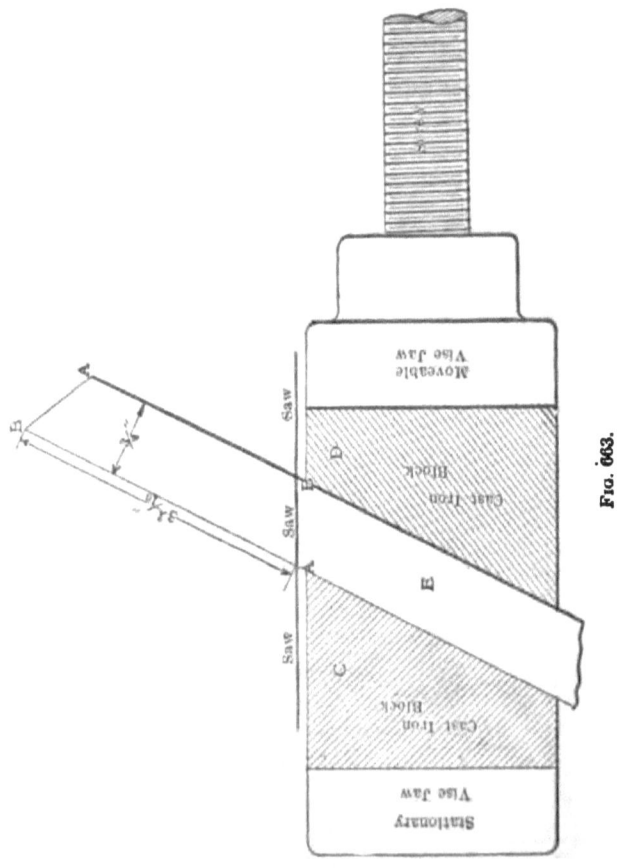

Fig. 663.

sometimes and at others the stripper would punch the bottom up as shown at *E*.

All this trouble was corrected by knurling the end of the punch as shown, and no further use was found for the bottom stripper. The knurled punch was made about three-thousandths smaller than the size required, then a light knurl was cut about $\frac{1}{8}$ inch long on the end and the sharp corner broken with emery-wheel cloth; the knurling expanded the punch end to the correct size. Very often before making the knurled punch, it was found necessary to chip cups out of the die, the friction was so great; but the knurling on the punch overcame the die friction, causing the cup to stick to the punch until it reached the stripper C.

CUTTING-OFF DIE BLANKS IN THE POWER HACK-SAW.

In machining die blanks used on a roll threading machine for threading small brass screws, the idea shown in Fig. 663 was found to be not only a time-saver but a steel-saver as well. The die blanks are bevelled at each end, as shown at X, which is a die blank, $3\frac{1}{16} \times 1 \times \frac{3}{4}$ inch, to fit the shoes of the machine, by which they were held when in use.

The idea, as shown in Fig. 663, is to saw the blanks at the required angle on the power hack-saw, allowing $\frac{1}{16}$ inch for finishing. This is done by the aid of the gray-iron blocks C and D. After each blank is sawed off, the bar E is turned bottom edge up and another blank is cut off, and so on until the bar is used up.

The blocks C and D are made the same height as the vise, and are used on other jobs besides the one illustrated here.

SQUARING UP A PUNCH PRESS.

Occasionally it is found necessary to overhaul and square up a punch press. When the bearing surfaces of the rams of the presses are found to be in fair condition, and if there are no tools suitable to replane them, scrape them to a good fit, then plane the seats for the punch shanks, so that they will be parallel with the bearings, and true up the lower faces of the ram. If the presses are intended to hold square-shank punches, they may be dressed up without difficulty.

A sweep *A* (see Fig. 664) should be made with a round shank *B*, and extreme care taken in the fitting, so that there will be no perceptible lost motion. This shank should then be securely fastened in the newly fitted ram in one of the presses and a micrometer clamped to the end of the sweep as shown. The bolster *C* must be planed up to an even thickness, and one side used as a surface plate to file and scrape the top bed on.

The micrometer thimble can now be run down and a reading taken from each of the four corners, care being taken that the

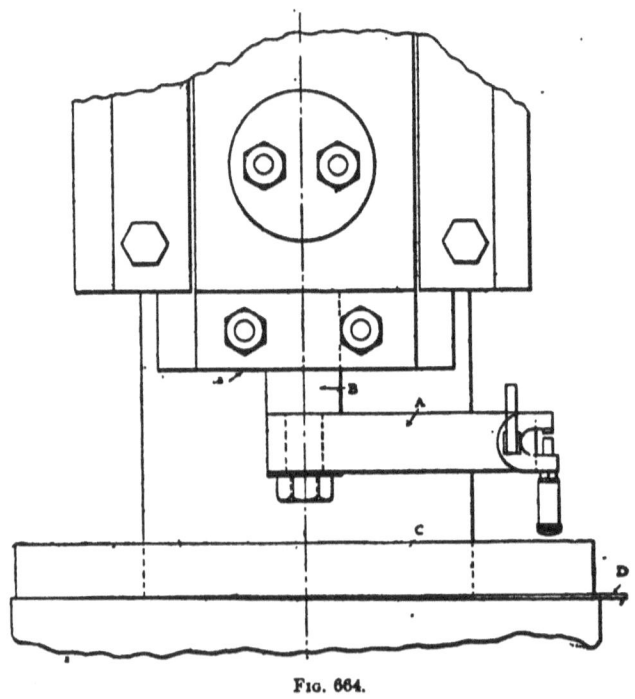

FIG. 664.

end of the thimble must just touch the bolster without any strain. Comparison of the reading showing that one end is higher than the other, this can be overcome by placing strips of tissue paper under the low end, as shown exaggerated at *D*. When the readings of the four corners are alike, the bolster can be removed and set up on the planer. The packing strips should be marked to correspond with similar marks on the corners of the bolster, and in setting up the planer, the strips from the opposite corners should be transposed and a cut taken over the top of the bolster. This will give the desired result.

This method should not be employed if the bolsters are used on different presses, and they should be plainly marked, preferably with the number of the press and the word "Front" on the side toward the operator.

IMPROVED METHOD OF BENDING TUBING.

The principal trouble in bending a tube is the tendency to buckle and wrinkle, but this method eliminates that trouble.

The tube, Figs. 665 and 666, which we will take as an example, is of brass, 1¾ inches diameter, with about ¹⁄₁₆-inch wall, and

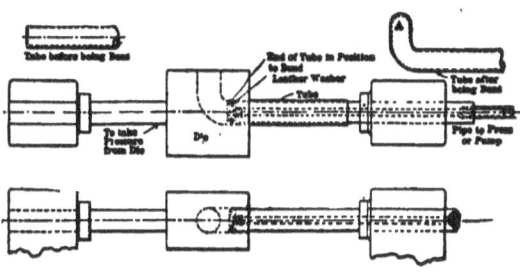

Figs. 665 to 668.

had to be bent at an angle of 90° without a wrinkle, crease, or mark of any kind; it also had to be round and of an exact diameter.

In order to get a tube with these requirements a die was made in halves, milled out to the shape of the bend wanted, hardened, and the cut polished very smooth.

The tube was laid in the lower half of the die, as shown in Figs. 667 and 668, with the end at the beginning of the curve; the upper half was then clamped in position, and a plunger pushed into the tube from the open end. Both the die and piston now having a support to withstand the pressure, the pump was turned on and the water pressure (6,000 pounds per square inch), besides keeping the tube round, pushed it around the curve in the die until it reached the required length.

The tube was then taken out and the round end A (the result of the pressure) was cut off, and we had a tube as smooth and round as the form of the die. The die must be well supplied with grease, as the friction on the sides is very great.

The drawings show the rig and tube as it is before and after it is bent; with this help no further explanation will be necessary in order thoroughly to understand the method.

THE FLATTENING OF PUNCHED DISKS.

Occasionally it is necessary to produce a lot of punched disks so that they will be perfectly flat and free from "buckles." To accomplish this satisfactorily adopt the method outlined and illustrated in the following:

First, we had a lot of round disks 8 inches in diameter and $\frac{1}{16}$ inch thick that had to be flat and not reduced in thickness. The stock was .040 carbon and the blanks had a 2-inch hole in them. Striking them two or three blows between a pair of flat dies had no effect upon them, so we then made a pair of dies like Fig. 669, the dies being concaved $\frac{1}{16}$ inch and the upper die convexed or "crowned" the same amount. The blanks were struck a good solid blow so as to dish them well, and they were then turned over and struck just hard enough to bring them back straight. It required a little practice to get the hang of the job, but after that no trouble was experienced.

The next job in this line was a lot of 7-inch disks .035 inch thick, made of spring copper. Although it was known that the disks must be flat, it was not expected that any trouble would be encountered in attaining the desired results. However, when the flattening stage was reached much trouble was encountered. The dies used on the $\frac{1}{16}$-inch disks would dish these blanks all right, but no kind of a blow would leave them flat enough. The trouble seemed to be in the spring of the metal, as the disks would have a "buckle" in them. Then a pair of dies like Fig. 670 were made, with two rings in them, projections on one face

PUNCHES, DIES, AND TOOLS. 437

and corresponding depressions on the other, great care being taken to turn out a first-class set of tools. This left the disk pretty good, and by working at them until the rings were gotten just large enough to take the slack out of the blank, they came through good and true. The rings also stiffened the blanks and kept them in the shape in which they left the flattening dies.

The next job in this line was a lot of square punchings that had to be brought to a given thickness, and they could be worked hot if that would bring them out any better. Having a pair of old flattening dies, they were annealed, and one planed out like Fig. 671, leaving the sides $A\ A$ ¼ inch high, and then grinding both the sides and the face of the die after hardening. The flattening was done in a light forging drop; the work was

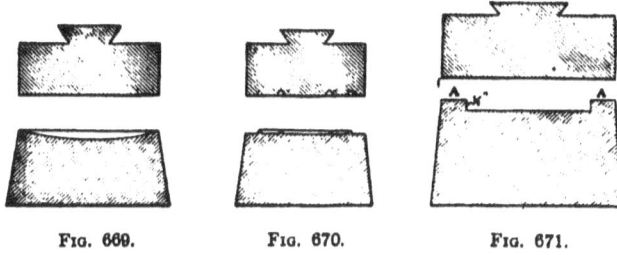

FIG. 669.   FIG. 670.   FIG. 671.

heated up to a dark red (or hot enough without raising scale) and struck one blow. The pieces came out of the dies flat and of an even thickness, but the cost of the work was more than was first estimated.

When the work to be flattened is very thin, there is no better way to proceed than to grind the dies good and true, lap them together with a little emery and oil, and then set them in the press so that when pieces of tissue paper are put under the corners the surface will "bite" all four corners alike. With a light power drop the cost of flattening thin press work should not be over six or eight cents per thousand. We know of a record of a press hand "flat-dropping" (as it is called) 44,000 pieces in a day of ten hours, using an ordinary rope drop press.

The trick of straightening a blank with a "buckle" in it is first to put in a crook that will take all the "buckle" up, and then take out the crook.

TABLE OF PUNCH AND DIE ALLOWANCES FOR ACCURATE WORK IN BLANKING AND PERFORATING TOOLS.

The following refers to the blanking, perforating, and forming of flat stock in the power press for parts of adding machines, cash registers, typewriters, etc., and its purpose is to give to the many what has been known to the few, and which has formerly not been tabulated for ready reference.

In this class of work it is generally desired to make two different kinds of cuts with the dies used. First, to leave the outside of the blank of a semi-smooth finish, with sharp corners, free from burrs and with the least amount of rounding on the cutting side. Second, to leave the holes and slots that are per-

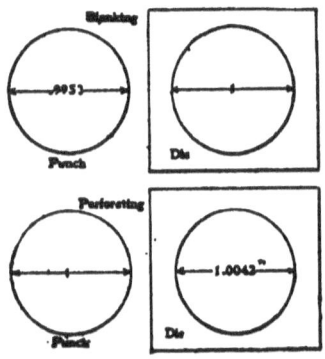

FIGS. 672 to 675.

forated in the parts as smooth and straight as possible, and true to size. The table given is the result of some years' experimenting on this class of work, and has stood the test of three years of use since it was compiled, and it has worked out to the entire satisfaction of those who have used it.

The die always governs the size of the work passing through it. The punch governs the size of the work that passes through. In blanking work the die is made to the size of the work wanted and the punch smaller. In perforating work the punch is made to size of work wanted and the die larger than the punch. The

PUNCHES, DIES, AND TOOLS.

clearance between the die and the punch governs the results obtained. It has been the aim in compiling the table for this class of work to put it in a simple form so that the average mechanic can make a die and be sure of the result.

Figs. 672, 673, 674, and 675 show the application of the table in determining the clearance for blanking or perforating hard rolled steel .060 inch thick. The clearance given in the table for this thickness of metal .0042, and the sketches show that for blanking to exactly 1-inch diameter this amount is deducted from the diameter of the punch, while for perforating the same amount is added to the diameter of the die. For a sliding fit make punch and die .00025 to .0005 inch larger; and for a driving fit make punch and die .0005 to .0015 inch smaller. The table follows:

TABLE OF ALLOWANCES FOR PUNCH AND DIE FOR DIFFERENT THICKNESSES AND MATERIALS.

Thickness of Stock. Inch.	Clearance for Brass and Soft Steel. Inch.	Clearance for Medium Rolled Steel. Inch.	Clearance for Hard Rolled Steel. Inch.
.010	.0005	.0006	.0007
.020	.001	.0012	.0014
.030	.0015	.0018	.0021
.040	.002	.0024	.0028
.050	.0025	.003	.0035
.060	.003	.0036	.0042
.070	.0035	.0042	.0049
.080	.004	.0048	.0056
.090	.0045	.0054	.0063
.100	.005	.006	.007
.110	.0055	.0066	.0077
.120	.006	.0072	.0084
.130	.0065	.0078	.0091
.140	.007	.0084	.0098
.150	.0075	.009	.0105
.160	.008	.0096	.0112
.170	.0085	.0102	.0119
.180	.009	.0108	.0126
.190	.0095	.0114	.0133
.200	.010	.012	.014

SECTION XVII.

Special and Novel Processes, Presses and Feeds for Working Sheet Metal in Dies.

SPECIAL COMPOUND PUNCHING AND FORMING DIE FOR SHEET-STEEL SHEAVE-PULLEY SIDES.

FIGS. 676 and 677 illustrate a novel arrangement of both press and punch with parts attached. Of the press it may be said that it not being of the inclined type, nor even having an open back, and it being desirable to remove the finished pieces by gravity, the plan of tilting the entire press sideways was adopted with perfect success in everything, except appearance, which was decidedly odd. New legs, giving a side inclination of 30° to the press, were provided in place of the usual supports. The centre of gravity of the upper portion of the press being low, it was found to stand safely upon new legs, and although the shorter one had on it by far the greater portion of the weight, no vibration or shakiness could be noticed while in action.

The special work for this press was the making of a small sheave-pulley side, of mild sheet steel, about 2 inches in diameter and .004 inch thick. These sheaves were produced in large quantities, about 40,000 sides per day being called for. They had previously been made by first blanking the side, then piercing the hole, and finally pressing into the desired shape in forming dies, the operations occupying three presses, each attended by an operator, as no automatic feeds had been employed for the work. After coming from the die the blanks were tumbled with sawdust as usual to remove any possible burr and the sharp edges, as well as to give a final semi-polish or brightening, and remove all traces of oil, which was plentifully used in the combined operation.

The design shown, while it was considered somewhat expen-

PUNCHES, DIES, AND TOOLS. 441

sive in its first cost, soon proved to be highly economical. Duplicate parts were made of those pieces which were likely to be broken or to wear out quickly, so that the press might be run

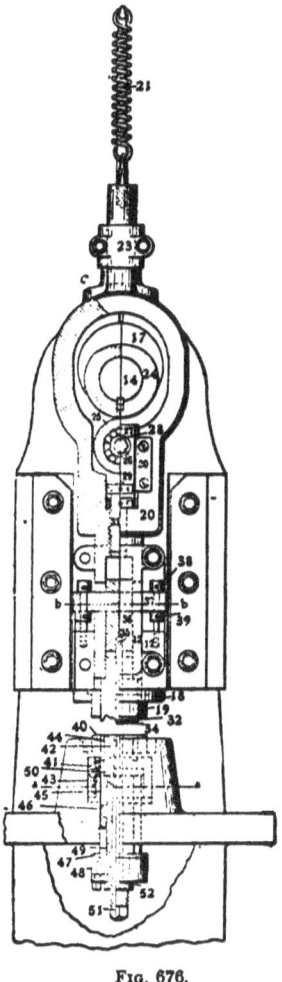

Fig. 676.

as uninterruptedly as possible, stopping only to make changes instead of repairs.

The alteration of the press, besides the inclined legs, consisted of a new special ram or slide, a new eccentric to move it, a special cam and cam-yoke, operating through the slide, with

442   PUNCHES, DIES, AND TOOLS.

a guide bracket on the top of the press, and a suitable heavy plate combining the functions of bolster, die-block, and support for the automatic roller feed, which was also added to it.

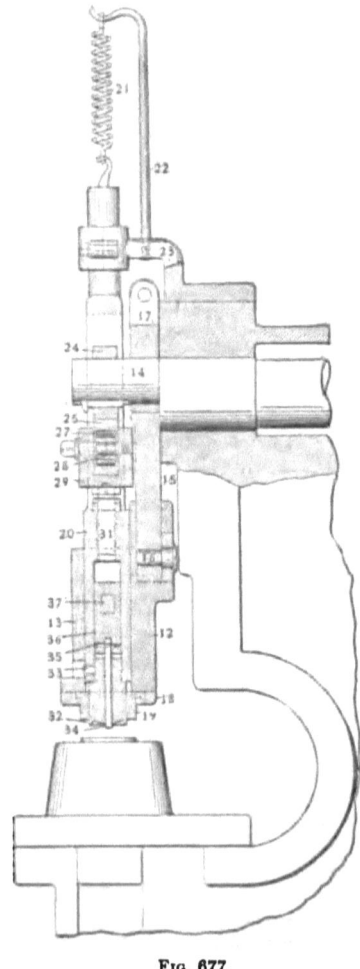

FIG. 677.

A detailed description of the drawing is given for those who may be interested. Figures represent the different parts. 12 is the slide, which is guided vertically in the press frame and carries a housing, 13, made in two parts for convenience of access, which form together a cylindrical bearing for the forming

PUNCHES, DIES, AND TOOLS. 443

plunger. The slide, 12, is connected with the crank, 14, by a pitman or connecting rod, 15, one end of the pitman being secured to the slide by a bolt, 16, while the other end encircles the eccentric, 17, the eccentric being adjustably clamped by the upper end of the pitman, so that any desired adjustment relative to the crank may be had. To the lower end of the slide, 12, by means of the ring, 18, is secured the pressure-plate, ·19. This plate serves also as a blanking die.

Within the housing is guided the slide, 20, which is held normally in an elevated position by the spring, 21, which connects the upper portion of the slide with a standard arising from a fixed portion of the press frame, a portion of which, 23, forms a guide for the upper end of the slide. A cam, 24, is keyed to the

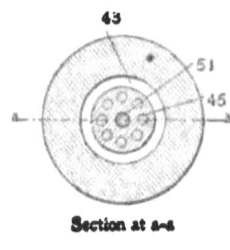

Fig. 678.

crank, 14, and operates the slide, 20, against the pressure of the spring by bearing on a hardened steel wheel, 25, carried on stud, 26. Anti-frictional rollers, 27, are provided, travelling upon a hardened-steel bushing, 28, fitted snugly on the stud. In order to obtain adjustment, this stud is carried in a yoke, 29, which is movable vertically in the slideway, 30; an adjusting screw, 31, is provided for varying the elevation of the yoke.

The upper forming die, 32, is fastened to the lower end of the slide, 20, by a set-screw, 33, passing through an opening in the front of the housing. A piercer or punch, 34, is provided, for punching the hole in the centre of the blank, which passes through a central opening in the former, 32, and has its upper end secured by a pin, 35, to the adjustable plunger, 36, carried in a recess in the slide, 12. An adjusting bar, 37, passes through the plunger, 36, and has its ends secured adjustably to the housing, 13, by means of the screws, 38, and lock-nuts, 39, in such a manner that the punch, 34, may be adjusted vertically with relation to the slide, 12.

To hold the strip during the forming operation, a lower pressure ring is provided at 40, in line with the upper ring, 19, which is supported upon a strong steel spring, 41, serving to hold the ring, 40, firmly up against the ring, 19, while the latter begins to descend, thereby clamping the stock under a powerful

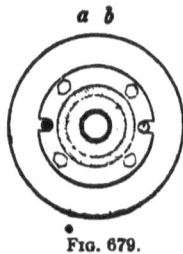

FIG. 679.

tension. The lower former, 42, rests solidly upon a ring, 43, which is firmly supported by the cast-iron die-block, resting in turn on the bed of the press.

In order to blank out or sever the pulley section from the strip after it has been formed, a lower blanking ring or punch, 44, is provided, encircling the lower frame, said ring being supported on a series of laterally adjustable steel pins. Two of these, one of which rests as shown at 45, the other being diamet-

FIG. 680.

rically opposite, are kept in contact with the punch by a coiled spring, 46, which causes the ring to act as a stripper, or means for forcing the blanked section out of the bottom former. The other pins rest on the steel washer, 47, which is supported by the nut, 48, the nut being split so that it can be securely clamped in any position.

PUNCHES, DIES, AND TOOLS. 445

The pins are of such length that there is a space between the bottom of the punch and the top of the pins, this space representing the amount of compression of the springs, 46. As the blanking ring, 44, is worn or ground away it may be adjusted vertically by the nut, 48, on the screw, 49. This pressure-plate, 19, also serves as a blanking die, the material being sheared between the edge of the plate or ring, 19, and the edge of the plate or ring, 44, as indicated in the sectional view of the punches and dies, Fig. 680. It is to be noted that the severing from the strip of sheet metal does not take place until the blank has finally been fully formed.

On account of the reciprocating motion of the punch, 44, there is a tendency of the former, 42, to work up out of the holder, which is overcome by having a sleeve, 50, fastened to the bottom of the former, 42, and threaded into the sleeve, 43.

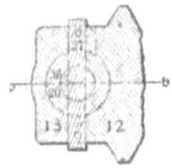

Section at b-b
FIG. 681.

In order that the hole in the centre may be punched by the piercer or punch, 34, at the same time the blank is cut, a round die is provided, which is supported and adjusted by the hollow screw rod, 51; this adjustable rod is threaded into the sleeve, 43, and held in proper position by the lock-nut, 52.

The operation is briefly described as follows: The strip of metal is placed or fed between the formers and the revolving of the crank forces the slides, 12 and 20, down in unison. As soon as the rings or plates, 19 and 40, come in contact with the work, the lower end is depressed and the formers, 32 and 42, begin to draw the metal into shape, the surrounding stock being in the meanwhile securely held by the pressure plates, 19 and 40. When the former, 32, has reached the limit of its downward stroke, and the blank is fully formed, the relative arrangement of the cams operating the slides is such that the slide, 12, and with it the pressure-plate, 40, and punch, 34, continues its downward movement. By this time, however, the springs, 46, have been compressed until the ring, 40, rests solidly on the pins, and the further downward movement of the pressure plate, 19, acting against stationary edge of ring, 40, shears off the metal, while at the same time the punch, 34, cuts out the centre hole. In the continued movement of the crank shaft, as soon as the slide, with the ring, 19, recedes, the pressure ring, 40, is forced upward by the springs, 46 and 41, out of the former, 42, when the cam, 24, has revolved so that the former, 32, will start downward, carrying with it the formed blank and strip, which, being moved along at the same time, should carry the formed blank out from under the tools, delivering it free from them and leaving the strip in readiness for the next operation.

A DIVIDING HEAD FOR A PUNCH PRESS.

In Figs. 682 to 685 are shown different views of an arrangement for application to a horizontal punch press, whereby boiled heads and similar flanged articles may be accurately spaced and punched without the tedious labor of previously laying out the

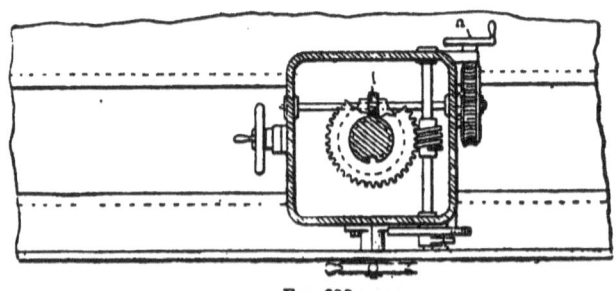

FIG. 682.

work by hand. The piece to be punched is assumed to have a central opening and by means of a stud or bolt extending up through this, and a crowfoot above, the work is securely fastened to the stiff vertical shaft h. The stand or carriage for the vertical shaft is movable along a horizontal side by means of a rack and pinion operated by the front hand-wheel, 7. The axis of the vertical shaft is thus always in line with the axis of the punch. This shaft has upon it a series of grooves which

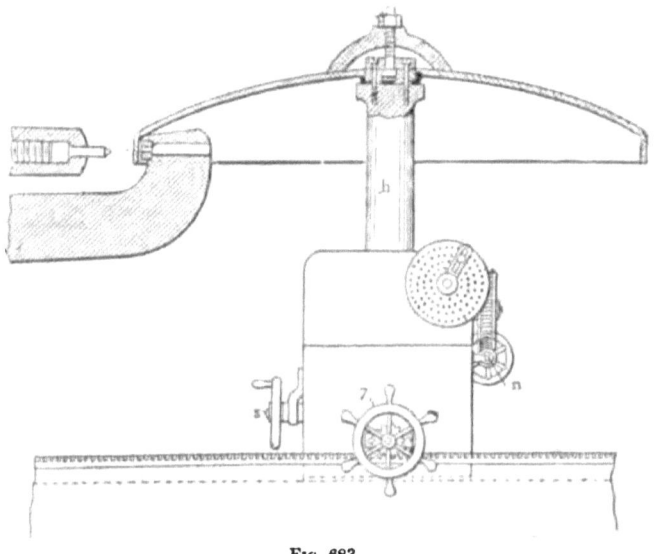

Fig. 683.

constitute a continuous rack for raising and lowering the shaft. In this circular rack meshes pinion l, which is operated through

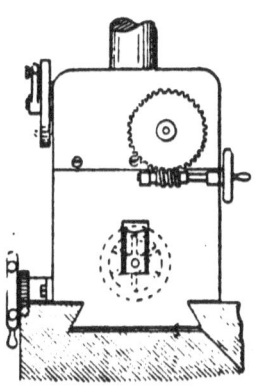

Fig. 684.

worm-wheel and worm by hand-wheel n. When the height of the shaft and the work is thus correctly adjusted, the step under

the end of the shaft is set to support the weight by the screw-operated wedge-block *r* and hand-wheel *s*. The graduated dial with worm and worm-wheel provide for the correct rotation of the work according to the number of holes to be punched. This

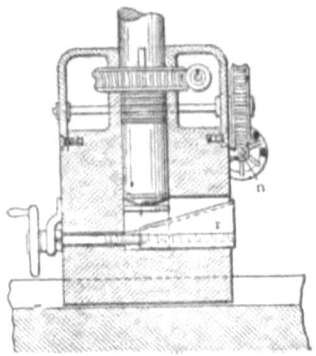

FIG. 685.

worm-wheel is splined to permit of the raising and lowering of the shaft as previously spoken of.

The arrangement here shown is, of course, equally applicable to vertical punches for dividing and punching flat circular disks, strainers, etc. It may also be applied to drills, either horizontal or vertical.

HYDRAULIC DRAWING-PRESS FOR HEAVY SHELLS AND FLANGES.

The engravings, Figs. 686 and 687, illustrate a powerful press which is designed for heavy drawing and other operations of a similar nature, and which is provided with an interesting arrangement of cylinders and piston for operating the blank-holder and the drawing punch.

The base of this press, which is shown at *A*, is mounted on a suitable foundation and provided with four heavy screws *B* which carry the head *C* in which the die *D* is secured. *C* is adjustable up and down by power, four nuts fitting the screws being operated for this purpose by means of gearing which is

PUNCHES, DIES, AND TOOLS. 449

driven in either direction, in the manner indicated, by the shaft above. The lower plate or bolster E carrying the blank-holder and adapted to slide up and down on the four screws is mounted

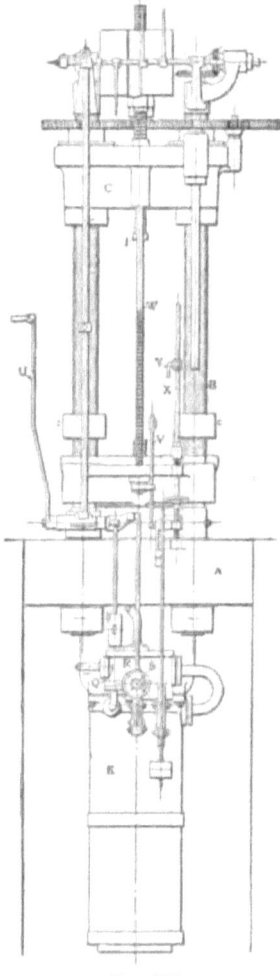

Fig. 686.

on a piston G in cylinder H. This piston is itself a cylinder in which is fitted another piston I, the latter being connected, as shown, with the drawing punch J. Below the cylinder H is cylinder K, which is fitted with a piston L connected by rod M with piston I.

450   PUNCHES, DIES, AND TOOLS.

At N is the reservoir for the liquid and at O a two-cylinder belt-driven pump for forcing the liquid through pipe P to the

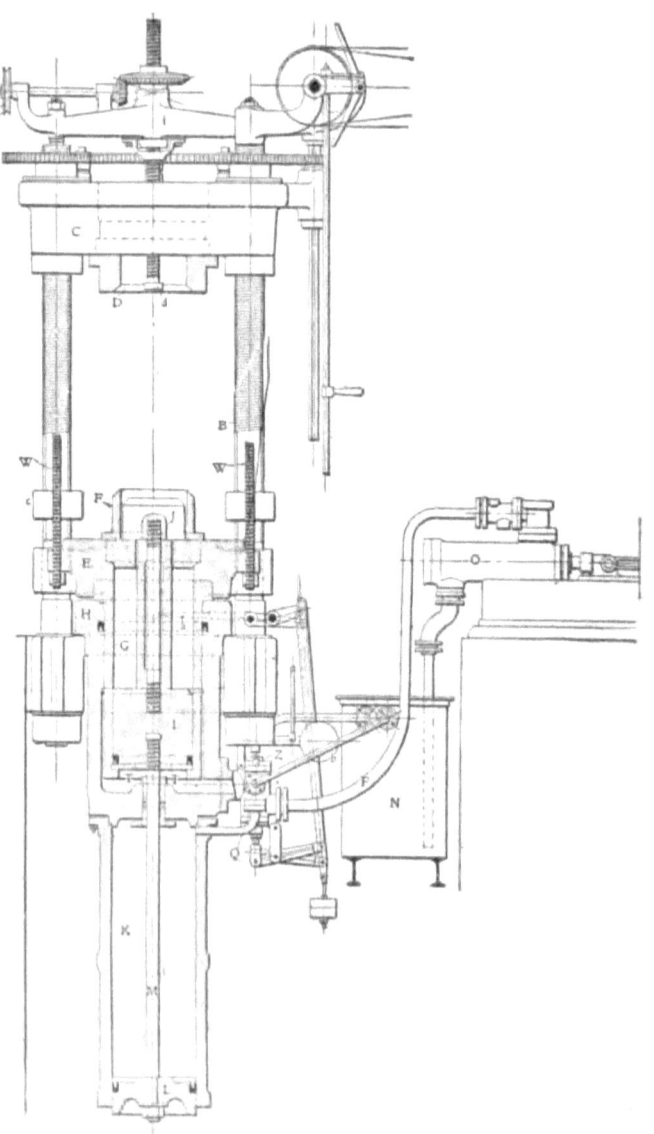

FIG. 687.

valve casing Q, and from there to the press. Two valves R and S are fitted in casing Q, the former controlling the supply and discharge of the pressure fluid to and from chamber T below pistons G and I, and being operated by the lever U, while the latter valve controls the supply and discharge of the fluids to and from cylinder K and is set by a hand-wheel, and operated automatically by a system of levers, these connecting with rod V operated by the adjustable dogs on rods W, and also connecting with rod X, which carries a tappet Y. At Z is a regulating valve, the function of which is to regulate the pressure upon piston L (called retractor piston), for which purpose it is operated by treadle A. As it serves at the same time as a safety valve for the retractor cylinder, it is fitted with a weighted lever b. By means of this valve Z, the blank-holder pressure is increased as desired without interrupting the operation of the machine.

The press being adjusted for the work in hand, and all three pistons being in their lowest position, the handle U is put in its mid-position and the pump started, the fluid then being forced directly through the valve casing Q and back to the reservoir. The controlling lever is then moved to the right and cylinder K begins to fill first (this being the case, however, only when the machine is first used) and all the pistons commence to rise. Piston L lifts the fluid in cylinder K and drives it up under the other pistons, thus accelerating their speed. The pistons rise until the blank to be drawn is gripped tight between holder F and die D, the bolster E then striking against the stop-nuts c which relieve the blank of any excess pressure. At the same time the tappet Y lifts the rod X and closes valve S, thus shutting off communication between cylinder K and chamber T, piston I continuing to rise under increased pressure due to automatic regulation of the pump.

As the punch draws the work to the desired depth in the die, the bottom of the piece strikes a stop rod d (which stop is adjustable through suitable gearing operated by a chain passing over a wheel e), frame f and rods W are lifted and valve S is thus opened, communication again being established between the chamber T and cylinder K. The lever U, at the same time, is put into mid-position and the pistons descend rapidly; the fluid again fills K, the remainder returning to reservoir N.

If, for any reason, the punch should stick in the work and prevent the pistons from descending, the controlling lever may be moved to the left; the valves then admit pressure fluid to cylinder *K* and the pistons are drawn back to their original position.

SIMPLE HYDRAULIC PRESS FOR HOBBING JEWELRY DIES.

Figs. 688 and 689 illustrate a hydraulic press which has been reduced to its simplest elements. The press proper, as shown in the engravings, consists of three principal parts only, the frame *A*, ram *B*, and platen *C;* the usual rods with nut, washers, etc., being dispensed with in favor of a construction which makes the base, the uprights, and the top a single piece, which is bored at

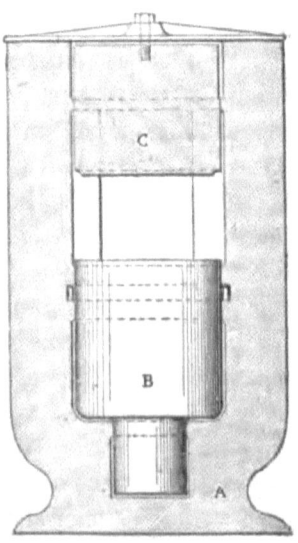

Fig. 688.

the top, as shown in the drawings, to receive the platen *C*, at the bottom where the ram *B* fits, and faced off on the upper surface of the openings for the platen *C* to form a seat, as shown in the

PUNCHES, DIES, AND TOOLS. 453

vertical section, Fig. 688. The plate on top of the upright is simply to hold the platen in place when pressure is removed from it. The stem on the lower part of the ram is turned to fit the bored seat in the frame by means of which the ram is accurately guided, and the drawing therefore shows absolutely all there is of the press, it being adapted for use in connection with

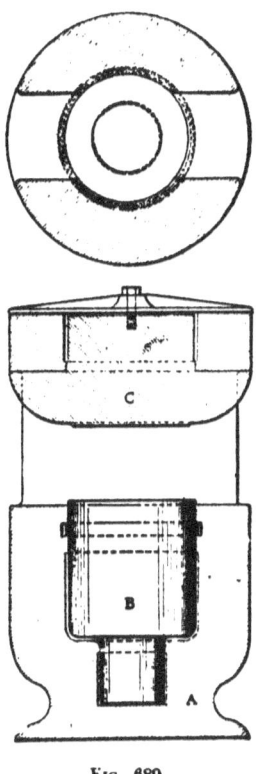

FIG. 689.

any suitable force pump. It is apparent, of course, that this construction secures great rigidity, and the press is to be used principally for hobbing dies used in jewelry manufacture, and for similar purposes where a hardened hob is by great pressure simply pressed into the die cold to form the impression. Charles Burroughs Company, Newark, N. J.

AN AUTOMATIC FEED FOR MEDAL EMBOSSING PRESS AND DIES.

In Section XIII., page 344, is given a description of the manner in which dies are made for the manufacture of small medallions of various styles. To make the description complete, we now show here what is an all-important part—that is, the automatic rack feed as made and used in conjunction with the dies.

Fig. 690 gives a front view of the upper and lower bolsters with the dies in the act of embossing a blank; and shows strip-

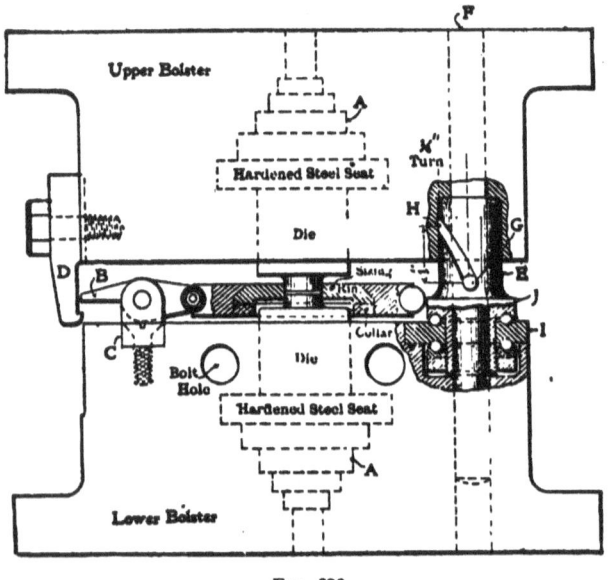

Fig. 690.

ping and sizing plates, a stripping lug and bracket (not shown except from one side) and the ball-bearing attachment for operating the rack feed.

As the reader is aware, the action of the embossing press is different from that of the ordinary power press, the lower die in the latter remaining stationary and the punch or die fastened to the ram moving downward, while in the embossing press the upper die remains stationary and the lower die moves upward.

PUNCHES, DIES, AND TOOLS. 455

The bolsters, as explained before, are steel castings, machined as per drawings, and fitted with hardened and ground-steel backings for the dies, as at *A*, these being firmly driven in from the face. The dies and the steel stripping and sizing plates have already been described.

The two stripping lugs, one of which is shown at *B*, are made in two pieces; they are tool-steel forgings and are held in bracket *C*—also of tool steel—which is let into the bolster and fastened with two countersunk screws. The two parts of each

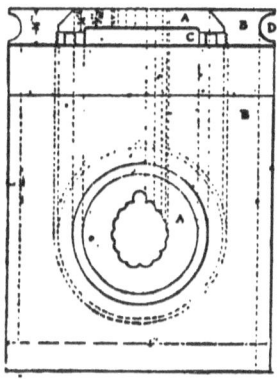

Fig. 690 a.

lug, which are made to fit together nicely, are connected with a spring on the outside, so that when coming in contact with the stripping dog *D*, on the upward stroke of the press, the lug will open just far enough to slip by the dog and then spring back and be in the position shown when the press has reached the end of its upward stroke; on the downward stroke the lug catches on the inside edge of the dog and forces the stripping and sizing plates down on the shoulder of the die, thus allowing the discharge of the embossed blank or medallion, which is accomplished, as explained before, by a sharp puff of air from a blower. The inner end of the lug is made to fit a groove in the stripping plate and is hardened. The extreme outside end where it comes in contact with the stripping dog is also hardened.

The sleeve *E*, for operating the feed attachment, is made of tool steel, bored slightly larger in the centre so that the bearing will be for about ¾ inch at each end, and turned as shown. The lower shoulder is grooved for a ball bearing and the end is threaded 24 pitch for two adjustable collars, the upper one of which is provided with a ball race and hardened. A spiral groove *H* wide enough to admit the roller *G* in pin *F*, is then

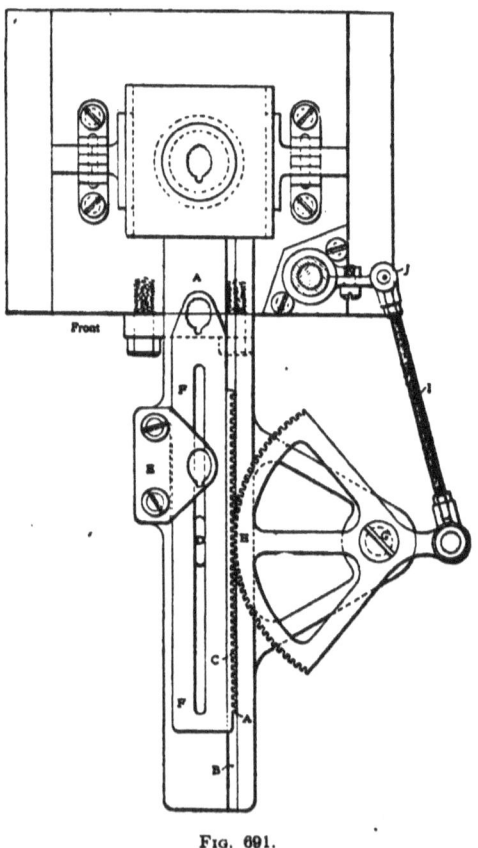

Fig. 691.

milled in the sleeve, the groove making about one-quarter turn in a distance of one inch, or whatever the full stroke of the press may be. After being hardened in whale oil, the sleeve is lapped to slide easily on the pin *F* (also hardened), which extends almost through both bolsters and is held in position by a set-screw (not shown) in the upper bolster.

PUNCHES, DIES, AND TOOLS. 457

The plate I is also made of tool steel and hardened; it is fastened to the bolster with two countersunk screws. Both sides of this plate are grooved for the balls J.

When the sleeve E was first made anything but satisfaction was met with, and everything was tried—fibre, etc.—to overcome the friction on the plate I during the quarter turn necessary. After breaking several plates, it was decided to try a ball-bearing as an experiment, and this greatly surprised the workers by working without any hitch or accident.

Fig. 691 represents a top view of the lower bolster with everything attached, and shows the slide mechanism as it would appear when the press has completed about half of its upward stroke.

The bracket A is made of cast iron, and the top surface and the end which are fastened to the bolster are planed. The groove B forms a guide for rack C and prevents undue friction or wear on the stop D (located just in the rear of the hopper E) which stop is so made that it may be dropped—by pressure on a spring lever on the under side of the bracket—below the surface of the bracket and so allow the easy removal of slide F, in case of accident, without having to disturb the hopper in any way. The bearing surface of the boss for segment H is milled off and drilled and tapped for a special large head-screw G on which H pivots. The bracket is fastened to the bolster with two cap screws, and has also two tapped holes for the screws which fasten the hopper in position.

The segment is made of brass, machined to the proper radius and cut 12 pitch. The end to which the connecting rod I is attached is fitted with a stud screwed in from beneath; this projects about $\frac{1}{4}$ inch above the surface (upper) and the eye of the connecting rod is slipped over it, so that the segment can be easily uncoupled and adjusted without pulling the whole thing apart.

The split collar J, attached to the sleeve and forming the connecting link with the feed, can be adjusted on the sleeve and fastened in any position.

The feed plate and rack are made of brass. The plate is usually a trifle thinner than the blank and is riveted to the rack, which has a slot milled the full length to receive the plate, and bring the same just level with the surface of the bracket. The slot, cut through the centre and almost the full length of the plates, regulates the extreme movement of the latter at both ends of the stroke, and is of great benefit in locating the hopper in the correct position.

The hopper is made of brass tubing of the same shape as the medallion, and is about 5 inches high in all. A slot, which is about $\frac{1}{2}$ inch wide at the top (to expedite the filling of the tube with the blanks to be embossed) gradually tapers to the same size as the neck of the medallion to insure the blank dropping in the hole in the feed plate on the backward movement of the latter. This tube is soldered into a brass plate, about $\frac{1}{8}$ inch thick, which in turn is soldered and riveted to what is called the "foot," also of brass, about $\frac{3}{8}$ inch thick and fastened with two screws to the bracket.

All the operator has to do after the different parts are adjusted is to lock the treadle of the press and keep the hopper filled with blanks, these being constantly removed by the slide or feed plate. The action of the feed is governed by the groove in the sleeve E, Fig. 690, as follows:

On the first movement of the press, which, as explained before, is upward, the roller operating in the groove in the sleeve causes the sleeve and its split collar to turn toward the rear, the connecting rod I, Fig. 691, operating the segment and causing that to move the feed plate toward the hopper; and on the completion of the upward stroke—the roller then reaching the bottom of the groove in the sleeve—the hole in the feed plate is brought directly under the hopper and allows the blank to fall in. On the return movement of the press, everything then necessarily working the opposite way, the feed plate and blank are carried toward the die, and as the stroke is completed, the blank then being directly over the sizing plate, drops through on to the face of the die. The hole in the feed plate when the press is at rest is always in this position. During the upward movement and while the feed plate is being moved toward the hopper the stripping lugs—as already described—slip by the dogs and spring back into position; and on the downward stroke the dogs

PUNCHES, DIES, AND TOOLS. 459

engage the ends of the lugs, the stripping and sizing plate is pressed down below the surface of the die, the blank is discharged as described, and the stripping lugs and plate return to normal position, as in Fig. 690. Each individual part is so timed and adjusted as to work in harmony with the others, and when finally this stage was reached after great trouble, it was decided that the limit had been reached.

RATCHET FEEDS FOR POWER PRESSES.

When an entirely new mechanism for feeding is required to be designed it is first necessary to be informed of the stroke of the press and the length of feed per stroke, besides the proportions and relations of the parts of the feed mechanisms. Should the press be provided with a feed gear, and such information as: What ratchet will give a certain feed? or *vice versa*, is required, the circumference of the feed rolls only is needed. For a feed gear of the type shown in Fig. 693 the circumference is used in the following formula:

$$\frac{\text{Cir. of feed rolls}}{\text{required feed}} = \text{teeth of ratchet.} \quad (a)$$

$$\frac{\text{Cir. of feed rolls}}{\text{teeth of ratchet}} = \text{required feed.} \quad (b)$$

In construction work we make use of this formula:

Req. feed $\times$ teeth of ratchet = cir. of feed rolls. (c)

The feed must start after the punch is stripped and stop before the punch re-enters the stock. One may save some trouble by laying out a diagram like Fig. 692, which shows the position of the plunger and feed-crank during one cycle. Beginning at A the punches pierce the stock, enter the die, strip and re-enter the stripper, the plunger stopping at the highest point A upon completion of the stroke. The length of this arc to be used for feeding as $a\,A\,b$, as the punches are then within the stripper and free from the stock. This shaded portion which represents the active travel of the feed-crank, is to be shifted around on the diagram so as to feed while the press-crank moves through $a\,A$

460 PUNCHES, DIES, AND TOOLS.

b. The adjustment for the number of teeth used on the ratchet is obtained by sliding the connecting rod in the slot, Fig. 693.

In Fig. 694 is shown a feed gear of somewhat different construction and better adapted for fine work. In this case the ad-

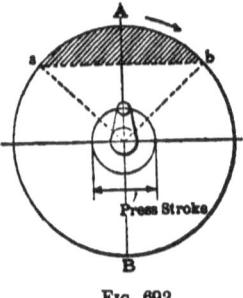

FIG. 692.

justment for the number of teeth is accomplished by adjusting the screw F under the pawl head and moving the stud in slot E. As the feed-crank turns from A to B the fork pushes the stud down, thus driving the ratchet in the direction of the ar-

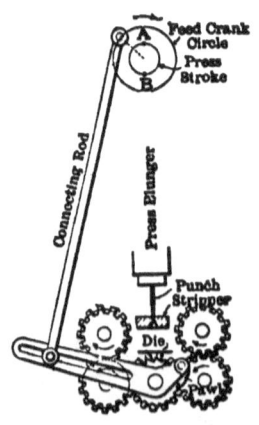

FIG. 693.

row. From B to D the fork rises and the lever is drawn back by the spring. The feed-crank is set 90° ahead of the press-crank.

Fig. 695 shows a substantial form for heavy work. The ten-

PUNCHES, DIES, AND TOOLS. 461

dency to run over is overcome by the application of a brake placed at some convenient place, generally upon the rolls and often against the face of one of the feed gears at the end of the press. This arrangement, which is quite simple, may be used to advantage in the perforating of heavy sheet metal. Adjustment for the number of ratchet teeth taken depends upon the adjustment of screw A and the position of the stud in the T-slot of the feed-crank. The slotted rod pushes the stud B down, and as

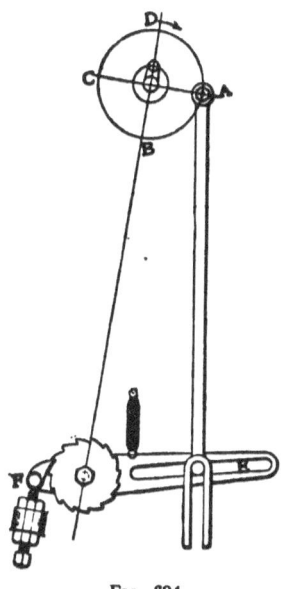

FIG. 694.

the rod rises the screw point strikes the stud and brings it back to its first position.

In the above attachments the motion of the ratchet is transmitted to the rolls by a gear or a train of gears, according to the direction in which the rolls are to turn and to the specific circumstances.

The illustrations, needless to say, are merely diagrams and are not necessarily in proportions that could be recommended for practical use. In Fig. 694 the construction of the adjustable stud is to be carefully considered, as the constant jamming of the fork is a serious factor of this otherwise simple design. In Fig. 696 considerable attention should be paid to the leverages,

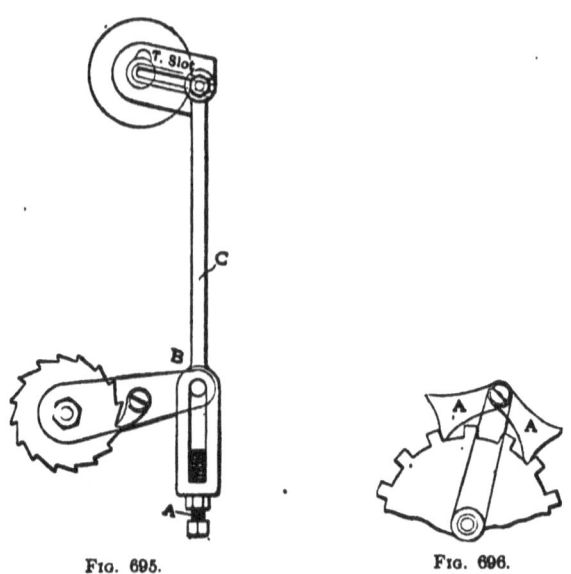

FIG. 695.   FIG. 696.

as the heavy sheets and the brake will offer high resistance. It is generally well to use large driven gears upon the ratchet shaft, which will help materially to reduce the starting strain.

THE RATCHET-PAWL FOR ROLL-FEEDS.

The pawl is by no means to be overlooked on account of its littleness. The general simplicity of the pawl permits us to dwell more upon its application and uses than its construction. Fig. 697 shows a ratchet-wheel turning about a centre pin. C is a vibrating lever carrying the pawl A. As the arm moves to the right the pawl lifts and slides over the points of the teeth; when it returns the pawl drops against the face of a tooth and carries the wheel with it. The idle pawl B is so placed as to fall behind each tooth as it passes under, preventing a reverse rotation.

In the case of a large number of fine teeth upon the wheel,

PUNCHES, DIES, AND TOOLS. 463

constant usage sometimes strips the faces and makes the feed uncertain. Now, considering it improbable that exactly opposite teeth will be spoiled, the arrangement shown at Fig. 698 may be employed. The pawls are of the same length and work together, so that if the ratchet is in good condition only half the strain is taken by each pawl and tooth, and when a tooth is

Fig. 697.

missed by one pawl the other is pretty sure of holding, thus making this extra pawl a safeguard.

Supposing that we have a ratchet with a certain number of teeth, and it is desirable to use spaces obtainable by using half teeth or a ratchet of twice as many teeth. Now rather than secure another wheel we may change one pawl, substituting in place of it a short one which will fall behind by one-half the

Fig. 698.

pitch, as in Fig. 699. These two pawls work alternately, each pushing the wheel ahead one-half of a tooth at a time. Similarly we might divide the space into thirds by using three pawls.

In feed mechanisms where it is desirable to run the wheel in either direction as upon shapers and planers, an arrangement like Fig. 696 is used. The ratchet has radial teeth, and the pawl, which is symmetrical, may occupy either position A or A'.

Fig. 699 shows another very convenient method of feeding. In this method the adjusting of the number of teeth per stroke is easily accomplished. The pawl *A* is made wide enough to run upon the shield *B*, which is movable about the centre. Noting

FIG. 699.

that the stroke is from A' to A in the direction of the arrow, it may be seen that there are only four teeth exposed, which, of course, will be the number used, and when A runs back to the position A' it again stops upon the shield. By making this shield longer it may serve to stop the feed entirely.

AUTOMATIC THROW-OUT FOR MISPLACED SHELLS.

The cuts in Fig. 700 show a very neat kink that is doing good service in at least one shop, and it should be adopted in a great many more where equally good results will be attained. The drawing, though not perfect in detail, shows enough to make plain the main point to which this description pertains. The presses are worked with dial plates. The shells, a sample of which is shown in Fig. 700, are of tin, and are fed into the plate by means of a hopper, which is supported by an arm extending from the top of the press. The theory of having the presses working on the self-feeding plan sounded all right, but in operation it was found that some of the shells worked into the plate upside down, or occasionally two at once. When they came down in this manner, and the plate moved around to the punches, instead of performing the operation, it either ruined the punches or stopped the press. As no positive escapement could be devised, this little kink was made, which would remove

PUNCHES, DIES, AND TOOLS. 465

all shells not in the right position. A is a piece of soft steel twisted at right angles, and is bolted to the gate of the press. B is a piece which swings on A, and is prevented from dropping by the stop C. D is a magnetized pin, and can be adjusted. E is a strip of tin which fits over pin D. On chute F is a strip of steep which comes directly over B. As chute F is always stationary, and A being bolted to the bed of the press, when A

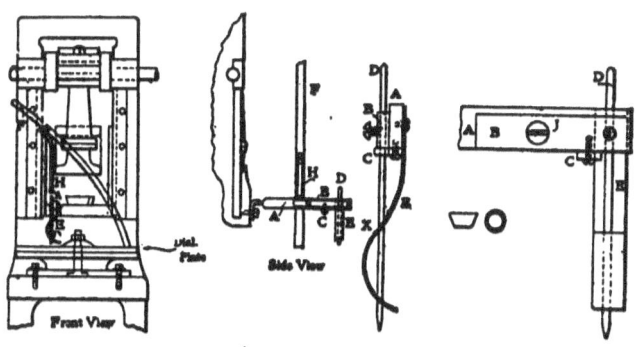

Fig. 700.

comes up a certain distance B comes in contact with H, and as B is movable on the screw J, it moves upward. As strip E also moves, on the screw, it moves upward on the pin, gradually nearing the point S. The pin is set so that it comes directly over the holes in the dial plate, and is also adjusted so that when the press comes to the time of drop the magnetized pin D will pick up any shell wrong side up, but will not interfere with the others.

OSCILLATING DIE-SLOTTING MACHINE.

The cutting of cavities in steel or any other metal for the making of dies, etc., is performed by certain well-understood methods, and the workman uses chisel, drill, miller, slotter, or file as the occasion may require. Any departure from these methods is a distinct novelty, and as the drawings in Figs. 701 and 702 suggest, there are many possibilities in the carving of metals into intricate forms. The machine shown in Fig. 701 is a patented one, and is neither slotter, shaper, nor miller, and is known as an oscillating slotting machine. The principle is that of a rocking tool fed down into the material and scooping out the chips by the toothed end of the formed cutter. An oscillating head is split to receive the tool shank and is driven by a

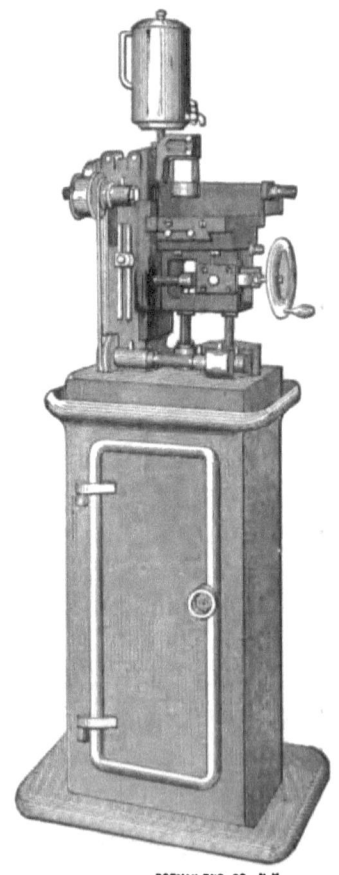

FIG. 701.

horizontal shaft at the top of the machine. The tool head, which has adjustment in any direction, and the feed to the cutter may be by hand or power, as shown.

Most mechanics will find the specimen of work seen in Fig. 702 equal to the machine in interest. Here is a block of square

PUNCHES, DIES, AND TOOLS. 467

steel, $2 \times 1\frac{1}{16} \times 1\frac{1}{16}$ inch, with sundry cleanly-cut holes in it. Three of these holes go clear through the steel block, and in finish and keenness of contour look like delicately-handled jobs with sharp broaches. But the square hole at the end changes shape when half through the chunk of steel and for the remainder of the way is hexagon, and the centre opening of the three at the other end of the block goes half-way only and has a semicircular bottom. The square hole is $\frac{5}{8}$ inch across the flats; the narrow

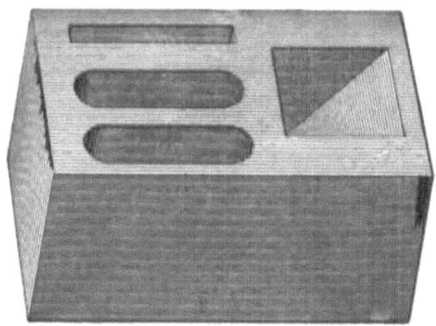

FIG. 702.

slot is $\frac{1}{8} \times \frac{7}{8}$ inch; the cut in the middle of the block is $\frac{1}{16} \times 1$ inch, and the slot with the half-round ends is $\frac{1}{4} \times \frac{7}{8}$ inch.

The square hole and the $\frac{1}{8} \times \frac{7}{8}$-inch slot in particular suggest the possibilities of the tool for irregular shapes, not to say awkward for other tools. A formed cutter can be made to suit any desired outline and worked through the metal. At this point may come an objection. It may seem from a cursory examination that the operation is rather a nibbling at the metal than any business-like carving of stock. The time taken for each cut, however, shown by the sample illustrated in Fig. 702 is approximately three minutes. This is a rapidity of action that in the working out of dies and similar operations puts the crude chisel and file another step to the rear.

INDEX.

A

ACCURATE automatic bending and shaping die, 74
    piercing and blanking, compound die for, 174
    sectional blanking punch and die, 155
Acid bath for hardening dies at low heats, 426
Agate wire-drawing dies, diamond, sapphire, and, 212
Aligning the punch and die for sub-press, 373
Armature blanking and piercing punch and die, 139
    disk and segment dies, 155
    disk, compound die for small, 172
    disk die with special stripper, compound, 177
    disks, special attachment for notching, 135
    punchings, slot index die for, 142
    segments and disk dies, 155
Arrangements, ingenious bending, 73
Art of drop forging, practical applications of, 391
Assembling cartridges in clips, 195
Attachment, automatic feed, for combination die, 90
    drawing die with blank holder, 269
    for notching armature disks, 135
Automatic bending and shaping, 74
    bending die, 67
    feed attachment for combination die, 90
    feed for medal embossing press and die, 454
    feed for punching, shearing, and drawing, 87

Automatic slide die for piercing, blanking, and drawing at one operation, 94
    throw-out for misplaced shells, 464

B

BAR steel, dies for drawing wire and, 211
Bath for hardening dies at low heats, 426
Beading dies, 302
Beading, expanding a double bead in brass cup, 305
    expansion punch and die for, 302
Bending an odd-shaped piece, 61
    and forming die, combination cutting off and, 119
    and forming die, cutting, 63
    and forming, five operations of, 80
    and forming die with automatic feed attachment, combination, 9
    and shaping die, accurate automatic, 74
    arrangement, ingenious, 73
    die, an automatic, 67
    dies, construction of, 13
    dies, copper clip, 19
    dies, piercing and blanking, 148
    dies, piercing and blanking and, 117
    dies, special and work, 30
    eyes of various shapes, dies for, 70
    in one die, blanking, drawing, and, 44
    novel and ingenious cutting dies, 33
    odd-shaped steel springs, hot-formed and, 83

INDEX.

Bending, piercing gold stock, cutting, 325
    tubing, improved method of, 435
Bicycle rim washers, cut-and-carry dies for, 97
Bit of jewelry die work, 318
Blanking and bending dies, 148
    and bending dies, piercing, 117
    and cutting dies, hardening, 425
    and drawing of rectangular shells, 273
    and drawing small shells, die for rapidly, 281
    and forming a sheet-metal roller, 107
    and forming die, combination, 263
    and perforating tools, table of allowances, 438
    and piercing a felt washer, 412
    and piercing punch and die, armature, 139
    compound die for accurate piercing and, 174
    die and punch, a double sectional, 157
    die for tool-steel blanks, double sectional, 166
    die with guide pins for punch, 115
    drawing, and bending in one die, 44
    drawing, and hole-cutting die with positive knockout, 279
    punch and die, accurate sectional, 155
    tools, gang and multiple, 114
Blanks, finding diameters of shells, 418
    for cup, finding size of, 422
    for cup leathers, 241
    for drawing into shells, sizing, 423
    for hat leathers, 242
    holders, attachments, drawing die with, 269
    in hack saw, power, cutting of die, 433

Bolster for holding round dies of different diameters, 427
Bolt dies, cast iron, 430
    machine, dies for, 408, 389
Bottom punch for sub-dies, making and finishing the, 377
Boxes, punch and die, for forming hinge springs for novelty, 77
Brass cup, expanding a double bead in a, 305
    stud, press tools for forming, 61
    tubes, punch and die for reducing, 291
Bullet jackets, die for nickel steel, 182
    jackets for 30-calibre cartridges, making cupro-steel, 192
Buttons with celluloid tops, die for finishing, 49

C

CAR wheels, making drawing dies of, 216
Carry die for bicycle rim washers—cut and, 97
Cartridge cases for quick-firing guns, drawing dies for, 186
    cases of quick-firing guns, dies for, 182
    cases, tools for making, 182
    in clips, assembling, 195
    making cupro-nickel bullet jackets for 30-calibre, 192
    shells, cupping the sheets for, 199
    dies for eighth drawing of, 204
    dies for fifth drawing of, 201
    dies for first drawing of, 201
    dies for fourth drawing of, 201
    dies for heading of, 206
    dies for ninth drawing of, 204
    dies for second drawing, 201
    dies for seventh drawing of, 204

INDEX. 471

Cartridge shells, dies for sixth drawing of, 203
    dies for tapering of, 208
    dies for tenth drawing of, 205
    dies for third drawing of, 201
    drawing and redrawing dies for, 186
    indenting the primer, 201
    other mechanical operations for, 209
    punches and dies for drawing brass, 183
    tools for making rifle, 182
Casting iron and steel with diamonds, making draw plates by, 214
Cast-iron bolt dies, 430
Celluloid tops, dies for finishing buttons with, 49
Chain-purse bodies, 104
Characteristic of drop-forgings, 391
Chemical tablets, dies for, 237
    making punches and dies for, 247
Chilled dies vs. tool-steel dies, 218
Chilled-iron wire-drawing dies, 213
Clips, assembling cartridges in, 195
Clips, forming and embossing die for spring, 122
Clock wheels, making a set of sub-dies for, 373
    work, sub-dies for, 361
Closing dies, 302
Coining of medals, 344
Cold rolled stock and sheet metal, lubricant for drawing, 425
    swaging process of pointing needles, 235
Coloring brass for laying out punchings, 425
Combination bending and forming die, 119
    blanking and forming die, 263
    cutting, drawing, and knurling die, 270
    die for punching, piercing, and splitting labels, 127

Combination shearing, piercing, bending, and forming die with automatic feed, 90
Composite punches and dies, 155
Compound armature disk die with special stripper, 177
    disk die for small disks, 169
    segment die, 162
    die for accurately piercing and blanking, 174
    die for small armature disks, 172
    dies, 155
    piercing die, 146
    punch and die for leather washers, 125
    punching and forming die for sheet-metal sheave pulleys, 440
    sub-die for punching an irregular piece, making an, 365
Construction of bending dies, 13
    of multiple punch for thin stock, 145
    of redrawing dies, economic, 290
    of spoon-making punches and dies, 313
    of sub-press dies, 361
Controlling screw blanks in thread-rolling machine, device for, 414
Copper clip bending dies, 19
    coloring brass for laying out punching, 425
Cupping the sheet for cartridge shells, 199
Cupro-nickel bullet jackets for 30-calibre cartridge making, 192
Curling die, 302
Curving dies for stove rims, 51
Cut-and-carry dies for bicycle rim washers, 97
    carry, and follow dies, 94
    dies, 254
    dies for double-acting press, 288
    finding the blanks for, 492
    leathers, making of tools for, 239
    leathers, table of blanks for, 241
Cutter plates, jewelry cutters and, 317

INDEX.

Cutters and cutter plates for jewelry work, 317
Cutting and bending dies, novel and ingenious, 33
   and drawing operation, lubricant used in first, 425
   bending and forming dies, 63
   bending, and piercing gold stock, 325
   dies, hardening, blanking, and, 425
   drawing and knurling dies, 270
   off, bending, and forming die, combination, 119
   off blanks for dies in power hack-saw, 433

D

DEPARTURE from established sub-press design, 379
Depth of U-leathers, 243
Design, departure from established sub-press, 379
   of sub-press dies, 361
Details of hornings and seaming operation, 308
Development of drawing processes, progress in, 182
Device for controlling screw-blanks in thread-rolling machine, 414
Diameters of shell blanks, finding, 418
Diamonds, making draw plates by casting iron and steel with, 214
   wire drawing dies of, 212
Dies, accurate bending and shaping, automatic, 74
   accurate sectional blanking punch and, 155
   allowances for accurate, table of punches and, 438
   and processes for making hydraulic leathers, 237
   and punch, double sectional blanking, 157
   and punches in holders, fastening, 428
   and their making, eyeglass strap punches and, 336

Dies and tools for making souvenir spoons, 354
   and tools for manufacturing spoons, 349
   armature blanking and piercing, 139
   armature disk and segment, 155
   at low heats, acid hardening bath for, 426
   automatic bending, 67
   automatic feed for embossing press and, 454
   beading, 302
   blanking, drawing, and bending in one, 44
   blanks, cutting off in power hack-saw, 433
   cast-iron bolt, 431
   chilled-iron wire-drawing, 213
   closing, 302
   combination blanking and forming, 263
   combination cutting-off, bending, and forming, 119
   combination shearing, piercing, bending, and forming, 90
   composite punches and, 155
   compound, 155
   compound armature segment, 162
   compound piercing, 146
   construction of bending, 13
   construction of spoon-making punches and, 313
   copper-slip bending, 19
   cupping, 254
   curling, 302
   cutting, bending, and forming, 63
   diamond, sapphire, and agate wire-drawing, 212
   dimensions of, table of drawing, 275
   drawing, 254
   drop, making of, 389
   economic construction of re-drawing, 290
   experience with thread-rolling, 415

INDEX. 473

Dies, eyeglass, 313
  feeds for working sheet metal in, 440
  finishing segments for sub-press, 375
  flanging, 254
  for accurate piercing and blanking, compound, 174
  armature punchings, 142
  beading, expansion punch and, 302
  bending an odd-shaped piece, 61
  bending eyes of various shapes, 70
  bending tubing, 435
  bicycle rim washers, cut and carry, 97
  blanking and drawing rectangular shells, 273
  blanking and piercing a felt washer, 412
  bolt machine, 387, 408
  cartridge cases, 182
  cartridge cases for quick-firing guns, 196
  cartridge cases of quick-firing guns, 182
  cartridge shells, indenting for primer, 202
  cartridge shells, other mechanical operations, 209
  chain purse bodies, 104
  chemical tablets, making punches and, 247
  circular and rectangular sheet-metal articles, 254
  clock wheels, making a set of sub, 373
  coining medals, 344
  combination cutting, drawing, and knurling, 270
  crimped box-corner fasteners, 42
  cupping the sheets for cartridge shells, 199
  cupro-nickel bullet jackets for 30-calibre cartridges, 192

Dies for cutting, bending, and piercing gold stock, 325
  cutting-off, bending, forming, and driving steel tacks, 57
  double acting cam, cupping, 288
  drawing a flanged cup, 277
  drawing a tin ferrule, punch and, 275
  drawing an odd-shaped piece, 258
  drawing and forming in a double-acting press, 263
  drawing brass cartridge shells, punches and, 183
  drawing central hub in heavy sheet-metal blank, 298
  drawing wire, 211
  eighth drawing of cartridge shells, 204
  expanding a double bead in a brass cup, 305
  eyeglass lens trial rims, making of, 329
  fifth drawing of cartridge shells, 201
  finishing buttons with celluloid tops, 49
  first drawing of cartridge shells, 201
  five operations, follow, 101
  five operations in bending and forming, 80
  flattening punched disks, 436
  forming a brass stud, 61
  forming corners of stoves and ranges, 53
  forming steel range bases, 25
  fourth drawing of cartridge shells, 201
  glove fasteners, striking up forming, 403
  heading cartridge shells, 206
  horning and seaming operation, 308
  jewelry, 318
  jewelry cutting, 317

474 INDEX

Dies for jewelry making, 321
  leather packing, 237
  leather washers, compound punch and, 125
  making a spring latch, 23
  making drop forgings, 392
  making German-silver forks, 353
  making joint ferrules, 25
  making rims for lenses, set of punches and, 333
  making shuttle carriers for sewing machines, 36
  paint tablets, 237
  perforating sheet metal, 151
  piercing, blanking, and drawing in one operation, 94
  pneumatic and steam hammer work, set of forging, 400
  punching an irregular piece, making a compound sub, 365
  punching and half-wiring operation, 307
  punching, forming, and cutting off pipe straps, 21
  punching four holes at right angles, 150
  punching holes in tool-steel parts, 131
  punching, piercing, and splitting labels, combination, 127
  rapidly blanking and drawing small shells, 281
  rectangular sheet-metal articles, 253
  re-drawing large shells, 266
  reducing brass tubing, punch and, 291
  reversing sheet-metal formed articles, 295
  rifle cartridges, 182
  second operation of cartridge shells, 201
  seventh drawing of cartridge shells, 204

Dies for sixth drawing of cartridge shells, 203
  small armature disks, 172
  small disks, compound armature disk, 169
  spring slip, forming and embossing, 122
  striking up number plates, 405
  sub-press, 372
  sub-press, grinding the upper, 372
  sub-press, making up the upper, 368
  sub-press work, substitute, 387
  tapering cartridge shells, 208
  tapering hollow screw tops, 310
  tenth drawing of cartridge shells, 205
  thin stock, a three-operation follow, 111
  third drawing of cartridge shells, 201
  tin nozzles, drawing and forming, 284
  tool-steel blanks, double sectional, 166
  tool-steel parts, sectional trimming, 160
  wire and bar steel, drawing, 211
  working gold-filled material, 313
gang or multiple blanking, 114
hardening, blanking, and cutting, 425
ingenious bending, 73
inside out shell drawing, 293
lens, trial and eyeglass, 313
locating and putting the parts of sub-press, 372
making a set of hinge, 19
making and finishing the bottom punch for sub, 377
making, jewelry, 313
making, methods of jewelry, 321

INDEX. 475

Dies, making the chuck for segmental pieces of sub-press, 374
making the filling in pieces for sub-press, 375
making the lower punch for sub-press, 371
for making the lower shedder for sub-press, 371
making the small punches for sub-press, 370
making thread-rolling dies, 416
methods of reproducing drop, 399
notching, 131
novel and ingenious cutting and bending, 33
number plate, 387
of car wheel, making drawing, 216
of different diameters, bolster for holding round, 427
of sub-press, aligning punches and, 373
piercing, blanking, and bending, 117–148
piercing, shearing, and bending, 28
presses for working sheet metal in, 440
punch for forming hinge springs in, 77
quintuple combination for drawing five shells, 254
re-drawing, 254
reducing, 254
reversing, 254
seaming, 302
sectional, 155
simple hydraulic press for hobbing jewelry, 452
simple wiring and its work, 304
sinking, 389
slotting machine, oscillating, for, 465
special bending and its work, 30
stamping small medallions in, 344
steam hammer, 389
steel wire-drawing, 214
stove rim curving, 51

Dies, stripping kink in drawing, 431
sub-press, accurately perforated blanks in, 361
sub-press clock work, 361
sub-press, construction of, 361
sub-press, design of, 361
sub-press, use of, 361
sub-press, watch work, 361
templet turning of punches and, 427
tool-steel dies, vs. chilled iron, 218
with automatic feed for punching, shearing, and drawing, 87
blank-holder attachment drawing, 269
guide pins for punch, blanking, 115
positive knock-out for notching, etc., 279
special attachment for notching armature disks, 135
special stripper, compound armature disk, 177
work, a bit of jewelry, 318
Dimensions of sub-presses—table of, 384
Disk, compound die for small armature, 172
die for small disks, compound armature, 169
die with special stripper, compound armature, 177
dies for flattening punched, 436
special attachment for notching armature, 135
Dividing head for punch press, 446
Double-acting cam press, cupping die for, 288
acting press, dies for drawing and forming in, 262
bead in brass cup, expanding a, 305
pointed tacks, dies for making and driving, 57
sectional blanking die and punch, 157

INDEX.

Double-acting sectional blanking die for tool-steel blanks, 166

Drawing a central hub in heavy sheet-metal blank, 298
   an odd-shaped cup, set of tools for, 258
   and forming dies for tin nozzles, 284
   and forming tools for double-acting press, 262
   and knurling die, a combination cutting, 270
   and re-drawing the cartridge shells, 186
   bar steel, pickling the stock for, 216
   brass cartridge shells, dies for, 183
   cartridge cases for quick-firing guns, 196
   cold-rolled stock and sheet steel, lubricant for, 425
   die dimensions, table of, 275
   die, stripping kink in, 431
   die with automatic feed for punching, shearing, and, 87
   die with blank-holder attachment, 269
   dies, 254
   dies, inside out shell, 243
   dies of car wheels, 216
   flanged cup, 277
   hole-cutting, die with positive knock-out, blanking, 279
   in one die, blanking, bending, and, 44
   into shells, sizing blanks for, 423
   lubricant used in first cutting and, 425
   lubricant used in re-, 424
   of rectangular shells, blanking and, 273
   of round and rectangular bar steel, 216
   press for heavy shells and flanges, hydraulic, 448
   processes, progress in development of, 182
   tin ferrule, punch and die for, 275

Drawing wire and bar steel, dies for, 211

Draw-plates, by casting steel and iron with diamonds, the making of, 214

Drop-die, methods of construction, 399

Drop-dies, making of, 389

Drop-forgings, 389.
   characteristic of, 391
   dies for making, 392
   dies, making set of, 394
   methods of, 393
   practical applications of the art, 391

E

Economic construction of redrawing dies, 290

Efficiency of leather packing, 237

Embossing die for spring clip, forming and, 122
   press and dies, automatic feed and medal, 454

Evolution and manufacture of pens, 221

Expanding a double bead in a brass cup, 305

Expansion punch and die for beading, 302

Eyeglass dies, 313
   lens rims, making dies for, 329
   strap punches and dies and their making, 336

Eyes, die for bending various shapes, 70

F

Fasteners, die for crimped box-corner, 42

Feed attachment, automatic for combination die, 90
   for medal embossing press and dies, automatic, 454
   for power press, rachet, 459
   for punching, shearing, and drawing, automatic, 87
   for working sheet metal in dies, 440
   punching without sub-press, fine power, 387

INDEX. 477

Feed ratchet pawl for roll, 462
Felt washer, blanking and piercing, 412
Ferrules, die for making joint, 25
Filling in pieces for sub-dies, 375
Fitting the parts of sub-press, locating and, 372
Five operations, follow die for, 101
    in bending and forming, 80
Flanged cups, drawing, 277
Flanges, hydraulic drawing press for heavy shells and, 448
Flanging dies, 254
Flattening of punched disks, die for, 436
Follow die for five operations, 101
    for thin stock, a three-operation, 111
Forgings, dies for making drop, 392
    dies, making sets of drop, 394
    drop, 389
    methods of drop, 393
Forming a brass stud, press tools for, 61
    a sheet-metal roller, blanking and, 107
    and bending odd-shaped steel springs, hot, 83
    and embossing die for spring clip, 122
    and shaping pens, 224
    corners for stoves and ranges, 53
    dies, 254
    dies, combination blanking and, 263
    dies, combination cutting off and bending, 119
    dies for glove fasteners, striking up, 403
    dies for pneumatic and steam hammer work, 400
    dies for sheet-metal sheave pulleys, punching and, 440
    dies for tin nozzles, drawing and, 284
    dies with automatic feed attachment, combination, 90
    five operations of bending and, 80

Forming sheet-metal shells, reversing, 295
    steel range bases, die for, 52
    tools for double-acting press, drawing and, 262
Four holes at right angles, punching, 150

G

GANG or multiple blanking dies and tools, 114
German-silver forks, making of, dies for, 353
Glove fasteners, striking up dies for, 403
Gold-filled material, working of, 313
Gold-pen making, processes for, 221
Gold stock, dies for cutting, bending, and piercing, 325
Grinding the upper die for sub-press, 372
Guns, dies for cartridge cases of quick-firing, 182
    drawing cartridge cases for quick-firing, 196

H

HACK saw, cutting off die blanks in power, 433
Half-wiring operation, punching and, 307
Hardening, blanking, and cutting dies, 426
    dies at low heats, acid bath for, 426
    pens, 226
Hat leathers, making of, 241
    table of blanks for, 242
Head for punch press, dividing, 446
Heavy sheet-metal blank, dies for drawing hub in, 298
Hinge dies, making a set of, 19
    springs, punch and die for forming, 77
Hobbing jewelry dies, simple hydraulic press for, 452
Holding round dies of different diameters, bolster for, 427
Hole-cutting die with positive knock-out, blanking, drawing, and, 219

Horning and seaming operation, details of, 308
Hot-forming and bending odd-shaped steel springs, 83
Hydraulic drawing press for heavy shells and flanges, 448
    leathers, dies and processes for making, 237
        making of cup, 239
        making of U-,
        practice to adopt for working, 244
        selection and preparation of leather for, 239
        three kinds of, 237
    press for hobbing jewelry dies, simple, 452

I

IMPROVED method of bending tubing, 435
Indexing sub-presses, 385
Ingenious bending arrangement, 73
Inside-out shell-drawing dies, 293
Irregular piece, making a compound sub-die for punching, 365

J

Jewelry cutters and cutter plates, 317
    die making, 313
        making, methods of, 321
        work, bit of, 318
    dies, simple hydraulic press for hobbing, 452
Joint ferrules, dies for making, 25

K

KINDS of hydraulic packing leathers, three, 237
Knurling die, combination cutting, drawing, and, 270

L

LABELS, combination die for punching, piercing, and splitting, 127
Large shells, making die for redrawing, 266
Latch, dies for making spring, 23
    needles, making steel spring and, 232

Leather for hydraulic washers, selection of, 239
    packing, efficiency of, 237
    washers, compound punch and die for, 125
Lens dies, 313
    trial rims, making dies for eyeglass, 329
Lenses, set of dies for making rims for, 333
Locating and fitting the parts of sub-press, 372
Lower punch for sub-press, making the, 371
    shedder for sub-press dies, making, 371
Lubricant for drawing cold-rolled stock and sheet steel, 425
    for first cutting and drawing operation, 425
    used in re-drawing cylindrical shells, 424

M

MACHINE, oscillating die slotting, 465
    releathering a, 245
Making a compound sub-die for punching an irregular piece, 365
    a set of sub-dies for clock-wheels, 373
    and finishing the bottom punch for sub-dies, 377
    cupro-nickel bullet jackets for 30-calibre cartridges, 192
    dies for drawing wire, 211
    draw plates by casting diamonds with iron and steel, 214
    of a sub-die and punch, 361
        cup leathers, 239
        German-silver forks, 353
        hat leathers, 241
        needles, 228
        paint tablets on punch press, 251
        sets of drop-forging dies, 394
        souvenir spoons, 354
        U-leathers, 242

INDEX.   479

Making punches and die for chemical tablets, 247
   the chuck for five segmental punches of sub-press, 374
   filling in pieces, 375
   lower lunch for sub-press, 371
   lower shedder for sub-press, 371
   small punches for sub-press, 370
   upper die for sub-press, 368
Manufacture methods of spoon making, 349
   of gold pens, 222
   needles, 231
   pins, 221
   sewing-machine needles, 233
   steel pens, 222
Mechanical operation on cartridge shells, other, 209
Medal dies, 313
   embossing press and dies, automatic feed for, 454
Medallions, stamping of small, 344
Medals, coining of, 344
Method of bending tubing, improved, 435
   drop forging, 393
   jewelry die-making, 321
   production, needles, 234
   reproducing drop dies, 399
   spoon-making and manufacturing, 349
Misplaced shells, automatic throwout for, 464
Multiple blanking tools for thin stock, construction of, 145
   tools, gang and, 114

N

NEEDLES, cold-swaging process of pointing, 235
   making of, 229
   manufacturing of sewing-machine, 233
   methods of production, 234
   steel spring and latch, 232

Nickel-steel bullet jackets, dies for, 182
Notching armature disks, special attachment for, 135
   dies, 131
Novel and ingenious cutting and bending dies, 33
   process of die making, 444
   of sheet-metal working, 440
Novelty boxes, punch and die for forming hinge springs for, 77
Number plate dies, 389
   plates, dies for striking up, 405

O

ODD-SHAPED pieces, bending of, 61
Oscillating die-slotting machine, 465

P

PAINT tablets, dies for, 237
   making of, on punch press, 251
Pawl for roll feeds, ratchet, 426
Pens, evolution and manufacture of, 221
   forming and shaping, 224
   hardening the, 226
   making, process for, 221
      preliminary processes of, 223
   manufacture of steel, 222
   slitting the, 227
   tempering the, 226
Perforating punches and dies, 131
   sheet metal, 151
   tools, table of allowances for blanking and, 438
Pickling stock for bar-steel drawing, 216
Piercing a felt washer, blanking and, 412
   and blanking, compound die for accurate, 174
   and splitting labels, combination die for punching, 127
   bending and forming with automatic feed attachment, 90
   blanking and bending dies, 148
      and drawing in one operation, automatic slide, 94

INDEX.

Piercing die, compound, 146
  gold stock, cutting, bending, and, 325
  punch and die, armature blanking and, 139
  punches and dies, 131
  shearing and bending dies, 28
Pipe straps, die for punching, forming, and cutting-off, 21
Plungers, steel, for sub-press, 378
Pneumatic and steam hammer work, forging dies for, 400
Pointing needles, cold-swaging process of, 235
Positive knock-out, blanking, drawing, and hole-cutting die with, 299
Power feed punching without sub-press, 387
  presses, ratchet feeds for, 459
Preliminary processes of pen making, 223
Press and dies, automatic feed for medal-embossing, 454
  cupping dies for a double-acting cam, 288
  dividing head for punch, 446
  for heavy shells and flanges, hydraulic drawing, 448
  hobbing jewelry dies, simple hydraulic, 452
  working sheet metal in dies, 440
  tools for forming a brass stud, 61
  for forming steel range bases, 52
  squaring a punch, 433
Presses, ratchet feeds for power, 454
Process of pin manufacture, 228
Processes for gold-pen making, 221
Progress in development of drawing processes, 182
Progressive sheet-metal working, 94
Proportions of sub-presses, 381
Punch, making of a sub-die and, 361
  press, dividing head for, 446
  making of paint tablets on, 251
  squaring, 433

Punched disks, dies for flattening of, 436
Punches and dies, allowances for accurate work, table of, 438
  and dies and their making, eyeglass strap, 336
  and dies, armature piercing and blanking, 139
  and dies for beading, expansion, 302
  and dies for drawing a tin ferrule, 275
  and dies for drawing brass cartridge shells, 183
  and dies for forming hinge springs, 77
  and dies for leather washers, 125
  and dies for making rims for lenses, set of, 333
  and dies for reducing brass tubing, 291
  and dies for sub-press, aligning the, 373
  and dies in holders, fastening, 428
  and dies, perforating, 131
  and dies, templet turning of, 427
  blanking dies with guide pins for, 115
  for blanking and piercing a felt washer, 412
  for sub-press, making the small, 370
  for thin metal, construction of, 145
  improved washer, 411
  piercing, 131
Punching an irregular piece, making a compound sub-die for, 365
  and forming die for steel sheave pulleys, 440
  and half-wiring operation, 307
  copper coloring brass for laying out, 425
  four holes at right angles, 150
  piercing, and splitting labels, combination die for, 127
  shearing, and drawing die with automatic feed for, 87

INDEX. 481

Punching slot index die for armature, 142
   small holes in tool-steel parts, 131
   without sub-press, fine power feed for, 387
Purse bodies, dies for chain, 104

Q

QUICK-FIRING guns, dies for cartridge cases for, 182
   drawing cartridge cases for, 196
Quintuple combination die for producing five drawing shells, 254

R

RANGE bases, dies for forming steel, 52
Rapidly blanking and drawing small shells, die for, 281
Ratchet feeds for power presses, 459
   pawl for roll feeds, 462
Rectangular bar steel, drawing of round and, 216
   sheet-metal articles, dies for circular and, 254
   shell, blanking and drawing of, 273
Re-drawing cartridge shells, drawing and, 186
   cylindrical shells, lubricants to use, 424
   dies, 254
   dies, economic construction of, 290
   large shells, die for, 266
Reducing die for brass tubes, 291
   dies, 254
Re-leathering a machine, 245
Reversing dies, 254
   formed sheet-metal shells, 295
Rifle cartridges, tools for making, 182
Right angles, punching four holes at, 150
Rims for lenses, set of dies for making, 333
Roll feeds, ratchet and pawl for, 462
Roller, blanking and forming a sheet-metal, 107

Round and rectangular bar steel, drawing of, 216

S

SAPPHIRE and agate wire-drawing dies, diamond, 212
Screw blanks in thread-rolling dies, device for controlling, 414
   tops, dies for tapering hollow, 310
Seaming dies, 302
   operation, details of horning and, 308
Sectional blanking die and punch, double, 157
   blanking die for tool-steel blanks, 166
   blanking punch and die, accurate, 155
   punches, 155
   trimming die for tool-steel parts, 160
Segment, armature-disk dies, 155
   die, compound armature, 162
   finishing the, for sub-press, 375
Set of punches and dies for making rims for trial lenses, 333
   of sub-dies for clock wheels, making a, 373
Sewing machine needles, manufacture of, 233
   machines, dies for making shuttle carriers for, 36
Shearing and bending die, piercing, 28
   and drawing, die with automatic feed for punching, 87
   piercing, bending, and forming with automatic feed, 90
Shedder for sub-press dies, making the lower, 371
Sheet metal, perforating, 151
   metal shells, reversing formed, 195
   steel, lubricant for drawing, 425
   steel sheave pulleys, punching and forming die for, 440
Shells and flanges, hydraulic drawing press for heavy, 448

INDEX.

Shells, automatic throw-out for misplaced, 464
   blanking and drawing rectangular, 273
   blanks, finding the diameters of, 418
   die for blanking and drawing small, rapidly, 281
   die for redrawing large, 266
   drawing dies, inside out, 293
   finding the blanks for drawing into, 423
   lubricant used in re-drawing cylindrical, 424
   punches and drawing brass cartridge, 183
   reversing sheet-metal formed, 295
Shuttle carriers, for sewing machines, dies for, 36
Simple wiring die with its work, 304
Slide die for piercing, blanking, and drawing, automatic, 94
Slitting pens in manufacture, 227
Slot index die for armature punchings, 142
Slotting machine, oscillating die, 465
Small armature disks, compound die for, 172
   holes in tool-steel parts, punching, 131
   medallions, stamping of, 344
   punches for sub-press, making the, 370
   shells, die for rapidly blanking and drawing, 281
Souvenir spoons, making of, 354
Special attachment for notching armature disks, 135
   bending die and its work, 30
Splitting labels, combination die for punching, etc., 227
Spoon making, manufacturing methods for, 349
Spoons, making of souvenir, 354
Spring clip, forming and embossing die for, 112
   latch, dies for making, 23
Springs, hot-forming and shaping of odd-shaped steel, 83

Springs, punch and die for forming hinge, 77
Squaring up a punch press, 437
Stamping of small medallions, 344
Steam hammer dies, 389
   work, forging dies for pneumatic and, 400
Steel bullet jackets, dies for nickel, 182
   drawing of round and rectangular bar, 216
   for sub-press plungers, 378
   pens, manufacture of, 222
   range bases, dies for forming, 52
   spring and latch needles, 232
   springs, hot-forming and bending of odd-shaped, 83
   wire drawing dies, 214
Stove-rim curving tools and dies, 51
Stoves and ranges, dies for forming corners and bases for, 53
Strap punches and dies and their making, eyeglass, 336
Straps, die for punching, forming, and cutting off pipe, 21
Striking number plates, dies for, 405
   up forming dies for glove fasteners, 403
Stripper, compound armature disk die with special, 177
Stripping kink in a drawing die, 43
Sub-press dies, 361
   accurately perforated blanks of irregular shape in, 361
   aligning the punches and dies of, 373
   and punches, making of, 361
   construction of, 361
   design, departure from established, 379
   design of, 361
   fine power feed for punching without, 381
   finishing segments for, 375
   for clock wheels, making a set of, 373
   for clock work, 361
   for watch work, 361
   grinding the upper die for, 372

INDEX. 483

Sub-press, indexing, 385
    locating and fitting the parts of, 372
    making and finishing the bottom punch for, 377
    making the chuck for segmental pieces for, 374
    making the filling-in pieces for, 375
    making the lower punch for, 371
    making the lower shedder for, 371
    making the upper die for, 368
    plungers, steel for, 378
    proportions of, 381
    punching an irregular piece, making a compound, 365
    use of, 361

T

Table of blanks for cup leathers, 241
    depth of U-leathers, 243
    diameters of blanks for U-leathers, 244
    dimensions of sub-presses, 384
    drawing die dimensions, 275
    punch and die allowances for accurate work, 438
Tacks, dies for making and driving double-pointed steel, 57
Tapering hollow screw tops, dies for, 310
Tempering steel pens, 226
Templet turning of punches and dies, 427
Thin metal, construction of multiple punch for, 145
    stock, a three-operation die for working, 111
Thread-rolling dies, experience with, 415
    making, 416
    machine, device for controlling screw blanks in, 414
Three kinds of hydraulic leathers, 237

Three-operation follow die for thin stock, 111
Throw-out for misplaced shells, automatic, 464
Tin ferrules, punch and die for drawing, 275
    nozzles, drawing and forming dies for, 284
Tool combinations for progressive sheet-metal working, 94
Tools for tapering hollow screw tops, 310
Tool-steel blanks, double sectional blanking die for, 166
    dies, chilled iron vs., 218
    parts, punching small holes in, 131
    parts, sectional trimming die for, 160
Trial rims, making dies for eyeglass lens, 329
Trimming die for tool-steel parts, sectional, 160
Tubing, improved method of bending, 435
Turning of punches and dies, templet, 427

U

U-leathers, making of, 242
Upper die for sub-press, making the, 368
Use of sub-press dies, 361

W

Washers, compound punch and die for leather, 125
    cut and carry dies for bicycle rim, 97
    punch, improved, 411
Watch work, sub-press dies for, 361
Wire and bar steel, drawing dies for drawing, 211
    drawing dies, diamond, 212
    dies, chilled iron, 213
    dies, steel, 214
Wiring die and its work, simple, 304
    dies, 302
    punching and half, 307
Working gold-filled material in dies and presses, 313

SCIENTIFIC AND PRACTICAL BOOKS

PUBLISHED BY

The Norman W. Henley Publishing Co.

152 Nassau Street, New York, U. S. A.

☞ Any of these books will be sent prepaid on receipt of price to any address in the world.
☞ We will send FREE to any address in the world our complete Catalogue of Scientific and Practical Books.

Appleton's Cyclopædia of Applied Mechanics
This is a dictionary of mechanical engineering and the mechanical arts, fully describing and illustrating upwards of ten thousand subjects, including agricultural machinery, wood, metal, stone, and leather working; mining, hydraulic, railway, marine, and military engineering; working in cotton, wool, and paper; steam, air, and gas engines, and other motors; lighting, heating, and ventilation; electrical, telegraphic, optical, horological, calculating, and other instruments; etc.
A magnificent set in three volumes, handsomely bound in half morocco, each volume containing over 900 large octavo pages, with nearly 8,000 engravings, including diagrammatic and sectional drawings, with full explanatory details. Price $12.00.

ASKINSON. Perfumes and Their Preparation. A Comprehensive Treatise on Perfumery
Containing complete directions for making handkerchief perfumes, smelling salts, sachets, fumigating pastils; preparations for the care of the skin, the mouth, the hair; cosmetics, hair dyes, and other toilet articles. 300 pages. 32 illustrations. 8vo. Cloth, $3.00.

BARR. Catechism on the Combustion of Coal and the Prevention of Smoke
A practical treatise for all interested in fuel economy and the suppression of smoke from stationary steam-boiler furnaces and from locomotives, 85 illustrations. 12mo. 349 pages. Cloth, $1.50.

BARROWS. Practical Pattern Making
This is the best treatise on pattern making that has appeared. There is a general introduction on pattern making as an art, followed by a section on material and tools, taking up subjects like lumber, varnish, hand tools, band saws, circular saws, etc. Then follows a section devoted to examples of wood patterns of different types, and one upon metal patterns. There is then a section upon pattern-shop mathematics and one upon cost, care, and invention. It is indispensable to every patternmaker. Cloth, $2.00.

BAUER. Marine Engines and Boilers: Their Design and Construction
A large practical work of 722 pages, 550 illustrations, and 17 folding plates for the use of students, engineers, and naval constructors.
Clearly written, thoroughly systematic, theoretically sound; while the character of its plans, drawings, tables, and statistics is without reproach. The illustrations are careful reproductions from actual working drawings, with some well-executed photographic views of completed engines and boilers. $9.00 net.

BENJAMIN. Modern Mechanism
A large octavo volume of 959 pages and containing over 1,000 illustrations dealing solely with the principal and most useful advances of the past few years. Issued under a title which exactly describes its contents—"MODERN MECHANISM." The most eminent experts have contributed to this volume, and the benefits to be derived from the result of their researches and scientific accomplishments are of incalculable value to the man seeking the highest and most advanced practice in Applied Mechanics. Bound in half morocco. $5.00.

BLACKALL. Air-Brake Catechism
This book is a complete study of the air-brake equipment, including the latest devices and inventions used. All parts of the air brake, their troubles and peculiarities, and a practical way to find and remedy them, are explained. This book contains over 1,500 questions with their answers, and is completely illustrated by engravings and two large Westinghouse air-brake educational charts, printed in colors. 312 pages. Handsomely bound in cloth. 20th edition, revised and enlarged. $2.00.

Publications of The Norman W. Henley Publishing Co.

BLACKALL. New York Air-Brake Catechism

This is a complete treatise on the New York Air-Brake and Air-Signalling Apparatus giving a detailed description of all the parts, their operation, troubles, and the methods of locating and remedying the same. It includes and fully describes and illustrates the plain triple valve, quick-action triple valve, duplex pumps, pump governor, brake valves, retaining valves, freight equipment, signal valve, signal reducing valve, and car discharge valve. 200 pages, fully illustrated. $1.00.

BOOTH AND KERSHAW. Smoke Prevention and Fuel Economy

As the title indicates, this book of 197 pages and 75 illustrations deals with the problem of complete combustion, which it treats from the chemical and mechanical standpoints, besides pointing out the economical and humanitarian aspects of the question. $2.50.

BOOTH. Steam Pipes: Their Design and Construction

A treatise on the principles of steam conveyance and means and materials employed in practice, to secure economy, efficiency, and safety. A book of 187 pages which should be in the possession of every engineer and contractor. $2.00.

BUCHETTI. Engine Tests and Boiler Efficiencies

This work fully describes and illustrates the method of testing the power of steam engines, turbine and explosive motors. The properties of steam and the evaporative power of fuels. Combustion of fuel and chimney draft; with formulas explained or practically computed. 255 pages; 179 illustrations. $3.00.

BYRON. Physics and Chemistry of Mining

For the use of all preparing for examinations in Mining or qualifying for Colliery Managers' Certificates. $2.00.

COCKIN. Practical Coal Mining

An important work, containing 428 pages and 213 illustrations, complete with practical details, which will intuitively impart to the reader, not only a general knowledge of the principles of coal mining, but also considerable insight into allied subjects, including chemistry, mechanics, steam and steam engines, and electricity. In elucidating the various divisions incorporated in this excellent work, the author has started at the task from the very inception, and has ignored all obsolete methods, excepting where they illustrate fixed principles or are in touch with the march of modern improvements. The treatise is positively up to date in every instance, and should be in the hands of every colliery engineer, geologist, mine operator, superintendent, foreman, and all others who are interested in or connected with the industry. $2.50.

FOWLER. Locomotive Breakdowns and Their Remedies

This work treats in full all kinds of accidents that are likely to happen to locomotive engines while on the road. The various parts of the locomotives are discussed, and every accident that can possibly happen, with the remedy to be applied, is given. The various types of compound locomotives are included, so that every engineer may post himself in regard to emergency work in connection with this class of engine.

For the railroad man, who is anxious to know what to do and how to do it under all the various circumstances that may arise in the performance of his duties, this book will be an invaluable assistant and guide. 250 pages, fully illustrated. $1.50.

FOWLER. Boiler Room Chart

An educational chart showing in isometric perspective the mechanisms belonging in a modern boiler-room. The equipment consists of water-tube boilers, ordinary grates and mechanical stokers, feed-water heaters and pumps. The various parts of the appliances are shown broken or removed, so that the internal construction is fully illustrated. Each part is given a reference number, and these, with the corresponding name, are given in a glossary printed at the sides. The chart, therefore, serves as a dictionary of the boiler-room, the names of more than two hundred parts being given on the list. 25 cents.

GRIMSHAW. Saw Filing and Management of Saws

A practical handbook on filing, gumming, swaging, hammering, and the brasing of band saws, the speed, work, and power to run circular saws, etc., etc. Fully illustrated. Cloth, $1.00.

GRIMSHAW. "Shop Kinks"

This book is entirely different from any other on machine-shop practice. It is not descriptive of universal or common shop usage, but shows special ways of doing work better, more cheaply, and more rapidly than usual, as done in fifty or more leading shops in Europe and America. Some of its over 500 items and 222 illustrations are contributed directly for its pages by eminent constructors; the rest has been gathered by the author in his thirty years' travel and experience. Fourth edition. Nearly 400 pages. Cloth, $2.50.

GRIMSHAW. Engine Runner's Catechism

Tells how to erect, adjust, and run the principal steam engines in the United States. Describes the principal features of various special and well-known makes of engines. Sixth edition. 336 pages. Fully illustrated. Cloth, $2.00.

Publications of The Norman W. Henley Publishing Co.

GRIMSHAW. Steam Engine Catechism

A series of direct practical answers to direct practical questions, mainly intended for young engineers and for examination questions. Nearly 1,000 questions with their answers. Fourteenth edition. 413 pages. Fully illustrated. Cloth, $2.00.

GRIMSHAW. Locomotive Catechism

This is a veritable encyclopædia of the locomotive, is entirely free from mathematics, and thoroughly up to date. It contains 1,600 questions with their answers. Twenty-fourth edition, greatly enlarged. Nearly 450 pages, over 200 illustrations, and 12 large folding plates. Cloth, $2.00.

HARRISON. Electric Wiring, Diagrams and Switchboards

A thorough treatise covering the subject in all its branches. Practical every-day problems in wiring are presented and the method of obtaining intelligent results clearly shown. 270 pages, 105 illustrations. $1.50.

Henley's Twentieth Century Book of Receipts, Formulas and Processes

Edited by G. D. Hiscox. A complete work giving ten thousand formulas which will be of value to the housewife, the painter, the carpenter, the metal worker, the farmer, the soap and candle maker, the photographer, the jeweller, the watchmaker, the electroplater, the electrotyper, the tanner, the mechanic, the engineer, and the manufacturer. 900 pages. $3.00.

Henley's Encyclopedia of Practical Engineering and Allied Trades

Edited by JOSEPH G. HORNER. The scope of this work is indicated by its title, as being both practical and encyclopædic in character. All the great sections of engineering practice and enterprise receive sound and concise treatment.
Complete in five volumes. Each volume contains 500 pages and 500 illustrations. Bound in half morocco. Price, $6.00 per volume, or $25.00 for the complete set of five volumes.

HISCOX. Gas, Gasoline, and Oil Engines

Every user of a gas engine needs this book. Simple, instructive, and right up to date. The only complete work on this important subject. Tells all about the running and management of gas engines. Full of general information about the new and popular motive power, its economy and ease of management. Also chapters on horseless vehicles, electric lighting, marine propulsion, etc. 450 pages Illustrated with 351 engravings. Fifteenth edition, revised, enlarged, and reset. $2.50.

HISCOX. Compressed Air in All Its Applications

This is the most complete book on the subject of Air that has ever been issued, and its thirty-five chapters include about every phase of the subject one can think of. Beginning with a history of the progress that has been made in this l ne, it takes up the properties of air, gives tables of its volume and weight, both dry and saturated, as well as numerous other conditions. Step by step the reader finds how it is used, the various methods of compression and apparatus employed, its use in transmitting power, air motors and their efficiency, and a host of other information in this connection. Pneumatic tools and their uses receive ample attention, as do the sand-blast, pneumatic tube transmission, and other applications, such as raising water, ice machines and liquid air, while the air brake and air signal also come in for their share. Taken as a whole it may be called an encyclopædia of compressed air. It is written by an expert, who, in its 825 pages, has dealt with the subject in a comprehensive manner, no phase of it being omitted. 545 illustrations, 820 pages. Price, $5.00.

HISCOX. Horseless Vehicles, Automobiles and Motor Cycles, Operated by Steam, Hydro-Carbon, Electric, and Pneumatic Motors

A practical treatise of 459 pages and 316 illustrations for Automobilists, Manufacturers, Capitalists, Investors, Promoters, and every one interested in the development, care, and use of the Automobile.
Nineteen chapters. Large 8vo. 316 illustrations. 460 pages. Cloth, $1.50.

HISCOX. Mechanical Movements, Powers, and Devices

This work of 400 pages contains 1,800 specially made illustrations with descriptive text. It is a Dictionary of Mechanical Movements, Powers, Devices, and Appliances, embracing an illustrated description of the greatest variety of Mechanical Movements and Devices in any language. A new work on illustrated Mechanics, Mechanical Movements and Devices, covering nearly the whole range of the practical and inventive field for the use of Machinists, Mechanics, Inventors, Engineers, Draughtsmen, Students, and all others interested in any way in the devising and operation of mechanical works of any kind. $3.00.

Publications of The Norman W. Henley Publishing Co.

HISCOX. Mechanical Appliances, Mechanical Movements and Novelties of Construction

The many editions through which the first volume of "Mechanical Movements" has passed are more than a sufficient encouragement to warrant the publication of a second volume of 400 pages, containing 1,000 larger and specially-made illustrations, which are more special in scope than those in the first volume, inasmuch as they deal with the peculiar requirements of the various arts and manufactures, and more detailed in their explanations, because of the greater complexity of the machinery illustrated and described. $3.00.

HISCOX. Modern Steam Engineering in Theory and Practice

This book has been specially prepared for the use of the modern steam engineer, the technical students, and all who desire the latest and most reliable information on steam and steam boilers, the machinery of power, the steam turbine, electric power and lighting plants, etc. 450 octavo pages, 400 detailed engravings. $3.00.

HORNER. Modern Milling Machines: Their Design, Construction and Operation

This work of 304 pages is fully illustrated and describes and illustrates the Milling Machine from its early conception to the present time. $4.00.

HORNER. Practical Metal Turning

A work covering the modern practice of machining metal parts in the lathe. Fully illustrated. $3.50.

HORNER. Tools for Machinists and Wood Workers, Including Instruments of Measurment

A practical work of 340 pages fully illustrated, giving a general description and classification of tools for machinists and woodworkers. $3.50.

Inventor's Manual; How to Make a Patent Pay

This is a book designed as a guide to inventors in perfecting their inventions, taking out their patents and disposing of them. 119 pages. Cloth, $1.00.

KRAUSS. Linear Perspective Self-Taught

The underlying principle by which objects may be correctly represented in perspective is clearly set forth in this book; everything relating to the subject is shown in suitable diagrams, accompanied by full explanations in the text. Price $2.50.

LE VAN. Safety Valves; Their History, Invention, and Calculation

Illustrated by 69 engravings. 151 pages. $1.50.

LEWES AND BRAME. Laboratory Note Book

A practical treatise prepared for the Chemical Student. 170 pages. Cloth, $1.00.

MATHOT. Modern Gas Engines and Producer Gas Plants

A practical treatise of 320 pages, fully illustrated by 175 detailed illustrations, setting forth the principles of gas engines and producer design, the selection and installation of an engine, conditions of perfect operation, producer-gas engines and their possibilities, the care of gas engines and producer-gas plants, with a chapter on volatile hydrocarbon and oil engines. $2.50.

MEINHARDT. Practical Lettering and Spacing

Shows a rapid and accurate method of becoming a good letterer with a little practice. Oblong. Paper cover. 60 cents.

PARSELL & WEED. Gas Engine Construction

A practical treatise describing the theory and principles of the action of gas engines of various types, and the design and construction of a half-horse-power gas engine, with illustrations of the work in actual progress, together with dimensioned working drawings giving clearly the sizes of the various details. Third edition, revised and enlarged. Twenty-five chapters. Large 8vo. Handsomely illustrated and bound. 300 pages. $2.50.

PERRIGO. Modern Machine Shop Construction, Equipment and Management

The only work published that describes the Modern Machine Shop or Manufacturing Plant from the time the grass is growing on the site intended for it until the finished product is shipped. By a careful study of its chapters the practical man may economically build, efficiently equip, and successfully manage the modern machine shop or manufacturing establishment. Just the book needed by those contemplating the erection of modern shop buildings, the rebuilding and reorganization of old ones, or the introduction of Modern Shop Methods, Time and Cost Systems. It is a book written and illustrated by a practical shop man for practical shop men who are too busy to read theories and want facts. It is the most complete all-around book of its kind ever published. 400 large quarto pages, 225 original and specially-made illustrations. $5.00.

Publications of The Norman W. Henley Publishing Co.

PERRIGO. Modern American Lathe Practice
A new book describing and illustrating the very latest practice in lathe and boring mill operations, as well as the construction of and latest developments in the manufacture of these important classes of machine tools. 300 pages, fully illustrated. $2.50.

REAGAN, JR. Electrical Engineers' and Students' Chart and Hand-Book of the Brush Arc Light System
Illustrated. Bound in cloth, with celluloid chart in pocket. 50 cents.

SAUNIER. Watchmaker's Hand-Book
Just issued, 7th edition. Contains 498 pages and is a workshop companion for those engaged in watchmaking and allied mechanical arts. 250 engravings and 14 plates. $3.00.

SLOANE. Electricity Simplified
The object of "Electricity Simplified" is to make the subject as plain as possible and to show what the modern conception of electricity is. 158 pages. Illustrated. Twelfth edition. $1.00.

SLOANE. How to Become a Successful Electrician
It is the ambition of thousands of young and old to become electrical engineers. Not every one is prepared to spend several thousand dollars upon a college course, even if the three of four years requisite are at their disposal. It is possible to become an electrical engineer without this sacrifice, and this work is designed to tell "How to Become a Successful Electrician" without the outlay usually spent in acquiring the profession. Twelfth edition. 189 pages. Illustrated. Cloth, $1.00.

SLOANE. Arithmetic of Electricity
A practical treatise on electrical calculations of all kinds, reduced to a series of rules, all of the simplest forms, and involving only ordinary arithmetic; each rule illustrated by one or more practical problems, with detailed solution of each one. Nineteenth edition. Illustrated. 138 pages. Cloth, $1.00.

SLOANE. Electrician's Handy Book
An up-to-date work covering the subject of practical electricity in all its branches, being intended for the every-day working electrician. The latest and best authority on all branches of applied electricity. Pocketbook size. Handsomely bound in leather, with title and edges in gold. 800 pages. 500 illustrations. Price, $3.50.

SLOANE. Electric Toy Making, Dynamo Building, and Electric Motor Construction
This work treats of the making at home of electrical toys, electrical apparatus, motors, dynamos, and instruments in general, and is designed to bring within the reach of young and old the manufacture of genuine and useful electrical appliances. Eighteenth edition. Fully illustrated. 140 pages. Cloth, $1.00.

SLOANE. Rubber Hand Stamps and the Manipulation of India Rubber
A practical treatise on the manufacture of all kinds of rubber articles. 146 pages. Second edition. Cloth. $1.00.

SLOANE. Liquid Air and the Liquefaction of Gases
Containing the full theory of the subject and giving the entire history of liquefaction of gases from the earliest times to the present. It shows how liquid air, like water, is carried hundreds of miles and is handled in open buckets. It tells what may be expected from it in the near future. 365 pages, with many illustrations. Handsomely bound in buckram. Second edition. $2.00.

SLOANE. Standard Electrical Dictionary
A practical handbook of reference, containing definitions of about 5,000 distinct words, terms, and phrases. An entirely new edition, brought up to date and greatly enlarged. Complete, concise, convenient. 682 pages. 393 illustrations. Handsomely bound in cloth. 8vo. $3.00.

STARBUCK. Modern Plumbing Illustrated
A comprehensive and up-to-date work illustrating and describing the Drainage and Ventilation of dwellings, apartments, and public buildings, etc. The very latest and most approved methods in all branches of sanitary installation are given. Adopted by the United States Government in its sanitary work in Cuba, Porto Rico, and the Philippines, and by the principal boards of health of the United States and Canada. The standard book for master plumbers, architects, builders, plumbing inspectors, boards of health, boards of plumbing examiners, and for the property owner, as well as for the workman and his apprentice. 300 pages. 50 full-page illustrations. $4.00.

USHER. The Modern Machinist
A practical treatise embracing the most approved methods of modern machine-shop practice, and the applications of recent improved appliances, tools, and devices for facilitating, duplicating, and expediting the construction of machines and their parts. A new book from cover to cover. Fifth edition. 257 engravings. 322 pages. Cloth, $2.50.

Publications of The Norman W. Henley Publishing Co.

VAN DERVOORT. Modern Machine Shop Tools; Their Construction, Operation, and Manipulation, Including Both Hand and Machine Tools

An entirely new and fully illustrated work of 555 pages and 673 illustrations, describing in every detail the construction, operation, and manipulation of both Hand and Machine Tools; being a work of practical instruction in all classes of machine-shop practice. Including chapters on filing, fitting, and scraping surfaces; on drills, reamers, taps, and dies; the lathe and its tools; planers, shapers, and their tools; milling machines and cutters; gear cutters and gear cutting; drilling machines and drill work; grinding machines and their work; hardening and tempering; gearing, belting, and transmission machinery; useful data and tables. Fourth edition. $4.00.

WALLIS-TAYLOR. Pocket Book of Refrigeration and Ice Making

This is one of the latest and most comprehensive reference books published on the subject of refrigeration and cold storage. It explains the properties and refrigerating effect of the different fluids in use, the management of refrigerating machinery and the construction and insulation of cold rooms, with their required pipe surface for different degrees of cold; freezing mixtures and non-freezing brines, temperatures of cold rooms for all kinds of provisions; cold-storage charges for all classes of goods, ice-making and storage of ice, data and memoranda for constant reference by refrigerating engineers, with nearly one hundred tables containing valuable references to every fact and condition required in the instalment and operation of a refrigerating plant. $1.50.

WOOD. Walschaert Locomotive Valve Gear

The only work issued treating of this subject of valve motion. 150 pages, illustrated. Cloth $1.50.

WOODWORTH. American Tool Making and Interchangeable Manufacturing

A practical treatise of 560 pages, containing 600 illustrations on the designing, constructing, use, and installation of tools, jigs, fixtures, devices, special appliances, sheet-metal working processes, automatic mechanisms, and labor-saving contrivances; together with their use in the lathe, milling machine, turret lathe, screw machine, boring mill, power press, drill, subpress, drop hammer, etc., for the working of metals, the production of interchangeable machine parts, and the manufacture of repetition articles of metal. $4.00

WOODWORTH. Dies, Their Construction and Use for the Modern Working of Sheet Metals

A complete treatise of 384 pages and 505 illustrations upon the designing, constructing, and use of tools, fixtures, and devices, together with the manner in which they should be used in the power press, for the cheap and rapid production of the great variety of sheet-metal articles now in use. It is designed as a guide to the production of sheet-metal parts at the minimum of cost with the maximum of output. The hardening and tempering of Press tools and the classes of work which may be produced to the best advantage by the use of dies in the Power press are fully treated.

The engravings show dies, press fixtures, and sheet-metal working devices, from the simplest to the most intricate, and the descriptions are so clear and practical that all metalworking mechanics will be able to understand how to design, construct and use them. $3.00.

WOODWORTH. Hardening, Tempering, Annealing, and Forging of Steel

A new book containing special directions for the successful hardening and tempering of all steel tools. Milling cutters, taps, thread dies, reamers, both solid and shell, hollow mills, punches and dies, and all kinds of sheet-metal working tools, shear blades, saws, fine cutlery and metal-cutting tools of all descriptions, as well as for all implements of steel, both large and small, the simplest and most satisfactory hardening and tempering processes are presented. The uses to which the leading brands of steel may be adapted are concisely presented, and their treatment for working under different conditions explained, as are also the special methods for the hardening and tempering of special brands. 320 pages. 250 illustrations. $2.50.

WOODWORTH. Punches, Dies and Tools for Manufacturing in Presses

A work of 500 pages, and illustrated by nearly 700 engravings, being an encyclopædia of die-making, punch-making, die-sinking, sheet-metal working, and making of special tools, subpresses, devices and mechanical combinations for punching, cutting, bending, forming, piercing, drawing, compressing, and assembling sheet-metal parts and also articles of other materials in machine tools. $4.00.

WRIGHT. Electric Furnaces and Their Industrial Application

This is a book which will prove of interest to many classes of people; the manufacturer who desires to know what product can be manufactured successfully in the electric furnace, the chemist who wishes to post himself on electro-chemistry, and the student of science who merely looks into the subject from curiosity. The book is not so scientific as to be of use only to the technologist, nor so unscientific as to suit only the tyro in electro-chemistry; it is a practical treatise of what has been done, and of what is being done, both experimentally and commercially, with the electric furnace. 288 pages. $3.00.

www.ingramcontent.com/pod-product-compliance
Lightning Source LLC
Chambersburg PA
CBHW031322230426
43670CB00006B/215